Handbook of Antennas for EMC

For a complete listing of the *Artech House Antenna Library*,
turn to the back of this book.

Handbook of Antennas for EMC

Thereza Macnamara

Artech House
Boston • London

Library of Congress Cataloging-in-Publication Data
Macnamara, Thereza.
Handbook of antennas for EMC/Thereza Macnamara.
p. cm.
Includes bibliographical references and index.
ISBN 0-89006-549-7
1. Radio–Antennas. 2. Electromagnetic compatibility. 3. Radio noise–Measurement. I. Title.
TK6565.A6M32 1995 95-30521
621.382'4–dc20 CIP

British Library Cataloguing in Publication Data
Macnamara, Thereza
Handbook of Antennas for EMC
I. Title
621.3824

ISBN 0-89006-549-7

685 Canton Street
Norwood, MA 02062

International Standard Book Number: 0-89006-549-7
Library of Congress Catalog Card Number: 95-30521

10 9 8 7 6 5 4 3 2 1

This book is dedicated to my daughter Louise,
my son David and my husband Tom

Contents

Preface

This handbook is intended as a background and reference book primarily for EMC engineers, but others seeking specific or general information on antennas, as well as circuit design engineers and quality assurance managers, will also find it very useful. The treatment is essentially practical but it aims to be authoritative and reliable.

Postgraduate students studying electromagnetic theory, antennas and propagation, and EMC topics will also find the book invaluable in understanding the concepts of EM radiation.

It explains the principles of antenna theory, without going into great mathematical detail. The reader is expected to have attained an academic level of about *Higher National Diploma* (HND) or have had some practical experience as a test engineer in the EMC field.

The required characteristics of an antenna, depend on the application. In general, for EMC applications, the larger the bandwidth, the fewer the number of antennas required to cover the total frequency range.

For emission testing, and radio monitoring site surveys, the antenna should be omnidirectional, and capable of receiving electric and magnetic field of any polarisation. Other practical considerations, such as using receivers which cover the same frequency ranges as the antennas may also influence the antenna selection. Portability, weight and size are obviously also very important considerations for site surveys.

For immunity and susceptibility testing, the antenna should also be broad band, but it has to capable of handling high power, since it has to produce high electric or magnetic fields.

For shielding or screening effectiveness measurements, the transmitting antenna, should be capable of handling high power, in order that the required dynamic range can be produced. Both receive and transmit antennas should also be highly directional so that only the field through the material under test is measured. However at the low end of the frequency range single element antennas cannot produce directional beams.

The book is written in such a way that each section in a chapter can be referred to independently, and the reader does not have to read a chapter from the beginning in order to find the definition of a particular symbol, for instance. The details of each symbol is given after most equations. Although this may seem repetitious to the reader, reading the book from the beginning, this approach is invaluable to the reader seeking to refer to a particular topic, and not wishing to spend time looking up the meaning of symbols used in equations and formulas. Each chapter is also written so as to be fairly self contained, although this was not always possible.

Reference to textbooks is given with the particular page or section number. This enables the reader to quickly access the particular topic, and the author considers this a welcome departure from references made to entire books, with the reader required to find the relevant section.

The book uses SI units throughout, and lists all the abbreviations and acronyms commonly used. The SI units avoids the use of the solidus (forward slash or division sign) and instead the use of negative indices is recommended. However, the solidus is commonly used in some cases, and has been retained in these cases.

Chapter 1 explains the main properties of an antenna in a qualitative manner, and is useful for the engineer not familiar with antennas or the jargon used.

Chapters 2 and 3 assist engineers in revision of the mathematics required for the theoretical side of engineering. These were requested by colleagues who felt that math textbooks were too complicated and gave too much irrelevant detail, that were not applicable to EM theory. These chapters are written in such a way that they can be omitted by those familiar with mathematics or by those who do not wish to study the theoretical aspects of antennas.

These next three chapters describe antennas that are classified according to frequency. They give the theoretical basis for the antennas, and also describe the practical aspects, so that engineers can deduce the electrical properties of an antenna by examination of the physical characteristics of the antenna.

Chapter 7 describes standard horns, which are used to calibrate antennas used for measurements. Antenna correction factors are also covered in this chapter.

Finally the appendices contain lists of acronyms, preferred scientific prefixes, scientific constants, conductivities, dielectric constants, conversion between dBmV and volts or watts, the periodic table and frequency wavebands.

Every care has been taken in the preparation of the manuscript. However the author would appreciate any comments on the topics covered or errors in the text be they typographical or factual.

Chapter 1
Introduction to Antennas

This chapter deals with antennas, without considering the theoretical aspects in any great depth, hence enabling *electromagnetic compatability* (EMC) engineers to understand the main characteristics of an antenna in a qualitative manner without recourse to the intricacies of Maxwell's equations. It also gives approximate formulas to enable engineers to obtain some key parameters if other parameters are known. The main features of an antenna that are required for EMC and radio-monitoring applications are also discussed.

An antenna is a device for radiating or receiving an electromagnetic radio signal. It transfers energy between a transmission line and free space; that is, it transforms a guided wave to a free space wave, or vice versa.

The term antenna is used as a generic term for both wire antennas such as dipoles and for aperture antennas such as horns and reflectors. However, its use is sometimes restricted to aperture antennas in the upper *radio frequency* (RF) and microwave regions, that is, above approximately 300 MHz, whereas the term aerial is used at the lower frequencies.

1.1 REQUIREMENTS OF AN ANTENNA FOR EMC

The important features of an antenna that are required depend on the intended application. For emission testing and radio-monitoring site surveys, the antenna should be omnidirectional, broadband, and capable of receiving electric and magnetic fields of any polarization. Other practical considerations, such as using receivers that cover the same frequency ranges as the antennas, may also influence the antenna selection. Portability, weight, and size are obviously also very important considerations for site surveys.

For immunity and susceptibility testing, the antenna should also be broadband, but it has to be capable of handling high power since it has to produce high electric or magnetic fields.

For shielding or screening effectiveness measurements, the transmitting antenna should be capable of handling high power and have high gain so that the required dynamic range can be produced. Both receiving and transmitting antennas should also have narrow beamwidths so that only the field through the material under test is measured.

In general, for EMC applications, the larger the bandwidth, the fewer the number of antennas required to cover the total frequency range.

1.1.1 Types of Antennas

Antennas can be broadly classified by the directions in which they radiate or receive electromagnetic radiation. They can be isotropic, omnidirectional, or directional.

An isotropic antenna is a hypothetical antenna that radiates uniformly in all directions so that the electric field at any point on a sphere (with the antenna at its center) has the same magnitude. Figure 1.1 depicts the radiation obtained from an isotropic antenna that, in fact, cannot be realized in practice, since in order to radiate uniformly in all directions an isotropic antenna would have to be a point source. The nearest approximation to an isotropic antenna is a Hertzian dipole, which is a dipole that is very small in

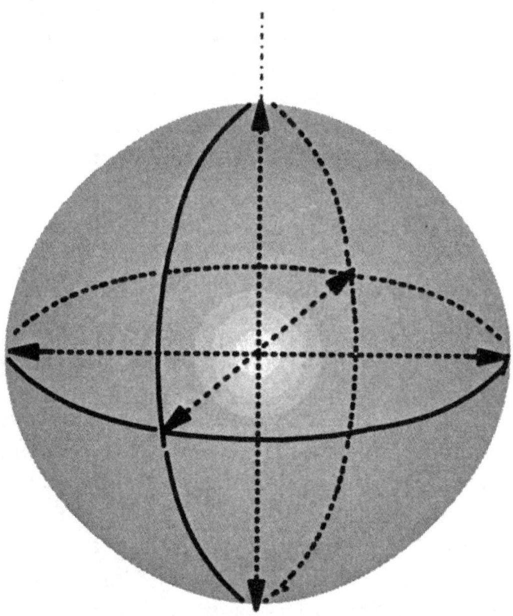

Figure 1.1 Radiation from an isotropic antenna.

terms of wavelength. The length of a Hertzian dipole is about one-hundredths of the wavelength at its operating frequency, and even in this case its pattern is not truly isotropic.

An omnidirectional antenna radiates uniformly in one plane. Examples of omnidirectional antennas are monopoles, dipoles, and biconicals. Figure 1.2 shows the radiation from a vertical dipole. The radiation is uniform in the horizontal plane, and the cross section of the radiation is in the form of a figure of eight in the vertical plane.

A directional antenna radiates most of its power in one particular direction. Examples of directional antennas are horns, reflector systems, log-periodics, and Yagis. Figure 1.3 shows the radiation from a reflector antenna. For a circularly symmetrical reflector the radiation pattern is the same in all planes.

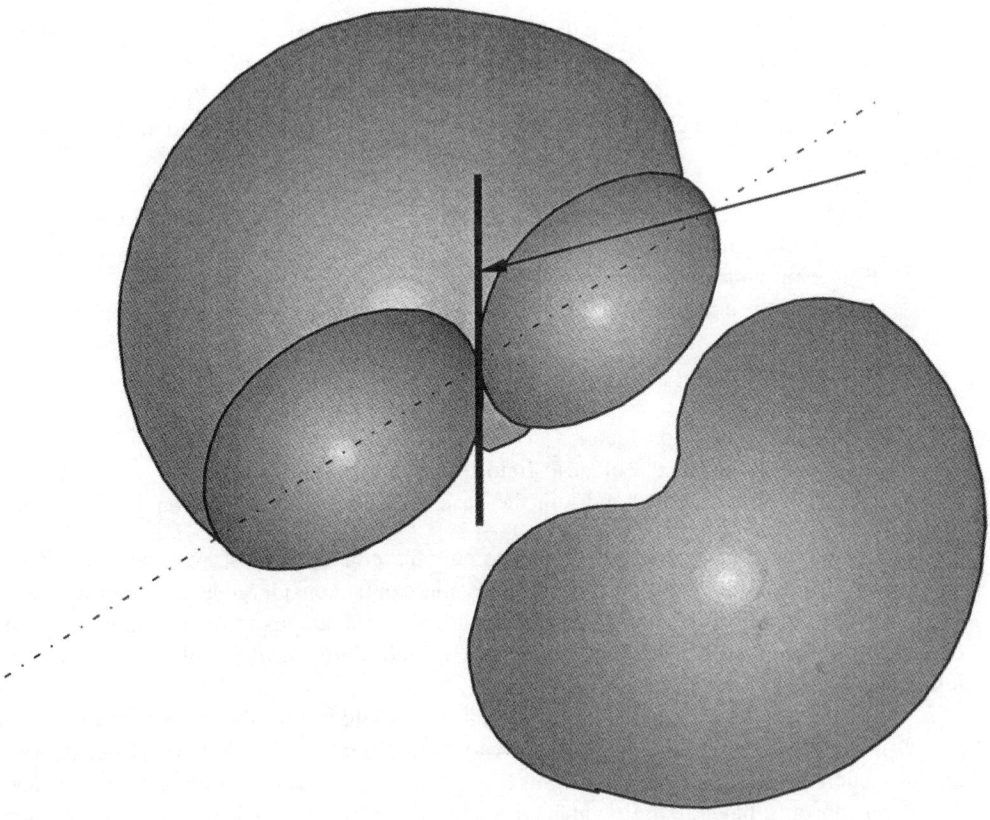

Figure 1.2 Radiation from a vertical dipole.

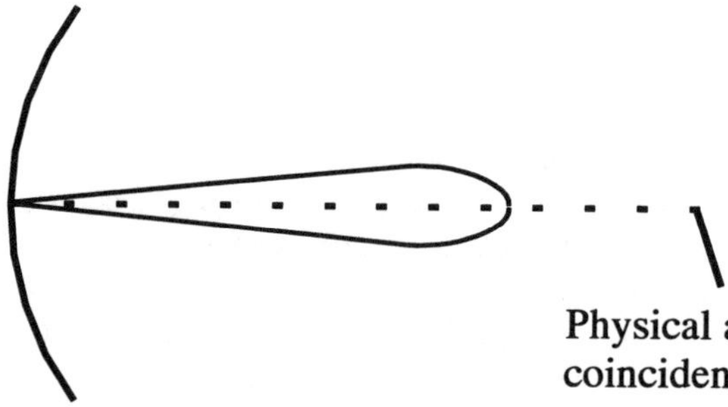

Figure 1.3 A directional reflector antenna.

1.2 MAIN CHARACTERISTICS OF AN ANTENNA

An antenna is chosen for use in a particular application according to its main physical and electrical characteristics. Moreover, the antenna must perform in a desired manner for the particular measurement system.

An antenna can be characterized by the following factors, not all of which apply to all antenna types:

1. Its radiation resistance;
2. Its radiation pattern;
3. The beamwidth and gain of its main lobe;
4. The position and magnitude of its sidelobes;
5. The magnitude of its back lobe;
6. Its bandwidth;
7. Its aperture;
8. Its antenna correction factor;
9. The polarization of the electric field that it transmits or receives;
10. The power that it can handle in the case of a transmitting antenna.

There are two principal planes in which the antenna characteristics are measured. These are known as the azimuth and elevation planes and can be considered as the horizontal and vertical planes, respectively, for land-based antennas. The angles in the azimuth plane are conventionally denoted by the Greek letter phi, ϕ, and in the elevation plane by the Greek letter theta, θ.

Some characteristics such as beamwidth and sidelobes are the same in both planes for symmetrical antennas such as circular waveguide horns and reflectors. Other characteristics such as the gain on boresight (i.e., where the azimuth and elevation planes intersect) can only have a single value. In general, for unsymmetrical antennas the characteristics are different in the two principal planes, with a gradual transition in the intervening region between these two planes.

1.2.1 Radiation Resistance

We can consider an antenna as a load that terminates the transmission line that feeds it. In the ideal case this load will have an impedance that is purely resistive; that is, the load will not have any reactive component, such as an inductance or capacitance. In practice, the impedance of an antenna is made up of a self-impedance and a mutual impedance [1, p. 29]. The self-impedance is the impedance that would be measured at the terminals of the antenna when it is in free space, given no other antennas or reflecting objects in the vicinity. The mutual impedance accounts for the coupling between the antenna and any other source. When the antenna is sufficiently isolated from other objects, this mutual impedance tends to zero. On the other hand, in some antennas such as the Yagi array, the operation depends on the mutual coupling between the driven element and the other parasitic passive elements. The self-impedance Z_a consists of a real part, which is the resistance R_a,and an imaginary part, which is the self-reactance X_a. In antennas such as the loop or dipole, the resistance and reactance are in series. The self-impedance is given by $Z_a = R_a + jX_a$.

The antenna resistance R_a accounts for the power absorbed by the antenna and appearing at its terminals. This power is mainly reradiated by the antenna, but a small amount is dissipated in the form of ohmic losses in the antenna structure. The resistance of the antenna can thus be visualized as a sum of a radiation resistance R_r and a loss (ohmic) resistance R_o. The ohmic power loss contributes to the loss of efficiency of the antenna and is of the order of fractions of a decibel [2, p. 39]. This makes it possible to set up an unambiguous network that is characteristic of the antenna itself [2, p. 39]. The rest of the power is radiated and is associated with a radiation resistance R_r. Thus the actual power P_r radiated by the antenna is given by

$$P_r = I^2 R_r \tag{1.1}$$

where I is the current.

At RF and lower frequencies, this concept of radiation resistance can be used because it is possible to define a unique driving point at which the antenna can be disassociated from the feed network and represented as a separate network. The magnitude of the impedances depend on the position of the driving point. In cases where the driving point is arbitrary, the equivalent circuit of the antenna can be either a series or parallel *inductance-capacitance-resistance* (LCR) network. In the latter case, the radiation and ohmic resistances appear in parallel [2, p. 39]. Radiation resistances vary from 73Ω for a thin resonant dipole to 0.01Ω for an HF notch antenna [1, p. 29].

1.2.1.1 Tuning Circuits

When the antenna has the same impedance as the transmission line that feeds it, the antenna is said to be matched to the line. When this occurs, maximum power is transferred from the transmission line to the antenna. In general, however, the impedance of the antenna is not the same as that of the transmission line. When the transmission line has a purely resistive impedance and the antenna has an impedance that contains a different resistive value as well as a reactive part, the optimum transfer of power can be achieved via the use of tuning circuits between the transmission line and the antenna. In general, these circuits consist of an LCR circuit in which the capacitance of the capacitor is altered in order to provide the maximum transfer of power.

1.2.2 Radiation patttern

The antenna is a reciprocal device; that is, it radiates or receives electromagnetic energy in the same way. Thus, although the radiation pattern is identified with an antenna that is transmitting power, the same properties would apply to the antenna even if it was receiving power. Any difference between the received and radiated powers can be attributed to the difference between the feed networks and the equipment associated with the receiver and the transmitter. The antenna radiates the greatest amount of power along its boresight and also receives power most efficiently in this direction.

The radiation pattern is a very important characteristic of an antenna. It facilitates a clearer understanding of the key features of an antenna that otherwise cannot be obtained from the textual technical description of an antenna. For instance, when the sidelobe level of an antenna is quoted, this level pertains to the near-in lobes (on either side of the main lobe), and when the average level of the two lobes is quoted, any asymmetry in the radiation pattern cannot be deduced. In textual technical descriptions, the level, position, and number of other sidelobes are rarely quoted. In EMC measurements these other lobes and the positions of nulls can be very important, and knowledge of their details can enable the engineer to determine if a low level of incoming signal is attributable to a null or if there genuinely is a low level in a particular direction.

The radiation pattern of an antenna is peculiar to the type of antenna and its electrical characteristics as well as its physical dimensions. It is measured at a constant distance in the far field of the antenna. The radiation pattern of an antenna is usually plotted in terms of relative power. The power at boresight, that is, at the position of maximum radiated power, is usually plotted at 0 dB; thus, the power at all other positions appear as a negative value. In other words, the radiated power is normalized to the power at boresight. If the power were plotted in linear units, the normalized power would be one at boresight. The main reason for using decibels instead of linear power is that the power at the nulls is often of the order of 10,000 times less than the power on boresight, which means that the scale would have to be very large in order to cover the whole range of power values. In addition, we would not be able to read difference in powers at the higher power levels since the gradient of the plot would be too small.

The radiation pattern is usually measured in the two principal planes, namely, the azimuth and the elevation planes. The radiated power is plotted against the angle that is made with the boresight direction.

If the antenna is not physically symmetrical about each of its principal planes, then one can also expect its radiation pattern in these planes to be unsymmetrical.

The radiation pattern can be plotted using rectangular/cartesian or polar coordinates. The rectangular plots can be read more accurately (since the angular scale can be expanded), but the polar plots give a more pictorial representation and are thus easier to visualize, like an analogue clock or *plan position indicator* (PPI) that is used in many radar sets.

1.2.2.1 Rectangular/Cartesian Plots

Rectangular/Cartesian plots are standard *x-y* plots where the axes are plotted at right angles to each other. The vertical or *y*-axis is used for the dependent variable, and the *x*-axis is used for the independent variable. The *y*-coordinate is called the ordinate, and the *x*-coordinate is called the abscissa. In a radiation plot the angle with respect to boresight is varied and the magnitude of the radiated power is measured; thus, the angle is the independent variable and the power radiated is the dependent variable. The magnitudes of the powers are the ordinates and the angles are the abscissae. It is important to remember that the radiated power is measured in the far field. The graph paper used to plot radiation patterns is not standard in the mathematical sense. The *x*- and *y*-axes are not numbered in ascending order from the origin of coordinates as in the case of standard graph paper. All values, whether negative or positive, are shown without a sign. A typical rectangular plot of an antenna radiation pattern is shown in Figure 1.4. The *y*-axis can show two sets of scales: one graduated from 0 dB to 4 dB, and another from 0 dB to 8 dB. Scales of 40 dB and 80 dB are obtained by multiplying the scales by ten. It should be noted that the zero is at the top and, thus, the numbers below should really be shown as negative values of -4 dB and -8 dB. Similarly the *x*-axis can show three sets of angular scales of 5, 30, and 180 deg on either side of the zero, representing the angles measured clockwise and counterclockwise from the boresight position and in standard mathematical convention denoted by positive and negative signs that are ommitted on radiation graph paper. Thus the radiation pattern can be plotted over 10, 60, or the whole 360 deg. The engineer or technician who performs the measurements usually encircles (or marks the scale being used by ticking) the relevant boxes in the key for each of the axes.

1.2.2.2 Polar Plots

In a polar plot the angles are plotted radially from boresight and the intensity or the power is plotted along the radius as shown in Figure 1.5. This gives a pictorial representation of the radiation pattern of the antenna and is easier to visualize than the rectangular plots. However, since the scale of the angular positions cannot be increased (i.e., they can only be plotted to scale from 0° to 360°), the accuracy cannot be increased as in the case of rectangular plots. The scale of the intensity or power, however, can be varied as in the

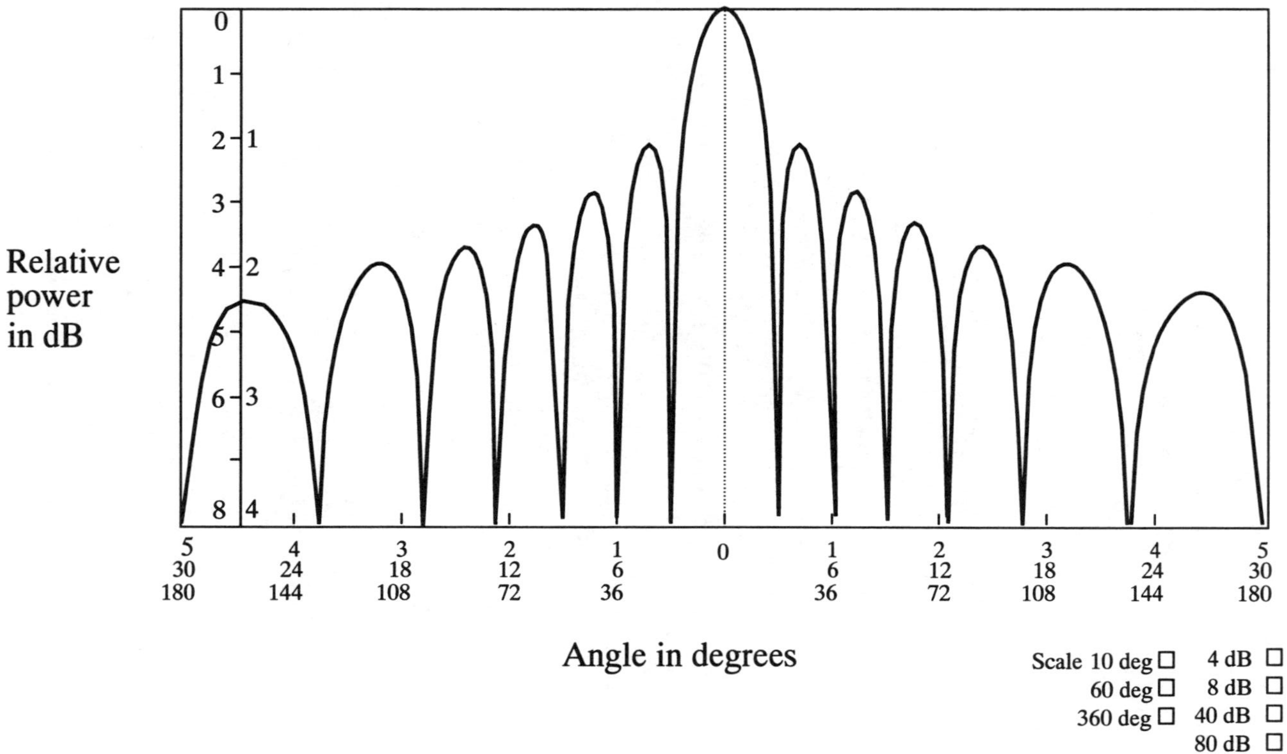

Figure 1.4 Rectangular plot of an antenna radiation pattern.

case of rectangular plots. On polar plots each circle represents a contour plot where the power has the same magnitude and is shown relative to the power at boresight. Since the power is in general a maximum value at boresight, these levels will always be less than the power at boresight and should thus be shown as negative values. However, they are usually written without a sign and should be assumed to be negative, contrary to standard arithmetic convention. Figure 1.5 shows levels from 0 dB to 25 dB in 5-dB steps. Some polar plots start at +10 dB as the maximum, go through the 0 dB, and then down to -30 dB or more. In almost all cases, the signs are omitted.

1.2.3 Main Lobe

The main lobe of an antenna is in the direction of maximum radiation, which is not to say that it is necessarily on the physical main axis of the antenna, but this is generally the case

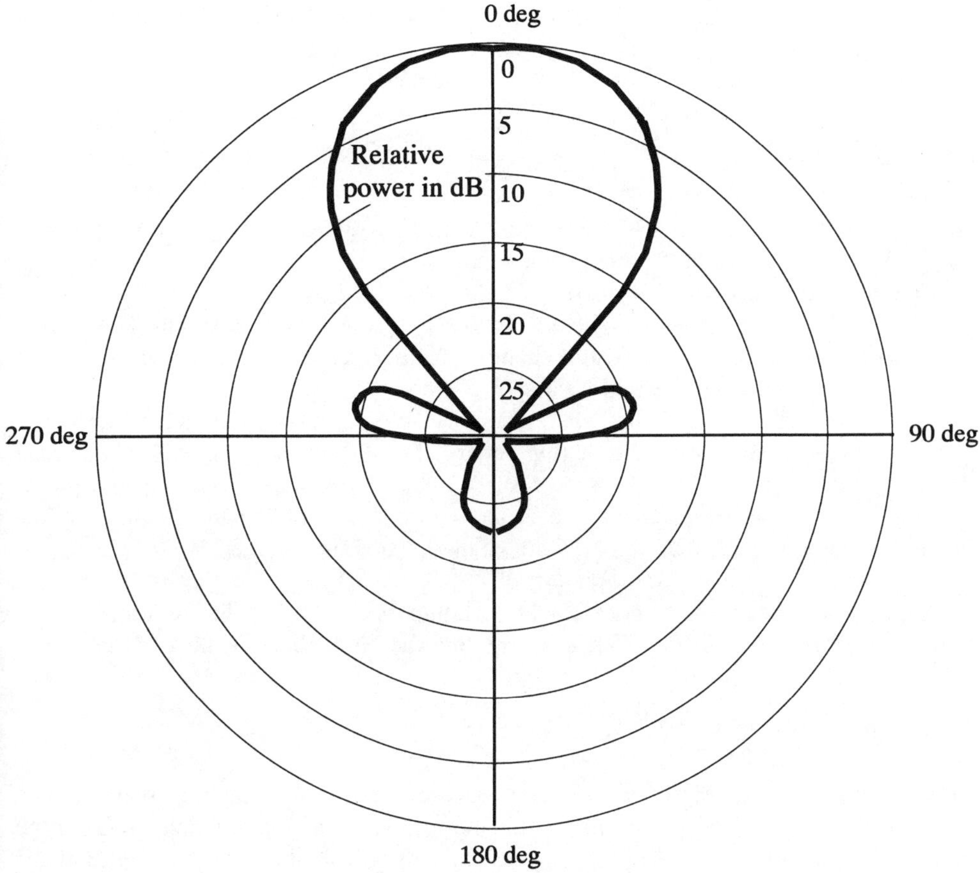

Figure 1.5 Polar plot of an antenna radiation pattern.

for most antennas used for EMC applications. The characteristics of an antenna such as the beamwidth and gain are related to the main lobe alone. The peak of the main lobe is called the boresight of the antenna, and the radiation pattern is often positioned so that its boresight coincides with the zero angular position of the graph, even when the antenna is not physically symmetrical. Figure 1.6 shows the radiation pattern of a symmetrical antenna.

1.2.3.1 Beamwidth—Half Power and 10 dB

The beamwidth only pertains to the main beam of the antenna and not the sidelobes. The beamwidth of an antenna is, in general, inversely proportional to its physical size; so the larger the antenna, the smaller is its beamwidth for the same frequency. If the antenna does not have the same dimensions in all planes, the plane containing the largest dimension will have the narrowest beamwidth. Thus for a rectangular horn, as shown in Figure 1.7, the beamwidth in the horizontal or azimuth plane AA′ is narrower than that in the vertical or elevation plane BB′. The planes AA′ and BB′ are also known as the *H* (magnetic) and *E* (electric) planes, respectively, for this type of antenna.

Definition of Beamwidth

The beamwidth of an antenna is commonly defined in two ways. The most well known definition is the 3-dB or half-power beamwidth, but the 10-dB beamwidth is also used, especially for antennas with very narrow beams.

The 3-dB or *half-power beamwidth* (HPBW) of an antenna is taken as the width in degrees (or sometimes in radians) at the points on either side of the main beam where the radiated power is half the maximum value.

In Figure 1.8 the 3-dB level, where the power is half that of the power at boresight, occurs at points K and L on the radiation pattern. The difference in degrees can be read off the *x*-axis as 25 deg, which is the HPBW. The 10-dB beamwidth is the width in degrees (or sometimes in radians) at the points on either side of the main beam where the radiated power is 10 dB, that is, one-tenth of the maximum value.

The points on the radiation pattern of Figure 1.8 where the power is 10 dB (i.e., one-tenth) below the power on boresight are at points M and N. The difference in degrees between these points is 44 deg. Thus, we say that the 10-dB beamwidth is 44 deg.

1.2.3.2 Boresight Directivity/Gain

The terms directivity (or directive gain) and gain are often used synonymously, but in fact they are not the same. The gain allows for the efficiency of the antenna, whereas the directivity does not. For instance, if we say that the linear directivity is 100 and the efficiency is 95%, then the gain of the antenna would be 95. In other words, the gain of

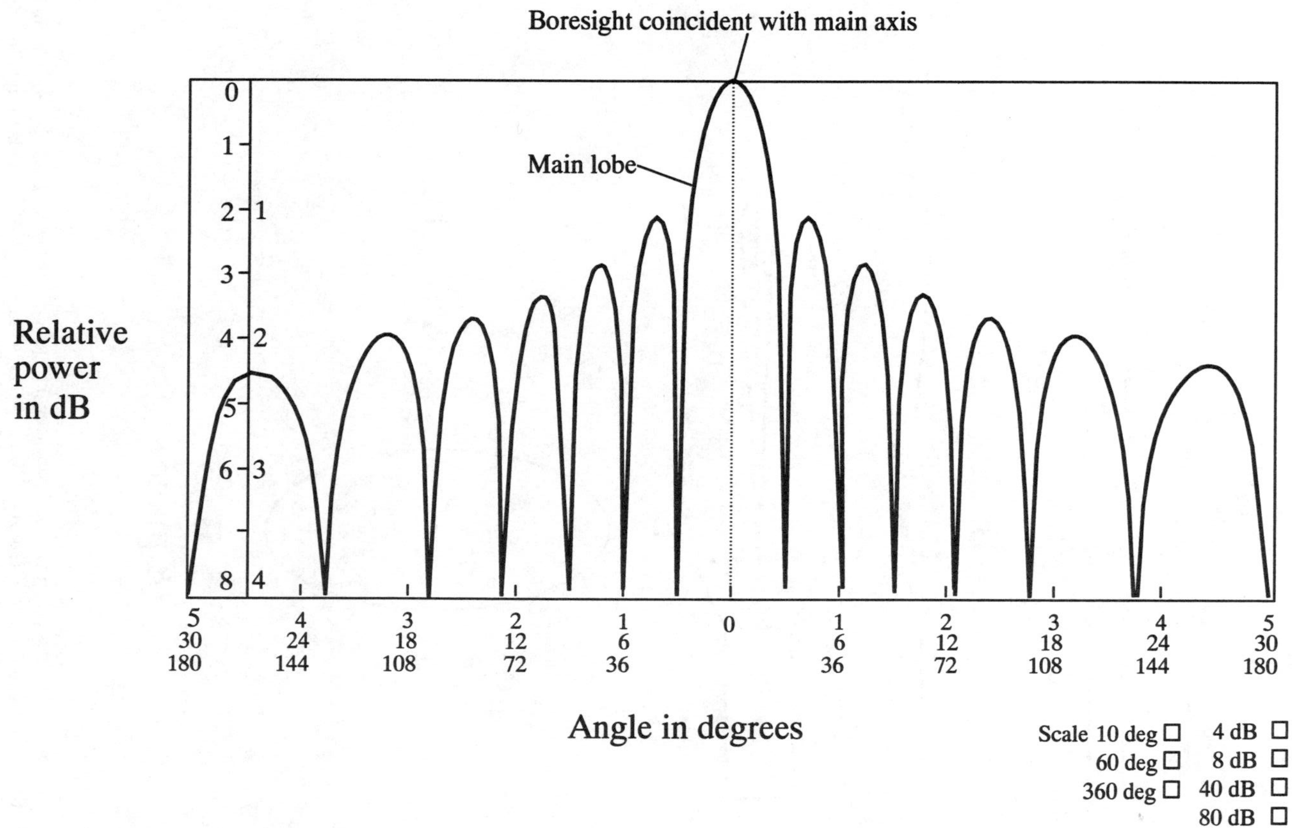

Figure 1.6 Rectangular plot of the radiation pattern of a symmetrical antenna.

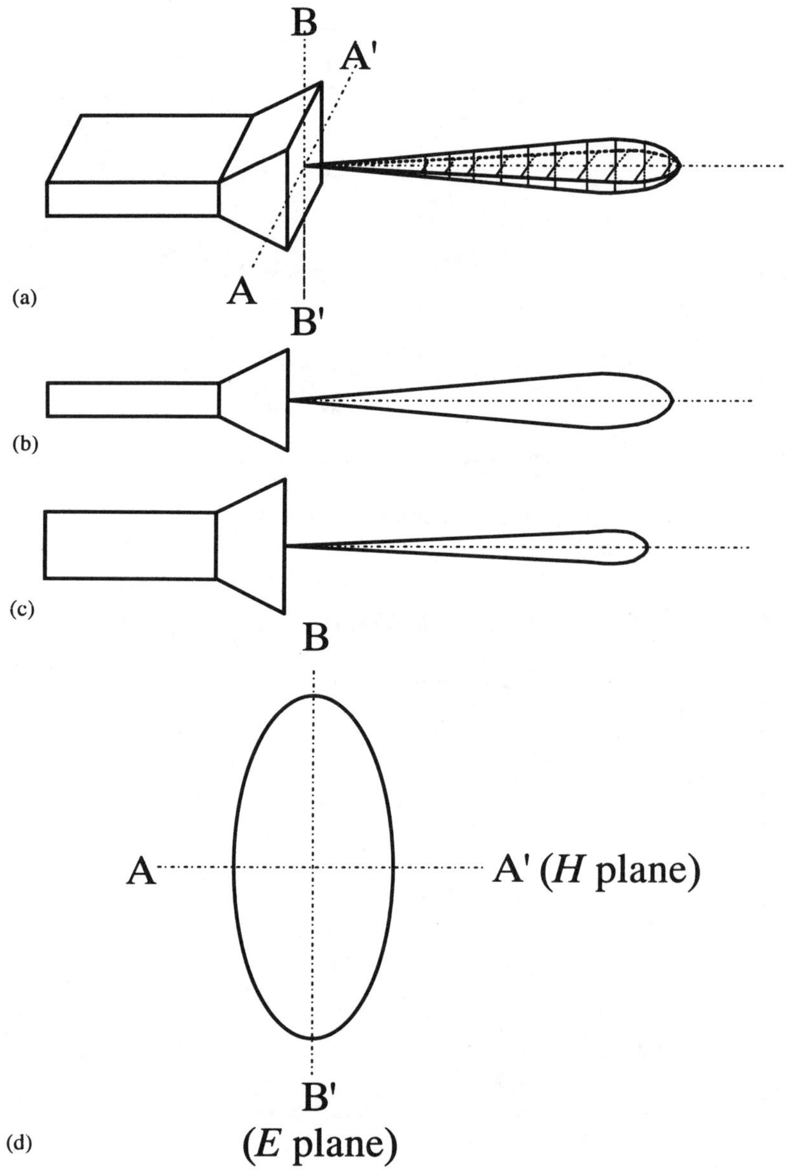

Figure 1.7 Main lobe of a rectangular pyramidal horn showing the beams in the principal planes. (a) main lobe; (b) side view showing the vertical or E plane beam width through BB′; (c) plan view showing the horizontal or H plane beam width through AA′; (d) front view of the beams.

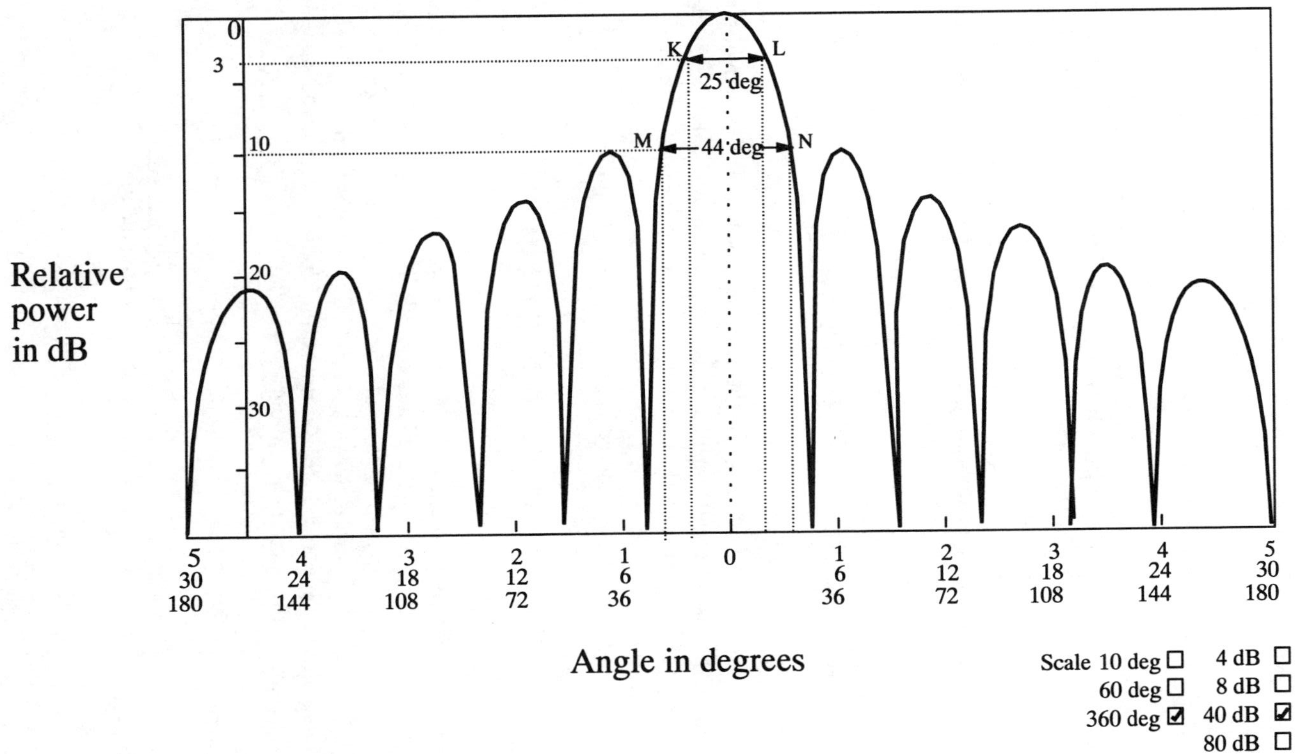

Figure 1.8 A radiation pattern showing the 3- and 10-dB beamwidths.

the antenna is the product of the directivity and the efficiency. In general, the gain of the antenna is that quoted in the specification, so we shall refer to this characteristic instead of the directivity.

The term gain is not the same as that of, for instance, an amplifier. An antenna is not usually an active device. We can think of an antenna as concentrating the electromagnetic energy in one particular direction, similar to the way in which the reflector of a torch concentrates the light from a light bulb into a narrow bright beam. The total light output is the same whether the reflector is present or not; but with the reflector, light is mainly concentrated in one direction, depending on the type and position of the reflector.

The *Institute of Electrical and Electronics Engineers* (IEEE) distinguishes between the directivity of an antenna and the directivity in free space. Both definitions are explained in a qualitative manner in this chapter.

Directive Gain of an Antenna

Directive gain is defined as 4π times the ratio of the radiation intensity in a given direction to the total power radiated by the antenna. The factor of 4π relates to the total number of angles (in solid angles or steradians) around a point, in three dimensions, just as 2 gives us the total number of angles (in radians) at a point in one plane. This can be considered as a measure of the ability of the antenna to concentrate the energy (that would normally be radiated over 4π solid angles or steradians) into a smaller number of angles.

Directive Gain in Physical Media

The IEEE definition of gain of an antenna relates the power radiated by the antenna to that radiated by an isotropic antenna (that radiates equally in all directions) and is quoted as a linear ratio or in *decibels referred to an isotropic* (dBi, "i" for isotropic). When we say that the gain of an antenna is, for instance, 20 dBi (100 in linear terms), we mean that an isotropic antenna would have to radiate 100 times more power to give the same intensity at the same distance as that particular directional antenna.

The gain G as a linear ratio is defined as

$$G = \frac{\{\text{power radiated on boresight}\}}{\{\text{power radiated by an isotropic antenna}\}} \tag{1.2}$$

The gain G_{dBi} expressed in decibels is defined as

$$G_{\text{dBi}} = 10 \log_{10}(G) \tag{1.3}$$

To convert the gain G_{dBi} expressed in decibles to the gain G as a linear ratio the following formula is used

$$G = 10^{(G_{dBi}/10)} \tag{1.4}$$

The gains of antennas vary between about 2 dBi for a dipole to around 70 dBi for a satellite ground-station antenna. These represent linear gain ratios of 1.5 and 10,000,000, respectively, compared with an isotropic antenna.

Relative and Absolute Gain

The radiation pattern of an antenna shows the power on boresight as 0 dB and the power in other directions as negative values. The gain in all directions is plotted relative to the gain on boresight. In order to find the absolute gain in any particular direction the gain on boresight must be known. If this gain is expressed in decibels, then this value can simply be added to the gain at any point to give the absolute gain. However, if the boresight gain is given as a linear ratio, then this has to be converted to decibels using the following formula

$$G_{dBi} = 10 \log_{10} \{\text{the gain as a linear ratio}\} \tag{1.5}$$

The absolute gain on boresight is measured by comparison with a standard gain antenna, which functions as a reference antenna whose gain is calculated or measured with a high degree of accuracy.

1.2.4 Sidelobes

The sidelobes are, strictly speaking, any of the maxima marked, for example, as A, B, C, D, E in Figure 1.9. However, in practice only the ''near-in'' lobes marked A are referred to as sidelobes. Sometimes the irregularities in the main beam of the radiation pattern mean that there are small peaks such as those marked F in Figure 1.9 (exaggerated for clarity) that could be mistaken for sidelobes. For this reason, the sidelobes are sometimes defined as the peaks, where the difference between the peak and an adjacent trough is at least 3 dB.

The sidelobes are characterized by their level below the boresight gain and their angular position relative to boresight.

1.2.4.1 Sidelobe Level

The sidelobe level is usually quoted as the level below the boresight gain. Since the radiation pattern is plotted with the boresight gain at 0 dB, the sidelobe level should be quoted in negative decibels. However, the sidelobe level is usually quoted as a positive quantity. The average value is taken where the two sidelobe levels are different. The absolute level of the sidelobe can only be calculated if the absolute boresight gain is known.

An aperture antenna, such as a reflector, can be illuminated by its feed in several ways. If we say that it is uniformly illuminated, we mean that the magnitude of the electric field across its aperture has a constant value and the phase at each point is the same. This type of illumination produces a radiation pattern with the narrowest beamwidth but is accompanied by one of the highest level of sidelobe. Adjusting the phase and/or magnitude of the illumination can result in a reduction of the sidelobe level, but this usually also broadens the beamwidth.

For an antenna that is symmetrical about its main physical axis, the radiation pattern is in general also symmetrical. Thus, the level of the sidelobes on opposite sides of the main beam would be the same. If the antenna is not physically symmetrical about its principal axis, then we would expect its radiation pattern to also be unsymmetrical. The level of the sidelobes on either side of the main beam would also be different if the antenna was unsymmetrical.

The radiation pattern shown in Figure 1.9 is that of a symmetrical antenna and, thus, both of the main sidelobe levels are the same at a level of -14 dB.

1.2.4.2 Sidelobe Position

The angular position of the sidelobes, relative to the position of the main beam, is usually quoted in degrees rather than radians. For a physically symmetrical antenna, where the radiation pattern is not symmetrical, the average of the two angular positions is taken as the sidelobe angular position.

Figure 1.9 shows the radiation pattern of an antenna that is symmetrical about its physical axis. The sidelobes are at equal angles of 43 deg off boresight.

1.2.5 Front-to-Back Ratio

The *front-to-back ratio* (F/B) is a measure of the ability of a directional antenna to concentrate the beam in the required forward direction. In linear terms, it is defined as the ratio of the maximum power in the main beam (boresight) to that in the back lobe. It is usually expressed in decibels, as the difference between the levels on boresight and at 180 deg off boresight. In Figure 1.10 the backlobe is shown at the 180-deg position, as being 35 dB below the level at boresight. In linear terms this means that the level of the back lobe is 3,162 times less than the level on boresight. In decibels, the front-to-back ratio is the difference of +35 dB in this case, but the sign is usually omitted.

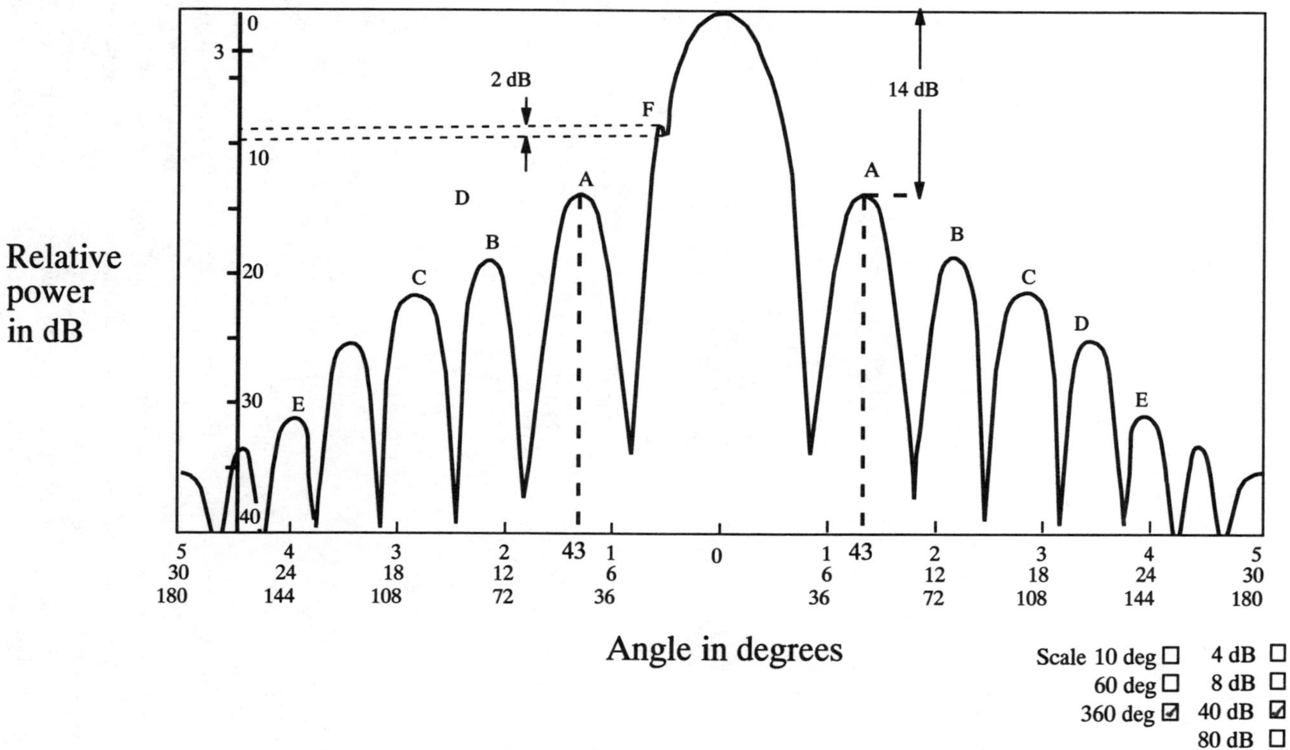

Figure 1.9 A radiation pattern showing the sidelobe levels and positions.

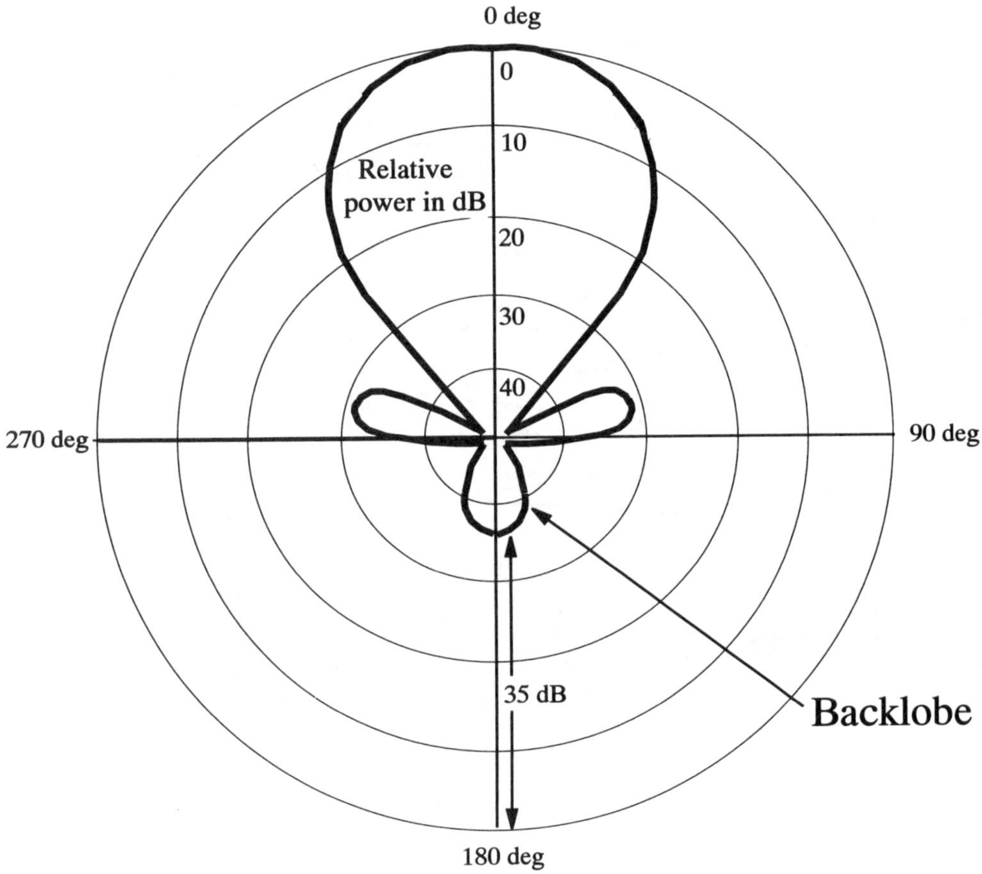

Figure 1.10 A radiation pattern showing the back lobe.

1.2.6 Bandwidth

The bandwidth of an antenna is a measure of its ability to radiate or receive different frequencies. The bandwidth is the range of frequencies that the antenna can receive (or radiate) with a power efficiency of 50% (0.5) or more or a voltage efficiency of 70.7%, (that is, ($\sqrt{0.5}$). A wide bandwidth is achieved at the expense of gain. In other words, if the antenna had the same size but a narrower bandwidth, its gain would be higher. The operating frequency range is specified by quoting the upper and lower frequencies, but the bandwidth is often quoted as a relative value.

Bandwidth is commonly expressed in one of two ways: as a percentage or as a fraction or multiple of an octave. (An octave is a band of frequencies between one frequency and the frequency that is double or half the first frequency; for instance, we have an octave between 10 MHz and 20 MHz.) When it is expressed as a percentage bandwidth, its center frequency should be quoted and the percentage expressed relative to this center frequency. When it is expressed in octaves, its lower and upper frequencies should also be quoted.

1.2.6.1 Percentage Bandwidth

When the bandwidth is expressed in terms of a percentage, it is defined by the relation

$$B_w = \left(\frac{\Delta f}{f}\right) \times 100 \tag{1.6}$$

where f is the frequency at the center of the band and Δf is the difference between the lowest and highest frequencies of operation. It is common to express the frequency characteristic of an antenna as its center frequency f and its total percentage bandwidth or as its center frequency f and half its total percentage bandwidth.

Example. An antenna can be used over the frequency range from 1 GHz to 2 GHz. Express this as a percentage bandwidth.

Solution. First we take the difference between the upper and lower frequencies, which is 1 GHz. The center frequency in this case is the average of 1 GHz and 2 GHz, that is, 1.5 GHz. Thus, Δf is 1 GHz and f is 1.5 GHz. Using (1.6) we get B_w = 66.7%.

The bandwidth of the antenna can be described as 66.7% at 1.5 GHz, or 1.5 GHz ± 33.3%, or 1.5 GHz ± 0.5 GHz.

1.2.6.2 Bandwidth as a Fraction or Multiple of an Octave

When the bandwidth is expressed in terms of a fraction or multiple of an octave, it is defined by the relation

$$B_w = \log_2\left(\frac{f_{\text{high}}}{f_{\text{low}}}\right) \tag{1.7}$$

where f_{high} is the highest frequency of operation and f_{low} is the lowest frequency of operation.

Many calculators do not have logs to the base 2. In these cases, we can still obtain logs to this base, by using the formula

$$\log_2\left\{\frac{f_{\text{high}}}{f_{\text{low}}}\right\} = \frac{\log_{10}\left\{\frac{f_{\text{high}}}{f_{\text{low}}}\right\}}{\log_{10}2} \tag{1.8}$$

Example. An antenna is quoted as having an operating frequency range from 500 MHz to 9 GHz. What is its frequency bandwidth in octaves?

Solution. The ratio of the upper frequency to its lower frequency is 18.

If we do not have a calculator that can give us logs to a base of 2 and we want to find the $\log_2 18$, we must divide $\log_{10} 18$ by $\log_{10} 2$. The result is 4.17 octaves to two decimal places.

1.2.7 Aperture Size

The aperture of an antenna governs the size of its beamwidth. In general, the larger the aperture, the narrower the beamwidth, and the higher is the gain at a given frequency.

The aperture size can be defined in two ways: either in terms of the actual physical size, in meters or feet, or in terms of wavelength. For instance, if we say that an antenna has an aperture of two wavelengths, then its actual physical size depends on its operating frequency. At a frequency of 1 GHz, the physical aperture is 60 cm; whereas at 10 GHz, its physical aperture is only 6 cm. It is more meaningful to refer to an antenna size in terms of its operating wavelength when the antenna is narrowband or single frequency because its beamwidth and gain are directly related to the aperture in terms of its operating wavelength. In this case we have to calculate its wavelength to find its physical dimensions. However, in the case of broadband antennas, its physical size is more appropriate because there are a range of operating frequencies.

1.2.8 Antenna Correction Factors

When an EMC engineer performs a site survey for instance, he/she wants to measure the maximum electric field strength or intensity at a particular location. However, the receiver reading is displayed in terms of volts or decibels (or more usually, microvolts, μV, or decibels referred to one microvolt, dBμV). The engineer must convert this value to the electric field intensity that is actually present at the position at which the antenna has been placed. In order to obtain this value the engineer has recourse to a set of tables or graphs showing the relationship between the receiver reading and the electric field intensity for that particular antenna, covering its operating frequency range. This correction factor is sometimes called the antenna factor.

The IEEE dictionary defines the antenna correction factor as a term or factor that is applied to the reading of the receiver so that the reading can be converted to the field strength in either volts per meter (electric field strength) or amperes per meter (magnetic field strength). This factor takes into account:

1. The effective height/length of the antenna;
2. The loss in the balun matching network between the antenna and the balun;
3. The loss due to the mismatch between the balun matching network and the transmission line that connects the balun to the receiver;

4. The loss due to the total length of the transmission line cables between the antenna and the receiver.

To be more accurate, the antenna factor should also take into account the proximity of the antenna to the ground and the frequency of measurement. The unit of antenna factor is m^{-1} in linear terms, and it is the ratio of the electric field strength at the antenna and the voltage reading of the receiver. Thus the antenna factor A_f is given by

$$A_f = \frac{E}{V} \tag{1.9}$$

However, we are usually given the antenna factor in decibels, and the gain of the antenna is also given in decibels. Hence, we must add the antenna factor to the gain of the antenna. Then we must convert this value A_{dB} in decibels to a linear value using the formula

$$A_f = 10^{(A_{dB}/20)} \tag{1.10}$$

This linear value A_f is the number by which we must multiply the receiver reading to obtain the true electric field at the antenna location.

In some cases, the antenna correction factor includes the gain of the antenna.

Example. The voltage measured at a receiver is 2 volts, and the antenna correction factor is 17 dB. What is the electric field strength at the antenna location? Assume that the gain at the antenna is 0 dBi.

Solution. Adding the gain of the antenna and the antenna correction factor gives us +17 dB. This means that if we were to ignore this factor, the electric field at the antenna location would be 17 dB less than the true value of the electric field at this point. So we must add this value to the reading of the receiver reading. We first convert the value in decibels to a linear value and then multiply the receiver reading by this linear value in order to calculate the true magnitude of the electric field at the position of the antenna.

Using (1.10), 17 dB converts to 7.08. We multiply 2 volts by 7.08 and get 14.16, which is the true value of the electric field at the antenna location in volts per meter.

Note that if the gain of the antenna is a high negative value that is greater than the antenna factor, we would get a negative value (-) by adding the gain of the antenna and the antenna factor. In this case we would have to divide the reading of the receiver by the linear equivalent of A_{dB}.

Refer to Section 7.3 for more details and another definition of antenna factor.

1.2.9 Polarization

Although this could apply equally to magnetic or electric polarization, it is used almost exclusively to describe the shape and orientation of the locus of the extremity of the electric field vector as it varies with time at a fixed point in space. This locus could be a straight line, an ellipse, or a circle.

1.2.9.1 Linear Polarization

In the case of linear polarization, the electric field varies sinusoidally in one plane, as depicted in Figure 1.11, and shows the variation with time at a fixed point in space. It is important to realize that a similar variation with distance also exists. Figure (1.11) shows the case for vertical polarization. Note that the extremity of the electric field vector at any fixed point in space is a straight line with a maximum value that is equal to twice the amplitude of the sine curve that depicts the variation of the electric field with time.

Horizontal polarization is shown in Figure 1.12. The electric field can also be polarized at any other angle between 0 and 90 deg to the horizontal. In general, however, the only other commonly used angle is 45 deg, which is known as slant polarization.

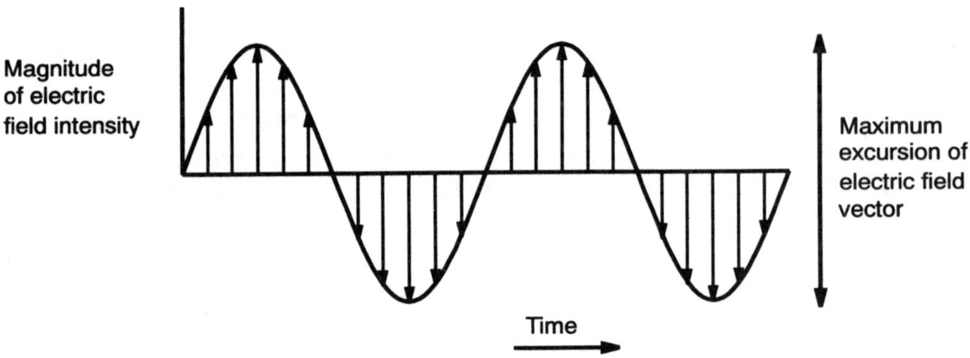

Figure 1.11 Variation of the electric field with time at a fixed point in space for vertical polarization.

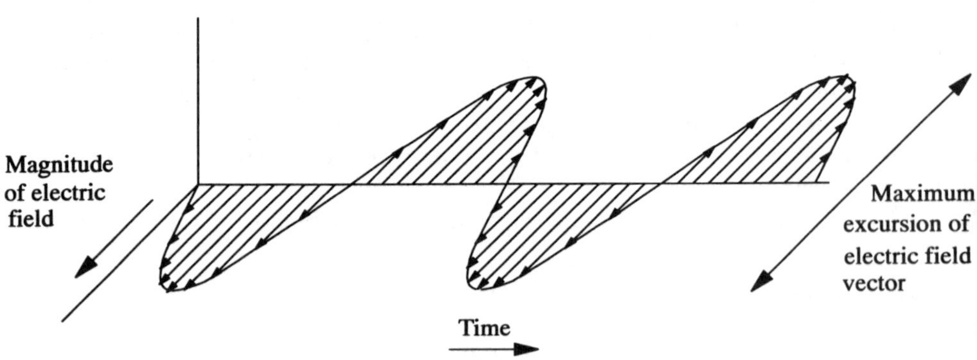

Figure 1.12 Variation of the electric field with time at a fixed point in space for horizontal polarization.

The polarization of a receiving antenna must match that of the incident radiation in order to detect the maximum field. If the angles are not the same, only the component that is parallel to the plane of incident polarization will be detected. If we have a vertically polarized antenna and the incident radiation is slant polarized, the magnitude of its component in the vertical plane will be reduced by a factor of cos(45 deg). Thus, only $1/\sqrt{2}$ or 0.707 of the electric field will be received. Since the radiated power is proportional to the square of the electric field, the power received will only be half (0.707^2) of the incident power.

1.2.9.2 Elliptical or Circular Polarization

Circular polarization is a special case of elliptical polarization, and thus elliptical and circular polarization are considered together in this section.

There are two definitions of axial ratio that are used by antenna engineers. The simpler one is discussed here. The axial ratio ξ is sometimes defined as the ratio of the major-to-minor axis. When this ratio is equal to one, that is, the major and minor axes are equal, we have circular polarization; whereas when it is infinite, it means that the minor axis is zero, that is, we have linear polarization.

It should be noted that the axial ratio ξ is not the same as the mathematical definition of the ellipticity of an ellipse. The mathematical ellipticity E_l varies between 0 and 1. The ellipticity E_l and axial ratio are related by the equation

$$E_l = 1 - \frac{1}{\xi} \qquad (1.11)$$

Circular polarization is a difficult concept to visualize. The electric field vector at each point in space sweeps out a circle through each period in time. Circular polarization can be generated by two electric vectors of equal magnitude that are orthogonal (at right angles) and also in phase quadrature, that is, in the time domain they are 90 deg out of phase with each other. Consider the two electric vectors depicted in Figure 1.13(a), showing the variation of the electric field in two directions y and x, that are perpendicular to each other. At a time denoted by point A, the electric field is at a maximum, say of value one in the x-direction and zero in the y-direction. Thus the resultant is one in the x-direction, as shown in Figure 1.13(c). One-eighth of a period later, the electric fields in both the x- and y-directions are 0.707, as depicted in Figure 1.13(b). The resultant is (using Pythagoras) $\sqrt{(0.707^2 + 0.707^2)}$, which is equal to one, but note that this is at an angle of 45 deg to the resultant at A, which is at 45 deg to both the y- and x-directions. At intermediate positions between A and B the resultants are still one, but they occur at angles varying between 0 deg and 45 deg to the resultant at A. At C, a further one-eighth of a period later, the electric field is zero in the x-direction but has the maximum magnitude of one in the y-direction. Thus the resultant is now one in the y-direction, which is

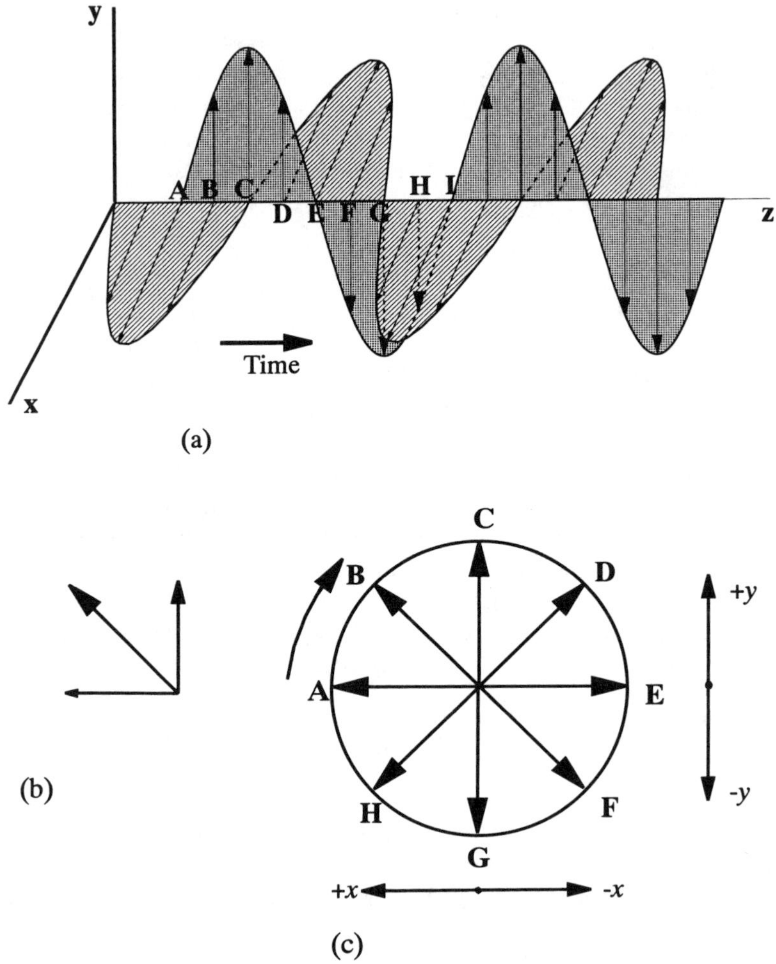

Figure 1.13 Circular polarization produced by two plane-polarized orthogonal waves in phase quadrature: (a) at right angles and in phase quadrature; (b) summation of vectors at time B; and (c) resultants of the two electric vectors over one period.

at right angles to the resultant vector at A. If we look at successive times, we can see that the resultant vector retains a value of one but presents successively increasing angles with the initial resultant vector at A until a whole period later at I when the resultant vector has a value and angle coincident with the resultant at A. We can thus see that in the time interval representing one period in time, the electric field has maintained its magnitude of one but has rotated through 360 deg, or one revolution.

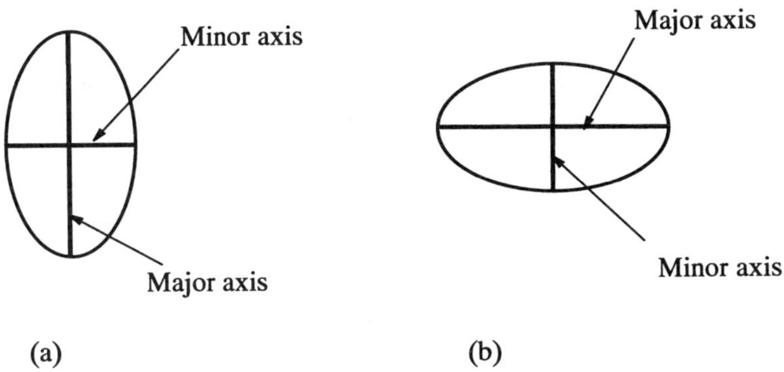

Figure 1.14 Two cases of elliptical polarization: (a) major axis vertical, and (b) major axis horizontal.

In the case of elliptical polarization, the electric field vectors may also be orthogonal (perpendicular to each other) and in phase quadrature, but they may not have the same magnitude. Elliptical polarization may also result if the orthogonal electric fields have the same magnitudes but a phase relationship that is not 90 deg or 270 deg. Other cases of elliptical polarization could be the result of orthogonal electric fields that are of unequal magnitudes and a difference in phase that could take any value between 0 deg and 360 deg. Figure 1.14 shows two examples of elliptical polarization. In Figure 1.14(a) the major axis is vertical, whereas in Figure 1.14(b) the major axis is horizontal. If the ratio of the major- to-minor axis in either of these cases is infinity, this would mean that the minor axis has a magnitude of zero and we would have linear polarization of vertical or horizontal orientation.

Sense of Polarization

In the example shown in Figure 1.13 the electric field vector rotates in a clockwise direction. We call this clockwise rotation of the electric field vector *right-hand circular polarization* (RHCP). If the resultant vector were to move in a counterclockwise direction, the wave would be a *left-hand circular polarized* (LHCP) wave. An LHCP wave would have resulted if either of the component electric field vectors had started half a period later. For instance, if the vertical electric field had commenced with the zero field at E, the resultant electric field would be horizontal in the negative x-direction (shown as E in Figure 1.13(c)). The vertical electric field, an eighth of a period later, would have been 0.707 in the positive y-direction, that is, upward. This would give a resultant of magnitude one at 45 deg between the negative x- and the positive y-directions (shown as D in Figure 1.13(c)). The resultant electric field has thus rotated in the counterclockwise direction.

1.2.10 Relationships Between the Key Parameters

The characteristics of an antenna are not always quoted by the manufacturers. In general though, we can usually deduce the unknown parameters if sufficient information is given. These formulas do not take into consideration losses in the receiving system and apply only to ideal conditions. The formulas also assume that the aperture of the antenna is very large in terms of the wavelength at the operating frequency.

1.2.10.1 Gain From Beamwidth

The gain of an antenna depends on the individual gains in the elevation and azimuth planes. The gains in these planes are inversely proportional to the half-power beamwidths. The narrower the beamwidth, the higher the gain. The rectangular aperture could be a horn or any other aperture. The formulas for both are given here for the sake of completeness. Although the EMC engineer will most probably only encounter horns in the general course of measurements, large aperture antennas are encountered in radio monitoring.

Rectangular Aperture With Uniform Illumination

For an uniformly illuminated aperture, the sidelobe level is 13 dB below the main beam. The linear gain for this type of rectangular aperture assuming 100% efficiency is given by [4, p. 14]

$$G = \frac{41300}{(\theta_1 \theta_2)} \tag{1.12}$$

where θ_1 and θ_2 are the half-power beamwidths in the two orthogonal planes.

Conical Log Spiral

The linear or numeric directivity, with respect to a circularly polarized isotropic source of a conical log spiral with a small cone angle can be calculated from the HPBWs in the orthogonal planes from the formula [5, p. 493]

$$D = \frac{32000}{(\theta_1 \phi_1)} \tag{1.13}$$

where θ_1 and ϕ_1 are the HPBWs in the orthogonal planes in degrees, and D is the linear directivity.

Rectangular Horn

In the case of a rectangular horn, the illumination across the aperture is not uniform and the efficiency is of the order of 60%. The variation is the same as that obtained for a rectangular waveguide. In the *H* plane, which is along the larger dimension *a*, the electric field has a half sinusoidal variation as shown in Figure 1.15 and the sidelobe level is 26 dB below the main beam if we assume that the phase is constant, as would be the case for a phase-compensated horn. The sidelobe level in the *E* plane or along the narrow dimension *b* is 13 dB. The linear gain of the horn on boresight is given by

$$G = \frac{31000}{(\theta_a \theta_b)} \tag{1.14}$$

where θ_E and θ_H are the half-power beamwidths in the *E* and *H* planes (i.e., along the *b* and *a* dimensions).

Circular Aperture With Uniform Illumination

For an antenna with a circular aperture and uniform illumination, the beamwidths are the same in all the planes and the sidelobe level is 18 dB below boresight. The linear gain *G* of a circular aperture antenna assuming 100% efficiency [3, p. 9] is given by

$$G = \frac{52000}{(\theta^2)} \tag{1.15}$$

where θ is the beamwidth in degrees in each of the planes.

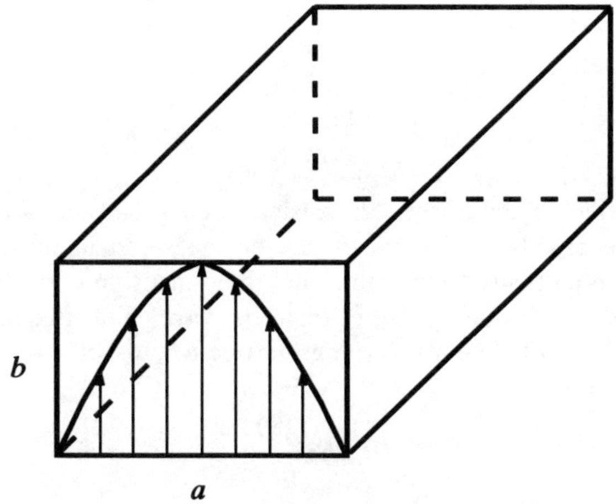

Figure 1.15 Electric field variation across a rectangular waveguide supporting the fundamental or dominant mode.

Circular Aperture With Nonuniform Illumination

If the circular aperture has nonuniform illumination such that there is a 10-dB taper across the aperture (i.e., the variation between the center and edge is 10 dB), then the sidelobe level is reduced to 26 dB below that of the main beam. In this case the efficiency of the antenna is also reduced to about 50% and the linear gain [3, p. 9] is given by

$$G = \frac{26000}{(\theta^2)} \tag{1.16}$$

1.2.10.2 Beamwidth From Aperture Size and Shape

The beamwidth of an antenna is different in each plane except for antennas with circular symmetry about boresight. In general, the larger the dimension in a particular plane, the narrower the beamwidth in that plane. Even in the case of circular symmetrical reflectors, the beamwidths are not necessarily the same in all planes if the illumination is different in each plane.

Beamwidth of a Rectangular Horn

For a rectangular horn, the broad dimension is known as the *H* plane and the electric field in this plane is not uniform. It has a variation, as shown in Figure 1.15, that produces a sidelobe level of 26 dB below that of the main beam if the aperture is phase compensated and, hence, the phase across it is constant. Because this dimension is larger than the *E* plane or narrow dimension, the beamwidth θ_H in this plane is smaller. The HPBW [3, p. 9] is given by

$$\theta_H = \frac{67\lambda}{a} \tag{1.17}$$

where λ is the wavelength and a is the dimension of the broadside of the waveguide. The wavelength and the broadside dimension can be measured in meters, centimeters, or inches as long as the units used for both λ and the dimension a are the same.

In most cases, the frequency rather than the wavelength is quoted. It is therefore more convenient to use a formula involving the frequency such as

$$\theta_H = \frac{20100}{aF} \tag{1.18}$$

where F is the frequency in megahertz and, in this case, the dimension a must be in meters.

For the E plane (the narrow dimension b of the waveguide) the sidelobes are only 13 dB below that of the main beam and the beamwidth is wider than that in the H plane. The equivalent formula for the beamwidth is given by [3, p. 9]

$$\theta_E = \frac{56\lambda}{b} \tag{1.19}$$

where λ is the wavelength and b is the narrow dimension of the waveguide. The wavelength and the broadside dimension can be measured in meters, centimeters, or inches as long as the units used for both λ and the dimension b are the same.

In most cases, the frequency rather than the wavelength is quoted. It is therefore more convenient to use a formula involving the frequency such as

$$\theta_E = \frac{16800}{bF} \tag{1.20}$$

where F is the frequency in megahertz and b is in meters.

Example. A rectangular waveguide horn operating at 10 GHz has an aperture of 36 cm by 20 cm and is phase compensated so that the phase is constant across the aperture. Calculate the beamwidths in each of its principal planes.

Solution. The principal planes referred to in this case are the E and H planes. The dimensions a and b will be 0.36m and 0.2m, respectively, and the frequency is 10000 MHz. Using (1.18) we get the H plane beamwidth as

$$\theta_H = \frac{20100}{0.36 \times 10000}$$

which gives a beamwidth of 5.6 deg in the H plane.

Similarly using (1.20) gives an E plane beamwidth of 8.4 deg. Note that these values are for ideal antennas and do not allow for efficiencies of less than 60%. In practice, the efficiency would be less than 60% and the measurement would introduce further errors, so the actual beamwidths obtained could be about 1.5 times the calculated values.

1.2.10.3 Gain From Aperture Size and Shape

The gain of an antenna depends on the shape and size of the aperture as well as the electric field distribution across the radiating aperture. The electric field distribution is commonly referred to as aperture illumination.

Rectangular Horn

The maximum linear gain is on boresight and is where the *E* and *H* planes intersect. For a rectangular horn the illumination across the aperture is not uniform, that is, neither the phase nor the magnitude is constant. However, the phase can be compensated by the use of a dielectric lens, for instance, so that it is constant. The efficiency of the horn is reduced to about 60% [3, p. 9], but this nonuniform illumination produces lower sidelobe levels that are typically of the order of 26 dB below the main beam. In addition, it should be noted that the broadband ridged waveguide horns, in common usage, will not have as high a gain as a normal smooth-walled rectangular pyramid horn. The ridged waveguide horn sacrifices the higher gain for its wider bandwidth characteristic.

For a horn of dimensions *a* and *b*, the linear gain *G* is given by [3, p. 9]

$$G = \frac{8ab}{\lambda^2} \tag{1.21}$$

where λ is the wavelength.

Note that the units used for dimensions *a* and *b* can be expressed in any form, that is, metric or inches, as long as the same units are also used for the wavelength.

In most cases, the frequency rather than the wavelength is quoted. It is therefore more convenient to use a formula involving the frequency. Moreover, instead of calculating the gain in terms of linear gain, it could be calculated in dBi (decibels referred to isotropic) as

$$G_{\text{dBi}} = 10 \log_{10} \left\{ \frac{abF^2}{11250} \right\} \tag{1.22}$$

where *a* and *b* are the dimensions of the aperture of the waveguide in meters and *F* is the frequency in megahertz.

If the dimensions of the waveguide are given in centimeters and the frequency in megahertz, the gain in decibels referred to isotropic can be calculated from the formula

$$G_{\text{dBi}} = 10 \log_{10} \left\{ \frac{abF^2}{1.125} \right\} \tag{1.23}$$

Rectangular Aperture With Uniform Illumination

For a hypothetical uniformly illuminated aperture, the efficiency can approach 100 %, but the sidelobe level is also high; typically of the order of 18 dB below the level of the main beam.

The linear gain G for a uniformly illuminated aperture with 100% efficiency and dimensions a and b is given by [3, p. 9]

$$G = \frac{16ab}{\lambda^2} \tag{1.24}$$

Note that the units used for the dimensions a and b can be expressed in any units as long as the same units are also used for the wavelength.

In most cases, the frequency rather than the wavelength is quoted. It is therefore more convenient to use a formula that involves the frequency. Instead of calculating the gain in terms of linear gain, we could also calculate it in decibels referred to an isotropic as

$$G_{\text{dBi}} = 10 \log_{10} \left\{ \frac{16abF^2}{90000} \right\} \tag{1.25}$$

where a and b are the dimensions of the aperture of the waveguide in meters and F is the frequency in megahertz.

If the dimensions of the waveguide are given in centimeters and the frequency in megahertz, the gain in decibels referred to an isotropic can be calculated from the formula

$$G_{\text{dBi}} = 10 \log_{10} \left\{ \frac{16abF^2}{9} \right\} \tag{1.26}$$

Circular Aperture

The circular aperture could be a parabola or any other aperture, such as a circular horn. The formulas for both are given here for the sake of completeness, although the EMC engineer will most probably only encounter parabolas in the general course of measurements.

Circular Aperture With Uniform Illumination

For a hypothetical aperture with uniform illumination, the efficiency may be as high as 100%, but the sidelobe level is of the order of 18 dB below the level of the main beam [3, p. 9].

The linear gain G for a uniformly illuminated aperture with 100% efficiency and diameter a is given by

$$G = \frac{10a^2}{\lambda^2} \tag{1.27}$$

Note that the units used for the dimension a can be expressed in any form, metric or imperial, as long as the same unit is also used for the wavelength.

In most cases, the frequency rather than the wavelength is quoted. It is therefore more convenient to use a formula that involves the frequency. Moreover, instead of calculating the gain in terms of linear gain, it could be computed in decibels referred to isotropic as

$$G_{\text{dBi}} = 10 \log_{10} \left\{ \frac{(aF)^2}{9000} \right\} \tag{1.28}$$

where a is the diameter of the aperture in meters and F is the frequency in megahertz.

If the diameter of the aperture is given in centimeters and the frequency in megahertz, the gain in decibels referred to an isotropic can be calculated from the formula

$$G_{\text{dBi}} = 10 \log_{10} \left\{ 10 \frac{(aF)^2}{9} \right\} \tag{1.29}$$

Circular Aperture With Nonuniform Illumination

A circular aperture that has an illumination taper can be used to give a reduced sidelobe level. In this case the illumination at the center is higher than that at the edges. For instance, if we say that the taper is 10 dB, we mean that the electric field at the edges of the parabola is 10 dB ($10^{0.5}$ or 3.16 times) lower than that at the center. The sidelobe level is reduced to 26 dB below that of the main beam, but the efficiency is also reduced to the order of 50%.

The linear gain G for an uniformly illuminated aperture with 50% efficiency and diameter a is given by

$$G = \frac{5a^2}{\lambda^2} \tag{1.30}$$

Note that the units used for the dimension a can be expressed in any form, metric or imperial, as long as the same unit is also used for the wavelength.

In most cases, the frequency rather than the wavelength is quoted. It is therefore more convenient to use a formula that involves the frequency. In addition, instead of calculating the gain in terms of linear gain, it could be calculated in decibels referred to isotropic as

$$G_{\mathrm{dBi}} = 10\ \log_{10}\left\{\frac{(aF)^2}{18000}\right\} \tag{1.31}$$

where a is the diameter of the aperture in meters and F is the frequency in megahertz.

If the diameter of the aperture is given in centimeters and the frequency in megahertz, the gain in decibels referred to an isotropic can be calculated from the formula

$$G_{\mathrm{dBi}} = 10\ \log_{10}\left\{\frac{(aF)^2}{1.8}\right\} \tag{1.32}$$

REFERENCES

[1] Rudge A. W., K. Milne, A. D. Olver, and P. Knight, *The Handbook of Antenna Design, Vol.* 1, Stevenage, UK and New York: Peter Peregrinus Ltd., 1982.

[2] Silver, S., *Microwave Antenna Theory and Design*, MIT Radiation Lab. Series, New York: McGraw-Hill, 1949.

[3] Hill, J. E., *Antenna Designer's Guide*, Watkins-Johnson Company Tech-notes, Vol. 3, No. 4, Palo Alto, CA, July/August 1976.

[4] BS 6657 British Standard Guide to Prevention of Inadvertent Initiation of Electroexplosive Devices by Radio Frequency Radiation, 1986.

[5] Dyson, J. D., ''The Characteristics and Design of the Conical Log Spiral,'' *IEEE Trans. on Antennas and Propagation*, Vol. AP-13, July 1965, pp. 488–499.

Chapter 2
Basic Math for EMC Engineers

This chapter gives an overview of the mathematics required by engineers in the EMC field. It is not intended as a replacement for a mathematics textbook but as a revision course, with emphasis on the application of mathematical principles to the theoretical aspects of the book. The topics are also explained in a qualitative manner so that the concepts may be understood without recourse to rigorous mathematical treatments. If the reader intends to pursue EM theory, it is strongly recommended that this chapter be studied before proceeding to Chapter 3.

2.1 ANGLES

The angle between two lines is defined as the distance that one line must be rotated in order to be superimposed onto the other line. Angles can be measured in degrees or radians. In navigation and geography, angles are measured exclusively in degrees; whereas in mathematics and science/engineering, both degrees and radians are used. Radians can be converted to degrees and vice versa according to the equalities

$$2\pi \text{ radians} = 360 \text{ deg}$$

$$1 \text{ radian} = 180/\pi \text{ deg}$$

Since π is approximately equal to 3.142,

$$1 \text{ radian} = 57.3 \text{ deg}$$

2.1.1 Convention for Angles

There are two main conventions used for angles, namely, those used in (1) navigation and (2) mathematics. It is important to appreciate the difference between the two conventions

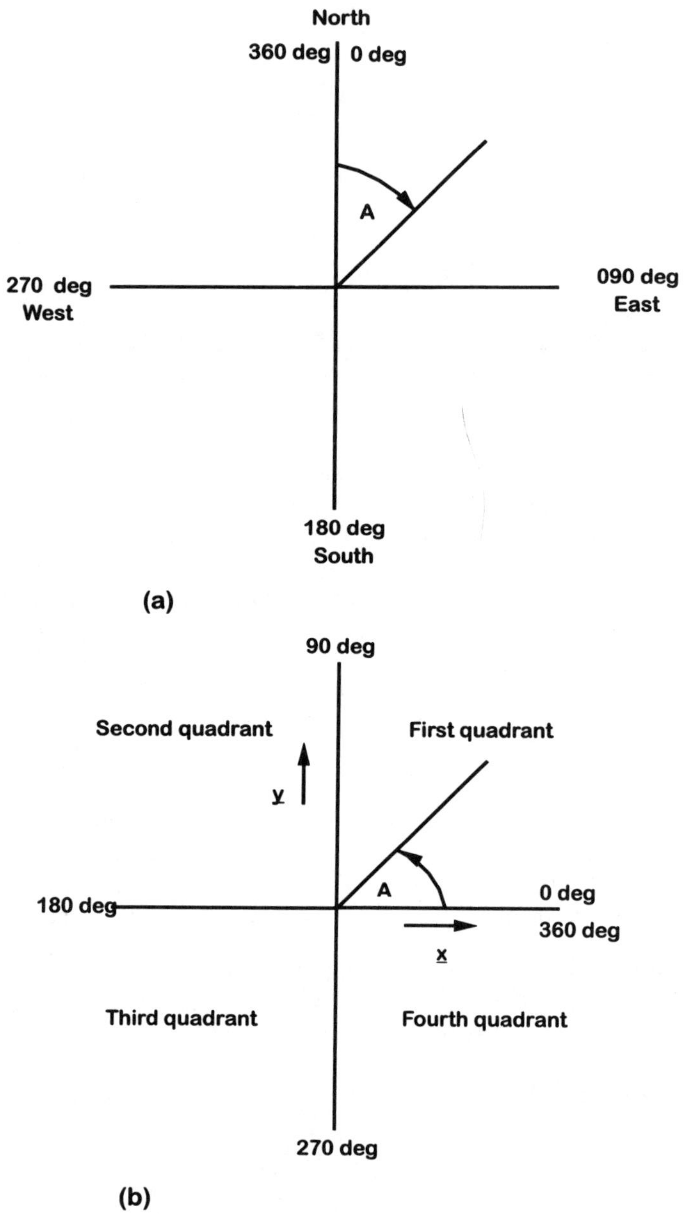

Figure 2.1 Convention for angles used for (a) navigation (bearings), and (b) mathematics.

since the mathematical convention is used for radiation patterns but the EMC engineer may encounter the navigational convention in the military field when dealing with emissions or susceptibility. To convert from mathematical angles D_m to navigational angles D_n, we use

$$D_n = 90 \text{ deg} - D_m \quad \text{for } 0 \text{ deg} < D_m < 90 \text{ deg} \tag{2.1}$$

(i.e., for mathematical angles between 0 deg and 90 deg) and

$$D_n = 450 \text{ deg} - D_m \quad \text{for } 90 \text{ deg} < D_m < 360 \text{ deg} \tag{2.2}$$

(i.e., for mathematical angles between 90 deg and 360 deg).

2.1.1.1 *Convention for Navigation*

In navigation, the angles are called bearings and are referenced to the Geographic North. The bearings are measured as positive angles (using three digits) in a clockwise direction from the Geographic North and vary from 0 deg to 360 deg. Thus, North is 000 deg or 360 deg, East is 090 deg, South is 180 deg, and West is 270 deg, as shown in Figure 2.1(a).

2.1.1.2 *Convention for Mathematics*

In mathematics, the angles are referenced to the positive x-axis, so when the angle A, shown in Figure 2.1(b), is 0 deg or 360 deg the direction is parallel to the x-axis; angle A = 90 deg is along the y-axis, A = 180 deg is along the negative x-axis, and A = 270 deg is along the negative y-axis. Mathematicians also use angles greater than 360 deg, so, for example, A = 450 deg is the same as 90 deg (along the y-axis).The angles are divided into four sectors called quadrants. The first quadrant contains angles between 0 deg and 90 deg, the second quadrant contains angles between 90 deg and 180 deg, the third quadrant contains angles between 180 deg and 270 deg, and the fourth quadrant contains angles between 270 deg and 360 deg. These are shown in Figure 2.1(b).

2.2 BASIC TRIGONOMETRY

The basic trigonometric functions are defined by the projections of lines onto the x- and y-axes, as shown in Figure 2.2(a). They can also be represented by the sides of a right-angled triangle, like the triangle depicted in Figure 2.2(b).

The variation of the trigonometric functions are plotted in Figure 2.3. We can see that sin A is positive in the first and second quadrants, cos A is positive in the first and last quadrants, and tan A is positive in the first and third quadrants. Summarizing, we can say

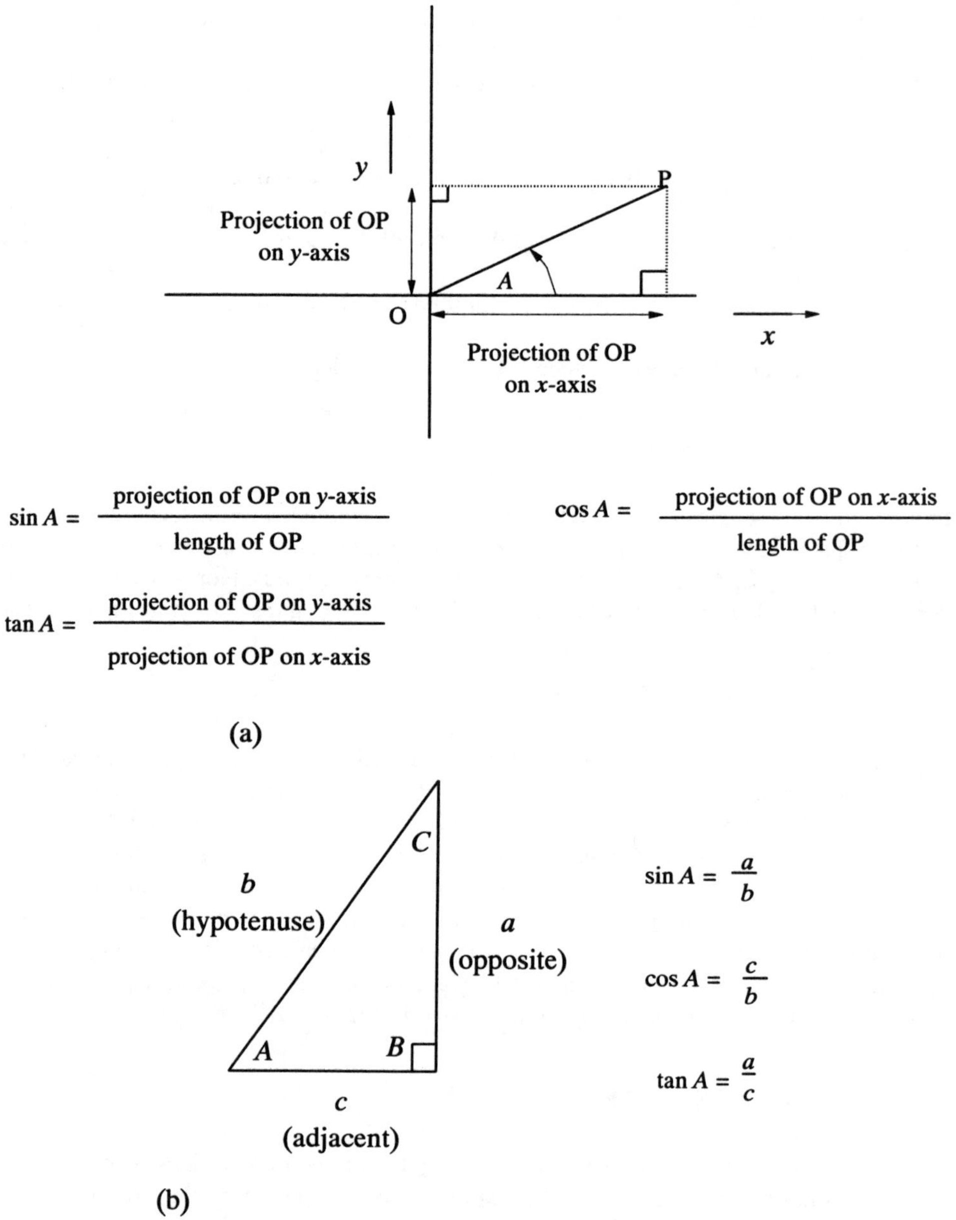

Figure 2.2 Basic trigonometric functions of an angle: (a) as projections; (b) in a right-angled triangle.

that all the functions are positive in the first quadrant, sin A is positive in the second quadrant, tan A is positive in the third quadrant, and cos A is positive in the last quadrant, as shown in Figure 2.3(d). The positive values of the trigonometric functions can be recalled by the mnemonic ''All Silly Tom Cats'' reading counterclockwise from zero in the direction of increasing angles or by the acronym ACTS reading clockwise from the first quadrant.

2.2.1 Reciprocal Trigonometric Functions

The reciprocal trigonometric functions are secant (sec), cosecant (cosec), and cotangent (cot) and are, respectively, defined as

$$\sec A = \frac{1}{\cos A} \tag{2.3}$$

$$\operatorname{cosec} A = \frac{1}{\sin A} \tag{2.4}$$

$$\cot A = \frac{1}{\tan A} \tag{2.5}$$

2.2.2 Inverse Trigonometric Functions

Inverse trigonometric functions are not the same as the reciprocal trigonometric functions and refer to the angles whose trigonometric functions are the stated values. On calculators and in program code they are often denoted by $\sin^{-1}$, asin, or arcsin in the case of sine functions, with similar annotation for the other functions. For example, if we know that the sine of an angle (sin A) is 0.5 and we want to calculate the angle, we would key in the number 0.5 on the calculator and then press the $\sin^{-1}$ or the asin key to get the result of 30 deg (assuming that the calculator is in degree mode).

2.2.2.1 Common Trigonometric Identities

Common trigonometric identities include

$$\sin^2 A + \cos^2 A = 1 \tag{2.6}$$

$$\sec^2 A = 1 + \tan^2 A \tag{2.7}$$

$$\operatorname{cosec}^2 A = 1 + \cot^2 A \tag{2.8}$$

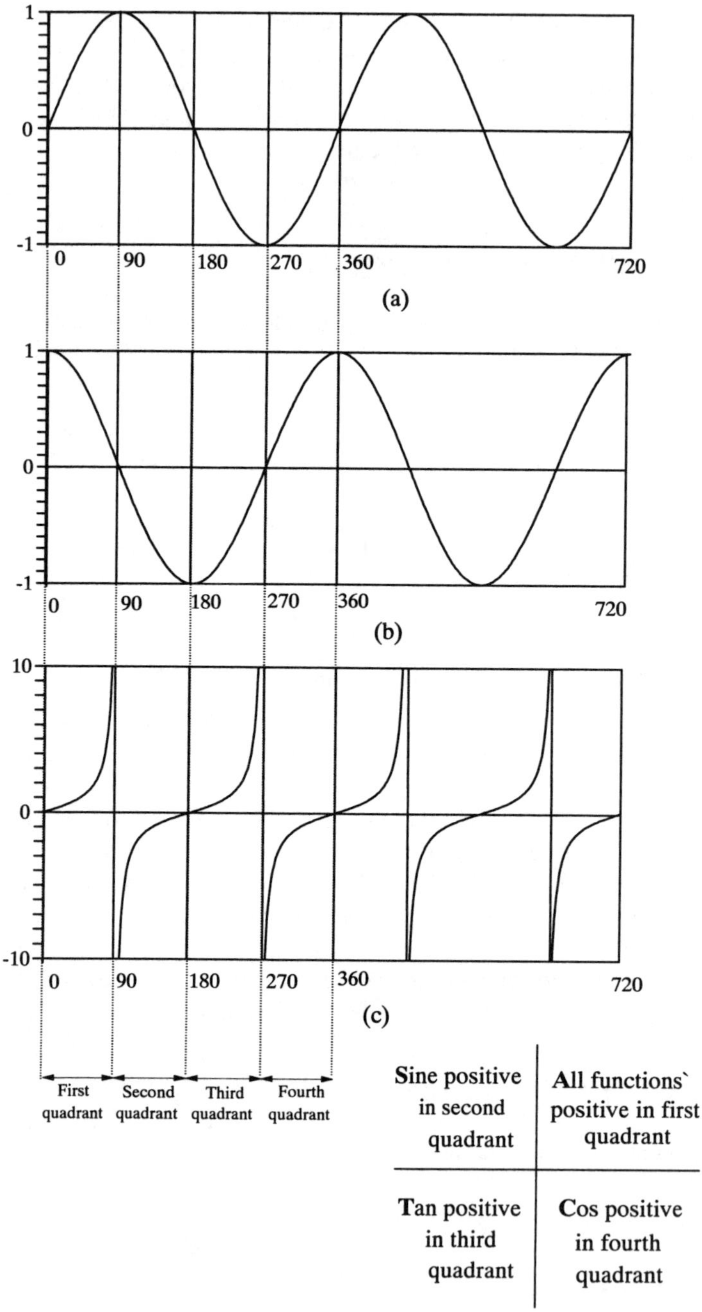

Figure 2.3 Variation of trigonometric functions with angle: (a) sine; (b) cosine; and (c) tangent.

Note that the powers of trigonometric functions are written, for example, as $\cos^2 A$ and $\cos^3 A$ rather than as $(\cos A)^2$ or $(\cos A)^3$.

2.2.2.2 Cosine Formula

The cosine formula is useful in calculating the angle in the case of any acute-angled triangle as shown in Figure 2.4(a) if the sides are known. Alternatively, the side opposite

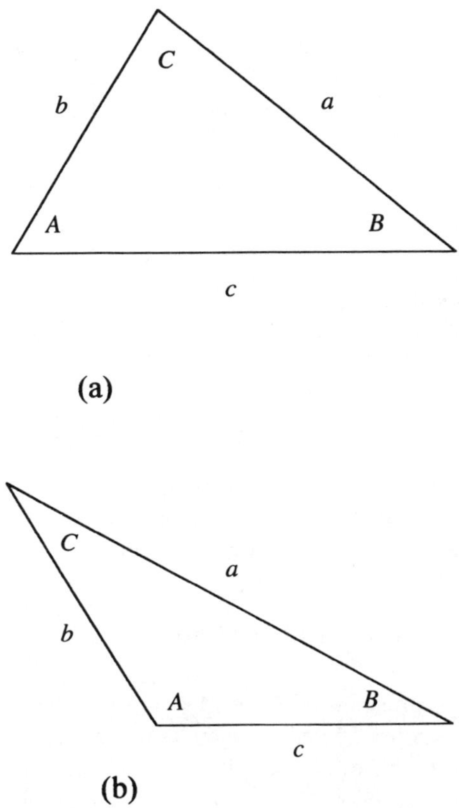

Sine formula

$$\frac{a}{\sin A} = \frac{b}{\sin B} = \frac{c}{\sin C}$$

Cosine formula

$$a^2 = b^2 + c^2 - 2bc \cos A$$

$$b^2 = a^2 + c^2 - 2ac \cos B$$

$$c^2 = a^2 + b^2 - 2ab \cos C$$

Figure 2.4 Sine and cosine formulas for acute-angled and obtuse-angled triangles.

a known angle can be calculated if the other two sides are known. The cosine formula is given by

$$a^2 = b^2 + c^2 - 2bc(\cos A) \tag{2.9}$$

Similar formulas for calculating b or c can be used as shown in Figure 2.4. The sides a and b may be inclined at an angle A, which is greater than 90 deg (in the case of an obtuse-angled triangle as shown in Figure 2.4(b)), but formula (2.9) still applies. In the case of angles between 90 deg and 180 deg, the cosine is negative, so the last term in (2.9) becomes positive.The cosine formula is useful in calculating the magnitude of vectors that are inclined at an angle of A. The lengths of the sides b and c are chosen to be the relative lengths of two vectors inclined at an angle A, and the resultant is represented by the length of the third side a.

2.2.2.3 The Sine Formula

The sine formula is given by

$$\frac{a}{\sin A} = \frac{b}{\sin B} = \frac{c}{\sin C} \tag{2.10}$$

The sides a, b, and c are the sides opposite angles A, B, and C as shown in Figure 2.4. This formula is useful when calculating the magnitudes of vectors and/or angles between vectors that are represented by the sides of any triangle.

2.3 POWERS, INDICES, AND LOGARITHMS

Powers, indices, and logarithms are all ways of expressing the same concept. When we have a number $A = x^3$ we say that x is raised to the power of 3. We could also say the index of the base x is 3 or that the logarithm of A to the base x is 3, which is written as $\log_x A = 3$. We use the standard convention of omitting the positive sign if the power is positive. If the power is negative, then the number is equal to the reciprocal of the positive power; that is, if $A = x^{-3}$, then we can say that it is also equal to $1/x^3$. We can take the logarithm of a number to any base. If we take it to a base of 10, it is called a common logarithm and is denoted by the roman expression $\log_{10}$, and when no base has been stated it is usually assumed to be a base of 10. It is also convenient to use logarithms to a base of e (where $e = 2.718$), and these are called natural or Naperian logarithms and are denoted by the roman expression ln. Logarithms are used because of the ease with which numbers can be multiplied and divided once they are converted into logarithms. The rules that apply to operations of multiplying, dividing, and squaring of indices are described in the following sections.

2.3.1 Multiplication of Numbers

Consider two numbers $A = x^2$ and $B = x^3$. If we rewrite them in their full forms ($A = x$ multiplied by itself twice and $B = x$ multiplied by itself three times) and then take their product, we get x multiplied by itself five times, that is, x^5. Note that we could get this result by adding the indices 2 and 3. The general rule for the multiplication of two of more numbers (that can be expressed as powers to the same base) is to sum their indices or powers, as is shown by

$$x^2 \times x^3 \times x^4 = x^{(2+3+4)} = x^9 \tag{2.11}$$

In the case of logs, when we multiply numbers, we add the logarithms; for example,

$$\log_x(AB) = \log_x A + \log_x B \tag{2.12}$$

The rule applies when the indices are positive or negative integers as well as fractions, as illustrated by

$$x^2 \times x^{-3.5} \times x^4 = x^{(2-3.5+4)} = x^{2.5} \tag{2.13}$$

Note that since negative indices can be written as the reciprocals of positive indices, (2.13) can also be rewritten as

$$x^2 \times \frac{1}{x^{3.5}} \times x^4 = x^{(2-3.5+4)} \tag{2.14}$$

Note that the index of the denominator is subtracted from the indices of the numerator. In the case of logs, the following rule applies

$$\log_x(A/B) = \log_x A - \log_x B \tag{2.15}$$

When we raise numbers to a power, we multiply the indices. This is demonstrated by the rule

$$(x^3)^4 = x^{3\times4} = x^{12} \tag{2.16}$$

In the case of logarithms we can say that the following rule applies

$$\log_x(A^2) = 2\log_x A \tag{2.17}$$

Logs were used to multiply and divide large numbers in the days before the availability of hand-held electronic calculators. Tables of logs and antilogs to the base ten (and base *e*) were available. Let us consider the two numbers 30 and 50. If we want to find their product we can multiply them in the normal way or, alternatively, put them in logarithmic form, add the logs together, and then find the antilog of their sum. The log of 30 is 1.477, and the log of 50 is 1.699, which gives the sum of 3.176. Taking the antilog of 3.176 we get the result 1,500. However, in engineering we use the *decibel* (dB) to express the relative magnitudes of two quantities such as transmitted and received powers. The ratios could be several orders of magnitude (10^6 or more), and decibels are a convenient form in which to write these large ratios. The decibel is defined as ten times the log of a number. Thus, 30 expressed in decibels is 14.77. Note that since $10 = 10^1$, the log of 10 is one but 10 expressed in decibels is 10. If we have a power P_1 that is ten times larger than another power P_2, then the ratio of P_1 to P_2 is 10 and in decibels we say that the ratio of P_1 to P_2 is +10 dB. However, if we consider the ratio of P_2 to P_1, we get 0.1 or 10^{-1}, which is -10 dB. Summarizing we can state that

$$10 \log \left(\frac{P_1}{P_2}\right) = -10 \log \left(\frac{P_2}{P_1}\right) \tag{2.18}$$

It is common practice to relate a power to 1 mW. In this case the power P_1 is divided by 1 mW and then expressed in *decibels referred to one milliwatt* (dBmW), which is usually shortened to dBm). For instance, if P_1 is 10W, the ratio of P_1 to 1 mW is 10,000 or 10^4; expressed in decibels, it would be +40. However, if we were to be told that a power of +40 dB existed we would not be able to state the absolute value of P_1 since all we could deduce would be that the relative power was 10,000 greater than another power level. To ascertain the actual value of P_1 we would have to specify that it has a value of +40 dBm, which this tells us that it is 10^4 greater than 1 mW, that is, that it is 10W.

If we had a level of -40 dBm we know that this power is 10^{-4} or 10^4 less than 1 mW and, therefore, is 0.1 μW. We will always get a negative number of decibels when we take the log of a number less than one. Of course, decibels are not just multiples of ten, they could be any number. It is useful to remember that the number in decibels is ten times the power to which 10 is raised. Thus, +3 dBm is $10^{+0.3}$ mW, which is 1.98 times greater than 1 mW, that is, 2 mW approximately. Thus, for every +3 dB the power doubles. Similarly for every -3 dB the power is halved. Approximate conversions between decibels referred to one milliwatt (dBm) and milliwatts (mW) are shown in Table 2.1. Accurate conversion tables are given in Appendix F.

2.3.2 Relationship Between dBm and dBμV

There is often confusion between decibels expressed as a power ratio (e.g., dBM) and decibels expressed as a voltage ratio (e.g., dBμV). The decibel was first introduced as 10

Table 2.1
Approximate Relationship Between dBm and mW

dBm	−15	−12	−10	−9	−6	−3	0	+3	+6	+9	+10
mW	0.03	0.06	0.1	0.13	0.25	0.5	1	2	4	8	10

times the log of a power ratio, so when it is used for voltages we must remember that the power P is proportional to the square of the voltage V ($P = V^2/R$) Thus the following relationship applies

$$10 \log \left(\frac{P_1}{P_2}\right) = 10 \log \left(\frac{V_1}{V_2}\right)^2 \tag{2.19}$$

Using the rule of (2.16) we can see that raising the ratio of voltages to a power of 2 results in the index being twice that of the ratio of the powers. Thus, the value in decibels will be 20 times the index of 10 for the voltage ratios but only 10 times the index of 10 for power ratios. In the case of voltage ratios, we can therefore say that for every +20 dB the voltage ratio increases by a factor of 10 and for every +6 dB the voltage ratio doubles. The level of voltages encountered in EMC are relatively low, so the voltages are usually related to 1 μV. Thus, in decibels they are expressed in dBμV. Table 2.2 gives an approximate conversion between decibels expressed as a voltage ratio (dBμV) and microvolts (μV). Accurate conversion tables are given in Appendix F.

Table 2.2
Approximate Relationship Between dBμV and μV

dBmV	−30	−24	−20	−18	−12	−6	0	+6	+12	+18	+20
μV	0.03	0.06	0.1	0.13	0.25	0.5	1	2	4	8	10

2.4 REAL AND COMPLEX NUMBERS

Consider a quadratic (an equation with a squared term or a product of two variables such as x^2 or xy) of the form

$$y = x^2 - 4 \tag{2.20}$$

We can plot the quadratic and obtain the curve shown in Figure 2.5(a). The solution(s) occurs when the curve crosses the x-axis (i.e., when $y = 0$). We can see that we have the two solutions or roots at $x = +2, -2$.

(a)

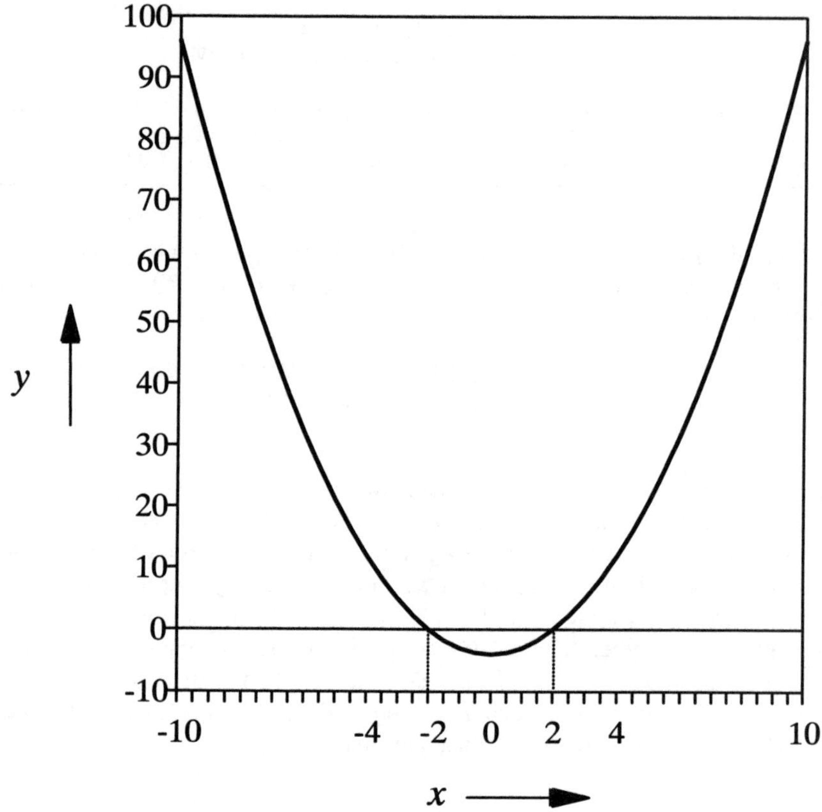

Figure 2.5 Graphs of quadratic functions: (a) $y = x^2 - 4$, showing real roots; and (b) $y = x^2 + 4$, showning no real roots.

We could also solve it analytically by factorization when $y = 0$ as

$$x^2 - 4 = 0$$

$$(x - 2)(x + 2) = 0$$

$$x = +2 \text{ or } -2$$

In either solution, we get real roots. Now let us consider the quadratic

$$y = x^2 + 4 \tag{2.21}$$

When we plot the quadratic, we can see from Figure 2.5(b), that it does not cross the

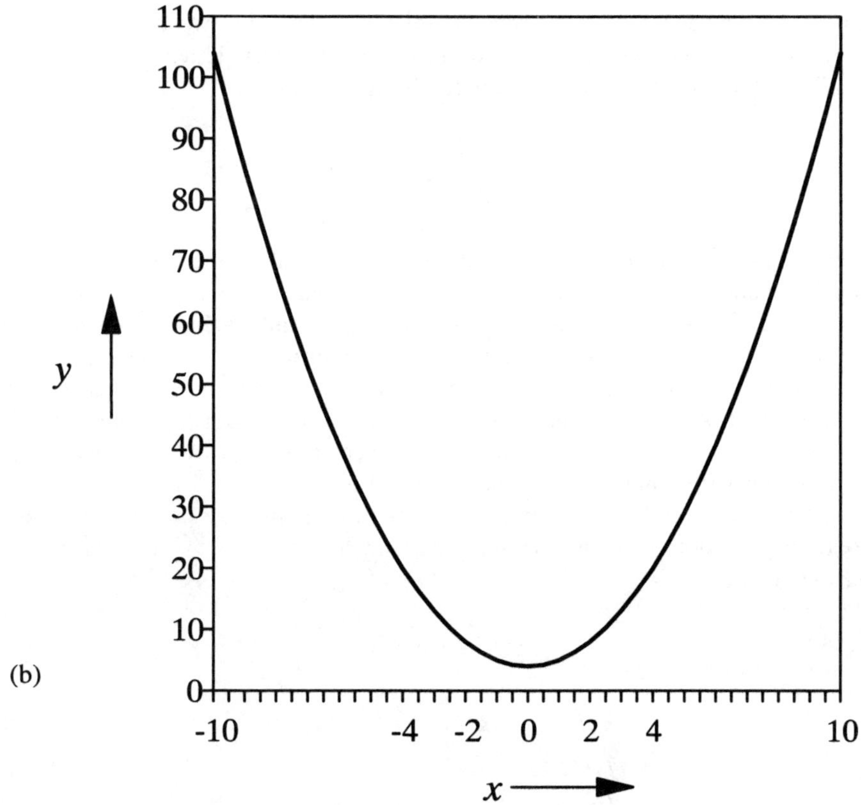

Figure 2.5. (continued)

x-axis. Thus, it does not have any real roots. We cannot factorize it, but we can solve it algebraically by using the formula for obtaining the roots of a quadratic expressed in the form $ax^2 + bx + c$. The roots of such an equation are given by

$$x = \frac{-b \pm \sqrt{b^2 - 4ac}}{2a} \tag{2.22}$$

where a is the coefficient of the x^2 term, b is the coefficient of the x term, and c is the constant (number). In (2.21) a is equal to 1, b is equal to zero, and c is 4. Inserting these values into (2.22) we get

$$x = \frac{0 \pm \sqrt{0 - 16}}{2} = \frac{\pm\sqrt{-16}}{2} \tag{2.23}$$

We cannot take the square root of a negative quantity (since the product of two negative or two positive numbers would always be positive); however, if we write -16 as the product of -1 and $+16$, then the solution can be written as

$$x = \frac{\pm 4\sqrt{-1}}{2} = \pm 2\sqrt{-1}$$

The square root of -1 is denoted by the letter *i* or *j*. In general, mathematicians tend to use *i* and physicists and engineers tend to use *j* (*j* is also called an operator sometimes). Thus, the roots of (2.21) are $+2j$ and $-2j$ and they are imaginary numbers. In order to depict these roots we define a complex plane that represents the real numbers on the *x*-axis and the imaginary numbers on the *y*-axis. This type of diagram is sometimes called an Argand diagram (after J. R. Argand) and is shown in Figure 2.6. In the quadratic equation (2.21) the roots do not have any real parts and, thus, appear on the *y*- (or imaginary) axis of Figure 2.6 where *x* (the magnitude of the real part) is zero.

In general, the solution of quadratics results in roots that are real or complex. The complex roots have both real and imaginary parts of the form $x + jy$. For example, if we consider the equation

$$y = x^2 + 4x + 5 \tag{2.24}$$

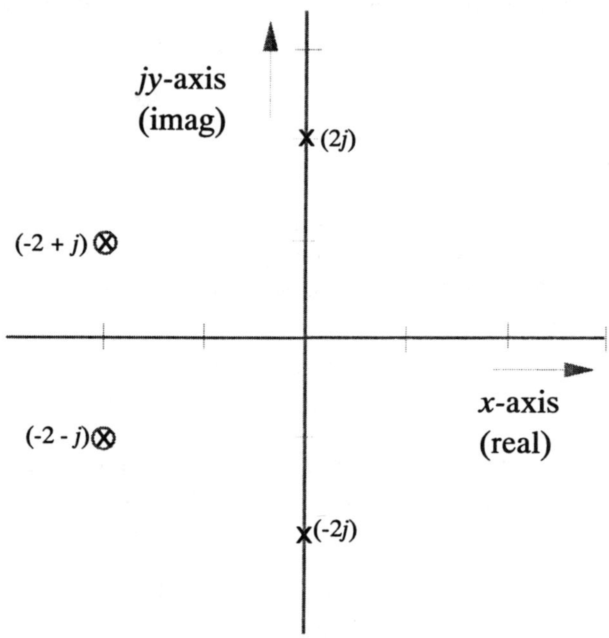

Figure 2.6 Argand diagram.

the roots occur when $y = 0$; and using (2.22) we get the following values for x

$$x = \frac{-4 \pm \sqrt{16 - 20}}{2}$$

$$x = -2 + j \text{ or } -2 - j$$

These are shown in the Argand diagram (Figure 2.6) by encircled crosses. In the case of higher order equations (such as a cubic of the form $y = ax^3 + bx^2 + x + c$), the roots could be a mixture of real and complex roots.

We can also represent complex numbers in polar form by a position vector (see Section 2.5 on scalars and vectors) of length $\boldsymbol{R}$, which makes an angle ϕ with the positive direction of the x-axis as depicted in Figure 2.7. The vector is uniquely described by the quantity $Re^{j\phi}$. The x coordinate (or abscissa) is given by $R \cos \phi$, and the y coordinate (or ordinate) on the imaginary axis is given by $jR \sin \phi$. The length of $\boldsymbol{R}$ represents the magnitude of the vector and is also called the modulus or absolute value. The modulus value is conventionally denoted by vertical lines on either side of $\boldsymbol{R}$ as $|\boldsymbol{R}|$. The modulus can be calculated by applying Pythagoras's theorem to give

$$|\boldsymbol{R}| = \sqrt{x^2 + y^2} \tag{2.25}$$

The angle ϕ that the vector $\boldsymbol{R}$ makes with the x-axis is called the argument and can be calculated by simple trigonometry since

$$\tan \phi = \frac{y}{x} \tag{2.26}$$

so ϕ is given by

$$\phi = \tan^{-1}\left(\frac{y}{x}\right) \tag{2.27}$$

Note that when $\phi = \pi/2$ or 90 deg there is no real part, only an imaginary part, and $\boldsymbol{R}$ is vertical or parallel to the y-axis. We can see this algebraically as well since according to the identity of L. Euler (1707–1783)

$$e^{\pm j\theta} = \cos \theta \pm j \sin \theta \tag{2.28}$$

$$Re^{j\pi/2} = R \cos \pi/2 + jR \sin \pi/2 \tag{2.29}$$

and $\cos \pi/2 = 0$ whereas $\sin \pi/2 = 1$; so

$$Re^{j\pi/2} = jR$$

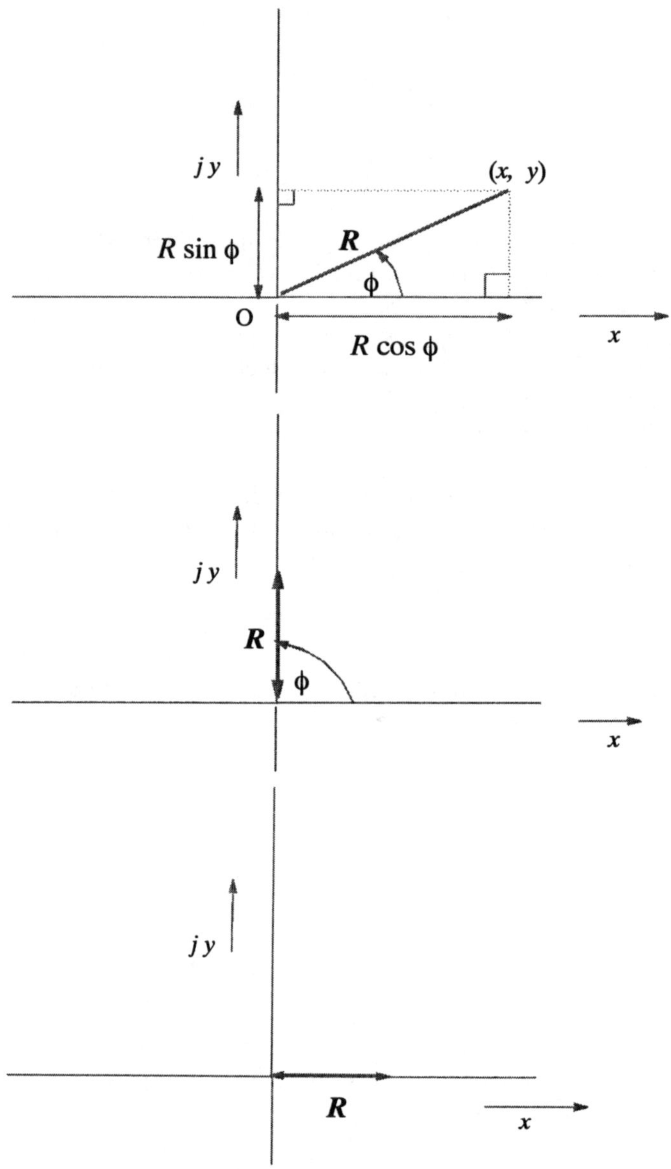

Figure 2.7 Polar representation of complex numbers: (a) In general ***R*** has real and imaginary parts. (b) When $\phi = 90$, ***R*** is imaginary. (c) When $\phi = 0$, ***R*** is real.

Similarly when $\phi = 0$ deg, $e^{j0} = 1$, so $Re^{j0} = R$. The evaluation of the position vector for other angles ϕ can be undertaken in a similar manner by calculating the trigonometric functions in the respective quadrants.

2.4.1 Addition of Complex Numbers

When we add complex numbers, we must add the real parts together and the imaginary parts to each other. We cannot add real parts to the imaginary parts. Thus, if we have two complex numbers P and Q, such that $P = 3 + 5j$ and $Q = 6 + 8j$, then $P + Q = 9 + 13j$. This property of complex numbers is utilized in what is known as stub matching of components to transmission lines.

If we have a load that has susceptance (1/reactance) of $+j$B in parallel with its conductance of $G_1 = 0.2$ S (siemens) and we want to connect it to a line whose characteristic impedance is 50Ω (i.e., a characteristic admittance of 0.2 S), we would get a standing wave since the load does not have the same impedance as the transmission line. This would result in a loss of power transfer, and is some cases could lead to voltage breakdown in high power systems. In order to match the load to the line, we would connect a stub, which is a short-circuited line, in parallel with the transmission line, as near to the load as possible. The conductance of the load can be varied by adjusting the position of the stub so that the conductance presented to the line equals 0.2S. The susceptance of the stub can be varied by adjusting its length so that its susceptance is equal and opposite to that of the load, that is, so that it is $-j$B. Thus the resultant susceptance is zero and the conductance is 0.2S. The load now presents an impedance that is the same as that of the transmission line; it is matched and, hence, maximum power transfer can be effected.

2.4.2 Complex Conjugate

The complex conjugate of a number has the same real part as the number, but its imaginary part has a sign opposite that of the number. For instance, if we have a complex number $S = (a + jb)$, its complex conjugate is $(a - jb)$ and is denoted by using an asterisk as a superscript after the letter (S^*). Conjugate matching is used to "match" two components that do not have purely resistive impedances.

2.5 SCALARS AND VECTORS

A scalar quantity is one that has magnitude but no direction, such as mass, resistance, time, and power. A vector quantity is one that has magnitude and direction, such as weight, force, current, magnetic field, and electric field.

2.5.1 Position Vector

A vector may be used to indicate the position of a point in space or in a plane, relative to another point (usually the origin of coordinates). The vector $\boldsymbol{R}$ in Figure 2.8(a) denotes the position of the point (x, y).

2.5.2 Vector Addition and Subtraction

Vectors can be added by using a parallelogram or triangle rule. The vectors are represented by scaled lines whose lengths are proportional to the relative magnitudes and whose

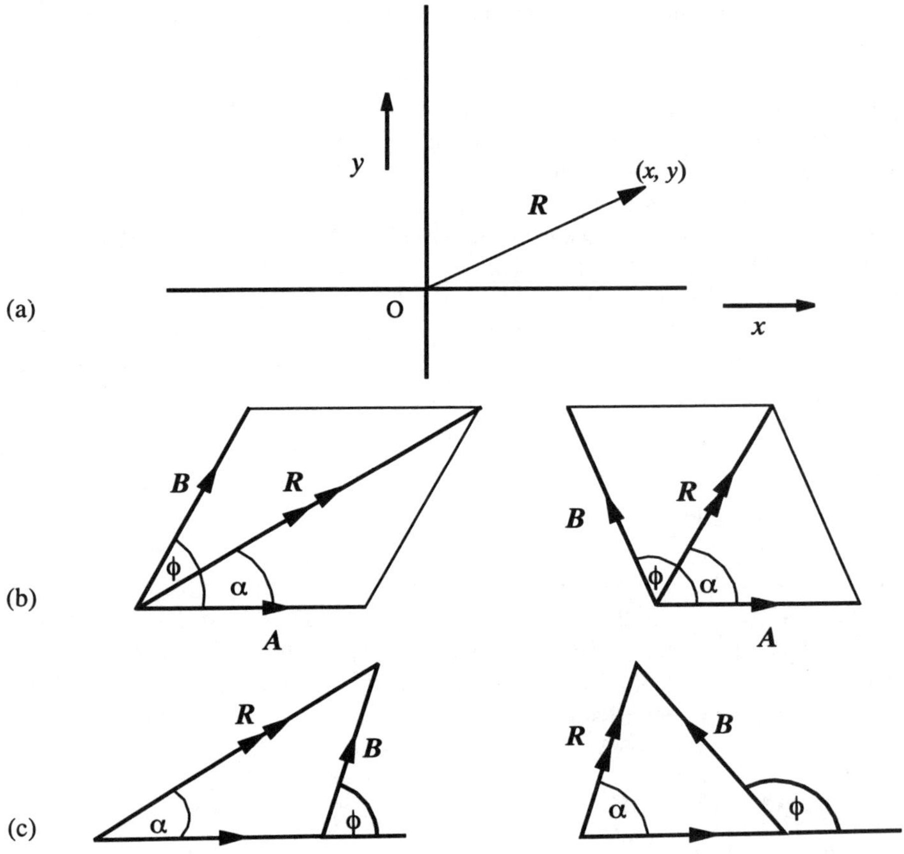

Figure 2.8 Rules for the addition of vectors. (a) $\boldsymbol{R}$ is the position vector for the point (x, y). (b) Parallelogram rule for vector addition. (c) Triangle rule for vector addition.

directions represent the directions in which the vectors act. Usually one vector is considered to be the reference vector and the second vector is drawn at an angle representing the relative angles between the two vectors. When more than two vectors are to be added, two are first chosen and summed and then the resultant is added to the third, and so on.

2.5.2.1 Parallelogram Rule

Consider two vectors ***A*** and ***B*** that represent the adjacent sides of the parallelogram shown in Figure 2.8(b); the lengths of the sides correspond to the relative magnitudes of the vectors, and the angle between them is the angle between the directions of the vectors. Note that their directions are both away from their common point of intersection. A parallelogram is constructed by drawing a line through the end of vector ***A*** parallel to vector ***B*** and a line through the end of vector ***B*** parallel to vector ***A***. The resultant is a vector ***R*** represented by the diagonal (through the point O) of the parallelogram and is denoted by the double arrow.

2.5.2.2 Triangle Rule

In this case the vectors to be added represent two sides of the triangle shown in Figure 2.8(c), but note that their directions are cyclic—the direction of vector ***A*** is toward their common point, whereas the direction of vector ***B*** is away from their common point. The resultant ***R*** of the two vectors is represented by the third side of the triangle, and its direction is anticyclic to the direction of the two component vectors. The length of the resultant is measured, and the appropriate scale is applied to calculate its magnitude. The angle that the resultant makes with either of the component vectors can be measured with a protractor.

A more accurate method of finding the magnitude and direction of the resultant is by calculation, using the cosine formula (2.9) to calculate the length of the resultant and then the sine formula (2.10) to calculate its direction.

2.5.3 Vector Multiplication

Vectors can be multiplied in two different ways. We can take their (1) scalar or dot product or (2) vector of cross product.

2.5.3.1 Scalar or Dot Product

The dot product of two vectors A and B is written with a dot between them and is defined as

$$\boldsymbol{A}\cdot\boldsymbol{B} = \boldsymbol{B}\cdot\boldsymbol{A} = AB\cos\theta = BA\cos\theta \tag{2.30}$$

where $\boldsymbol{A}$ and $\boldsymbol{B}$ (boldface) are vector quantities, A and B are scalars, and θ is the angle between the vectors $\boldsymbol{A}$ and $\boldsymbol{B}$. Note that the dot product of the two vectors is a scalar and the product of the magnitude of one vector and the resolute of the other (the component of the other one resolved) in the direction parallel to the first, as shown in Figure 2.9(a). It follows that when θ is zero, the two vectors are parallel and the dot product is equal to the product of the magnitudes of the two vectors. We can see that this is the case algebraically since cos θ is equal to 1 when θ is zero. When θ is 90 deg the two vectors are perpendicular and the resolute of one in the direction parallel to the other is zero, which results in the product also being zero. In physics we encounter many cases of the dot product of vectors. For instance, the work done is the dot product of the force and the distance moved. We may have an object, such as a train, that is constrained to travel in a straight line subjected to a force $\boldsymbol{F}$ that is applied at an angle θ as shown in Figure 2.9(b). The resolute of the force in the direction of motion of the train is $F \cos \theta$. When $\theta = 0$ deg, $\cos \theta = 1$ and the resolute has its maximum value of F; whereas when $\theta = 90$ deg, $\cos \theta = 0$ and the resolute has its minimum value of zero, that is, the force F is zero and thus the work done would also be zero.

2.5.3.2 *Vector or Cross Product*

The vector product of two vectors $\boldsymbol{A}$ and $\boldsymbol{B}$ is written with a cross (or multiplication sign) between them and is defined as

$$\boldsymbol{A} \times \boldsymbol{B} = \mathbf{k}AB \sin \theta = -\mathbf{k}BA \sin \theta \tag{2.31}$$

The cross product is the product of one vector and the resolute of the other in the direction perpendicular to the first and in the plane containing the two vectors. The resolute of $\boldsymbol{A}$ perpendicular to $\boldsymbol{B}$ is $A \sin \theta$ if θ is the angle between the vectors. The product is a vector whose direction is perpendicular to the plane containing the two vectors and governed by the right-hand rule. The vector product $\boldsymbol{A} \times \boldsymbol{B}$ is in the direction opposite that of the vector product $\boldsymbol{B} \times \boldsymbol{A}$. In Figure 2.9(c) we can see that if we take the vector product of $\boldsymbol{A} \times \boldsymbol{B}$ we move in an counterclockwise direction from A to B, and thus the resultant is in the upward direction; whereas if we take the vector product of $\boldsymbol{B} \times \boldsymbol{A}$ we move in a clockwise direction from $\boldsymbol{B}$ to $\boldsymbol{A}$, and thus the resultant is in the downward direction. If we say that $\boldsymbol{A} \times \boldsymbol{B}$ is positive, then $\boldsymbol{B} \times \boldsymbol{A}$ is negative. The unit vector $\mathbf{k}$ (see Chapter 3) is used to show that the resultant vector is in the direction of the z-axis, whereas $-\mathbf{k}$ denotes a vector in the direction of the negative z-axis. We encounter the vector product in the Poynting's power flux vector, which is the product of the electric field intensity $\boldsymbol{E}$ and the magnetic field intensity $\boldsymbol{H}$. Poynting's power flux vector gives us the power density (power per unit area) at a particular location and is given by

$$\boldsymbol{P}_{\mathrm{d}} = \boldsymbol{E} \times \boldsymbol{H} \tag{2.32}$$

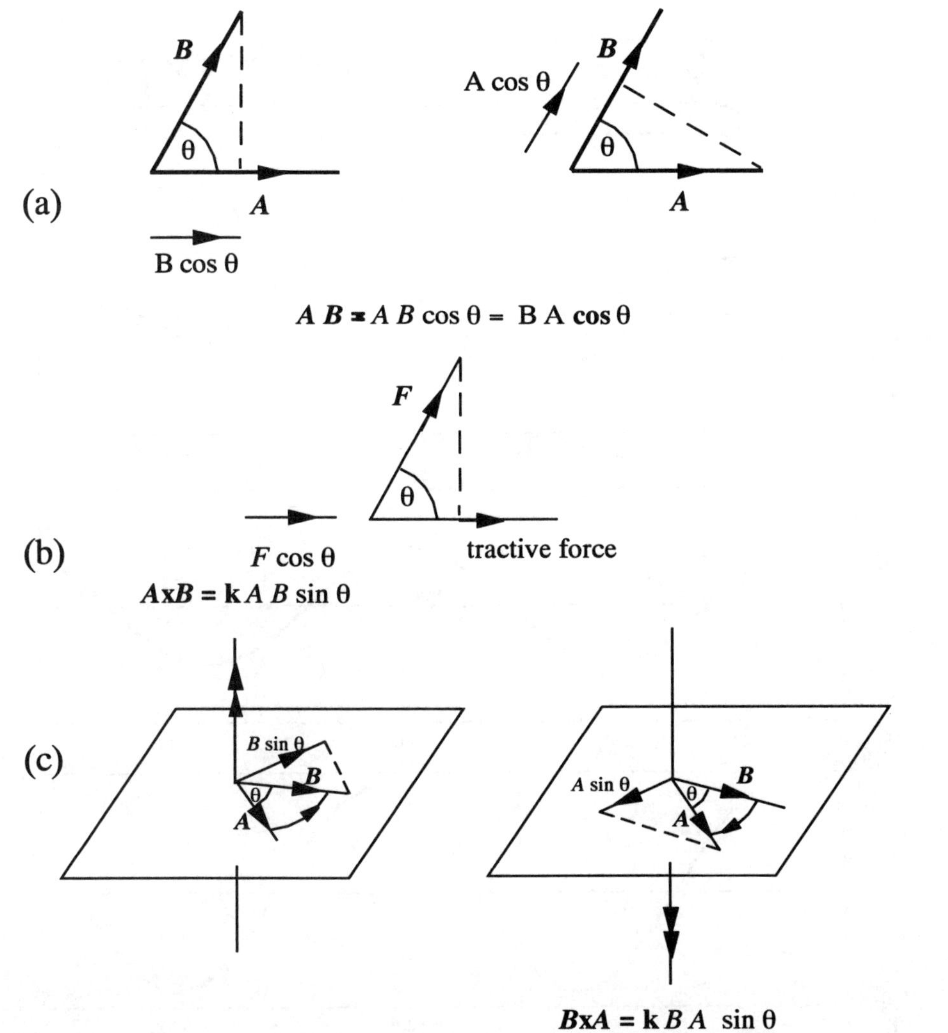

Figure 2.9 Scalar and vector product of vectors: (a) scalar or dot product; and (b,c) vector or cross product of vectors.

where $\boldsymbol{P}_d$ is in watts per square meters, W m^{-2}; $\boldsymbol{E}$ is in volts per meter, V m^{-1}; and $\boldsymbol{H}$ is in amperes per meter, A m^{-1}. $\boldsymbol{E}$ and $\boldsymbol{H}$ are often written in V/m and A/m, respectively.

2.5.4 Phasors

A phasor is a rotating vector that represents a quantity that rotates in time rather than in space at an angular velocity equal to the phase velocity of the quantity. In engineering we

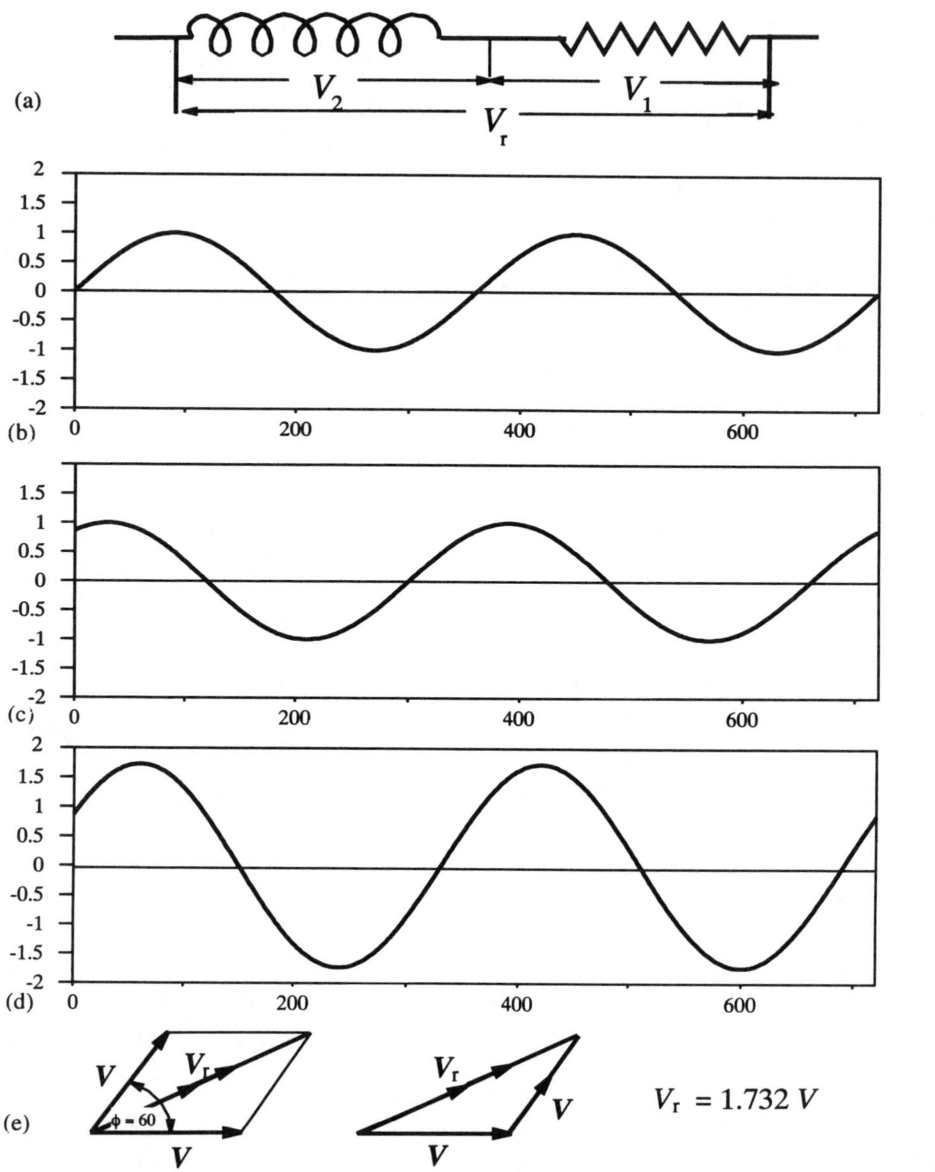

Figure 2.10 Addition of waveforms using algebraic addition and phasors: (a) circuit of an inductor in series with a resistor, (b) $V \sin \omega t$, (c) $V \sin(\omega t + \phi)$, (d) $V \sin \omega t + V \sin(\omega t + \phi)$, and (e) addition of waveforms using vectors as phasors.

use phasors to represent sinusoidal variations such as voltages, and the length of the phasor is proportional to the amplitude of the variation. If the voltage mains is at 50 Hz, one cycle takes 1/50 of a thousandth of a second (or 0.02 sec) and thus the phasor would take 0.02 sec to complete one revolution or 2π radians. In one second it would sweep out 50 revolutions or 100π rads, so we can say that the angular velocity, ω, is 100π rads/s (since $\omega = 2\pi f$). At any time t, the vector sweeps out an angle ωt. If we have more than one voltage at the same angular velocity but with relative phase differences, then we can depict both the phasors on the same diagram. Since both the vectors are rotating at the same velocity, the relative phase difference will be the same whether they are stationary or rotating; so we shall keep them stationary. If we have a sinusoidal variation as shown in Figure 2.10(b), then the length of the vector shows the amplitude or maximum value in each cycle. Consider the circuit of Figure 2.10(a), which shows an inductor in series with a resistor that is connected to a sinusoidally varying voltage source. We want to calculate the total voltage V_t across the inductor and resistor. The voltage waveforms across the inductor and resistor, respectively, are shown in Figure 2.10(b, c) with amplitudes of V_1 and V_2. In the case of a pure inductor the voltage V_2 would lead the voltage V_1 by exactly 90 deg, but in the general case we shall consider V_2 as leading V_1 by a phase of ϕ. These voltages can be represented on a phasor diagram by two straight lines that are scaled to represent the relative amplitudes of V_1 and V_2 and separated by an angle ϕ. The vector $\mathbf{V}_1$ is taken as the reference phasor and, since $\mathbf{V}_2$ leads $\mathbf{V}_1$ by a phase of ϕ degrees, $\mathbf{V}_2$ is drawn at an angle of $+\phi$. The resultant is found by using the laws of vector addition. Utilizing vectors as phasors is a very useful tool when analyzing the voltages and currents that are subjected to sinusoidal variations. We would normally find the resultant of two sinusoidal variations by adding the sine waves on a point-by-point basis. If we have $\mathbf{V}_1 = \mathbf{V}_2 = 1$ (each representing one unit) and ϕ equal to 60 deg we would get the waveform of Figure 2.10(d), which has an amplitude of 1.732, and a phase of 30 deg compared with the voltage across the resistor. In order to find the resultant using phasors, we can draw a scaled drawing as shown in Figure 2.10(e) and, by using the parallelogram or triangular rule, can directly measure the angle and length of the resultant (1.732 units). Alternatively, we could draw a sketch of the three vectors and then use the cosine rule to calculate the length of the resultant and the sine rule to calculate the angle of the resultant.

2.6 FOURIER ANALYSIS AND TRANSFORMS

Fourier analysis was first suggested by J. B. J. Fourier (1768–1830). He showed that a square wave could be decomposed into a sum of sine waves. Fourier transforms are used to transform a function from one domain to another, such as from the time to the frequency domains.

2.6.1 Fourier Analysis

Consider a square wave of frequency f ($= \omega/2\pi$) as shown in Figure 2.11(a). This wave has a period T equal to $2\pi/\omega$. If we add a sine wave of the same frequency and a sine wave of three times the frequency but one-third of the amplitude (sin $3\omega t$)/3, as shown in Figures 2.11(b, c), we get the resultant wave of Figure 2.11(d), which looks similar to the square wave but is not a very good approximation. If we add another sine wave of five times the frequency and one-fifth of the amplitude (sin $5\omega t$)/5, we get the plot of Figure 2.11(e). As we add more terms we get nearer to the shape of a square wave. To get an exact square wave function $f(\omega t)$ we have to add an infinite number of terms of the series given by the summation

$$f(\omega t) = \frac{4}{\pi} \sum_{n=1}^{\infty} \frac{\sin(2n - 1)\omega t}{(2n - 1)\omega t} \tag{2.33}$$

where ω is the angular frequency in radians per second.

Note that each term is an odd multiple of the frequency of the first term, sin ωt, and the amplitudes are reciprocals of the odd numbers. The first term is called the fundamental and the other terms are known as higher order harmonics. The second term has three times the frequency of the fundamental and is therefore known as the third harmonic. If we sum the first eight terms of the above series we get a fairly good approximation to the square wave. The resultant is shown in Figure 2.11(f) and is given by the sum

$$f(\omega t) = \frac{4}{\pi}\left(\sin \omega t + \frac{1}{3}\sin 3\omega t + \frac{1}{5}\sin 5\,\omega t + \cdots + \frac{1}{15}\sin 15\omega t\right) \tag{2.34}$$

Most periodic and many other nonperiodic waveforms can be decomposed into a summation of periodic sine and cosine waves.

2.6.2 Fourier Transforms

We have seen how the square wave can be decomposed into a summation of sine waves. The waves of Figure 2.11 show us the variation of the intensity or displacement with time, that is, the pattern in the time domain. If we were to look at the variation of the intensity with frequency for a sine wave we would get a single line at a frequency f (= $\omega/2\pi$) and the length of the line would be equal to the amplitude (the maximum value of the intensity). This type of plot, which shows us the variation in the frequency domain, is called the frequency spectrum and is shown in Figure 2.12(a). The Fourier transform has transformed a time domain plot into the frequency domain. In the case of the square wave, if we were to look at its spectrum we would get a series of lines with the slope of the amplitudes equal to -6 decibels per octave (the amplitude is halved each time the frequency is doubled). The time and frequency domain plots are shown in Figure 2.12(b). For a square pulse we would get the time and frequency domain plots of Figure 2.12(c). In antenna technology we use the Fourier transform to transform the illumination across an

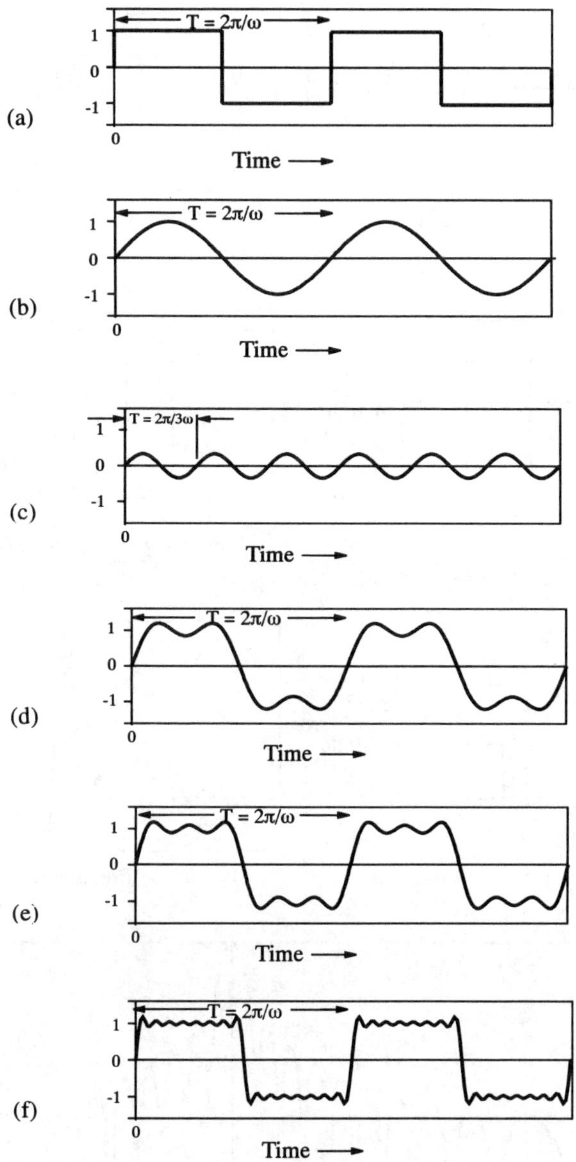

Figure 2.11 Fourier analysis of a square wave. (a) square wave of frequency $f = \omega/2\pi$ and amplitude 1, (b) sine wave of frequency $f = \omega/2\pi$ and amplitude 1, (c) sine wave of frequency $f = 3\omega/\pi$ and amplitude 1/3, (d) summation of the first two terms $\sin \omega t$ and $(\sin 3\omega/t)/3$, (e) summation of the first three terms $\sin \omega t + (\sin 3\ \omega t)/3 + (\sin 5\ \omega t)/5$, and (f) summation of the first eight terms.

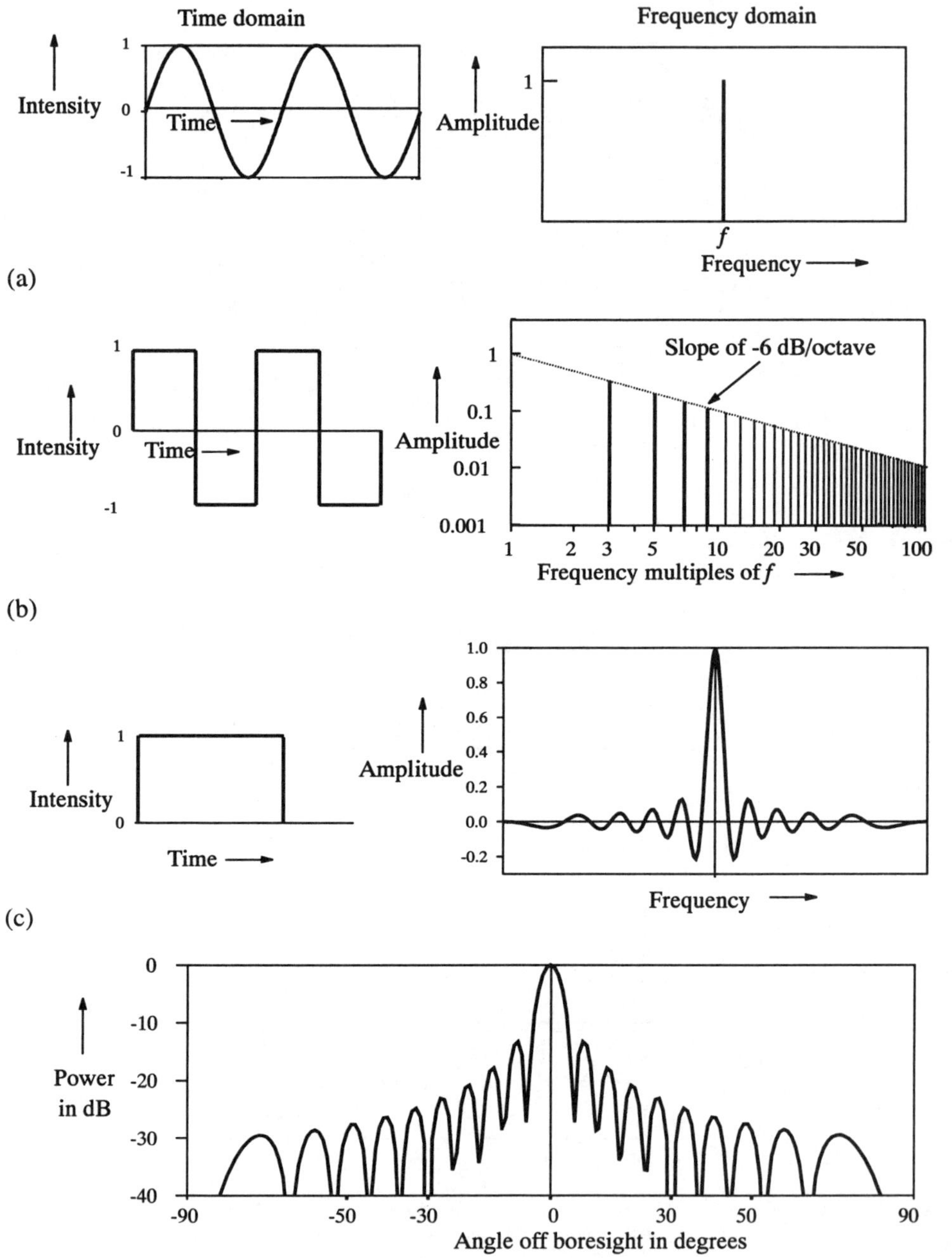

Figure 2.12 Fourier transform of: (a) a sine wave, (b) a square wave, (c) a square pulse, and (d) far field radiation pattern for uniform illumination.

antenna to the electric field variation in the far field. For example if we have a uniform illumination (constant electric field and no phase variation) across an antenna, it is the same as the square pulse shown in Figure 2.12(c) and its Fourier transform is given by

$$f(\theta) = \frac{\sin u}{u} \tag{2.35}$$

where $u = (\pi a \sin \theta)/\lambda$, a is the width of the aperture of the antenna, and θ is the angle measured from the normal to the aperture.

The radiation pattern that is the variation of power (proportional to the square of the electric field) with angle θ is given by

$$F(\theta) = \frac{\sin^2 u}{u^2} \tag{2.36}$$

The exact number, height, and position of the sidelobes depend upon the size of the electrical aperture a/λ, that is, the size in terms of the wavelength. In Figure 2.12(d) we have the radiation pattern for an antenna that is 10 wavelengths long. Note that the y scale shows the power in decibels and is therefore a logarithmic scale.

Other uses for Fourier transforms are in measuring the frequency content of transients and pulses so that their effect on equipment can be estimated.

2.7 PARAMETERS

We must distinguish between the strict definition of parameters used by mathematicians and the parameters used by engineers.

2.7.1 Mathematical Definition of Parameters

A parameter is defined by mathematicians as a third variable upon which the other two variables depend. For instance, instead of writing the equation of a circle as

$$x^2 + y^2 = a^2 \tag{2.37}$$

or

$$y = \sqrt{a^2 - x^2} \tag{2.38}$$

each of the variables can be written in terms of another variable θ.

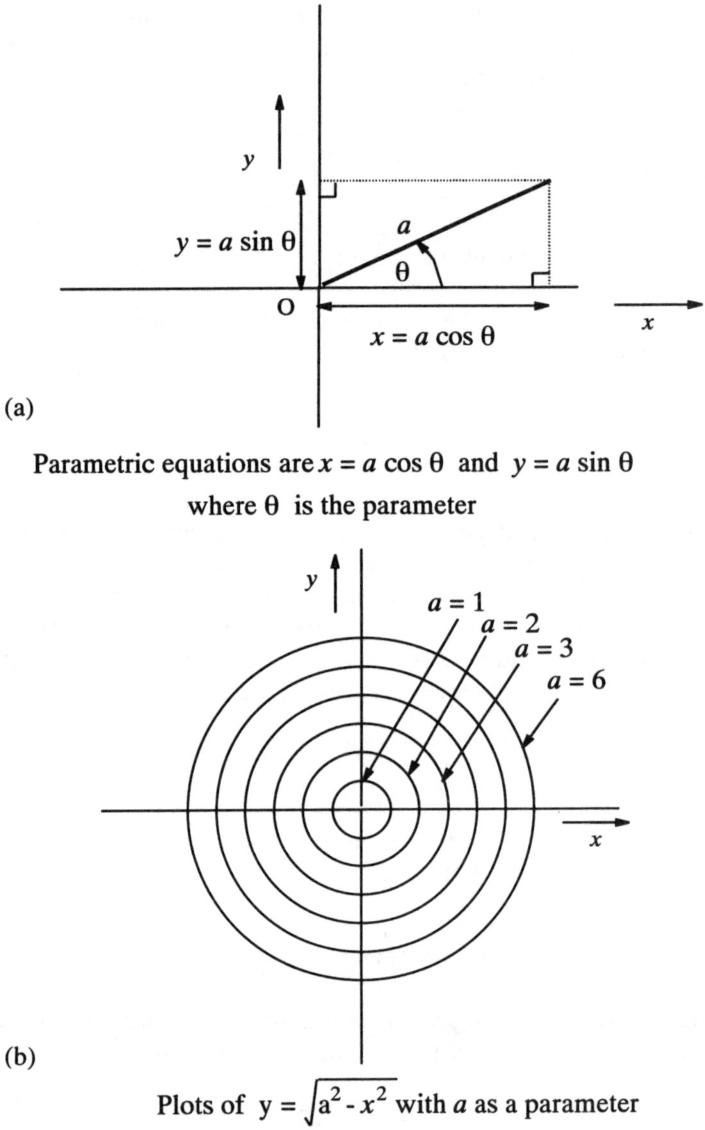

Figure 2.13 Parameters used by (a) mathematicians and (b) engineers.

$$x = a \cos \theta \tag{2.39}$$

$$y = a \sin \theta \tag{2.40}$$

Equations (2.39) and (2.40) are known as parametric equations, and θ is known as the parameter. These are depicted in Figure 2.13(a).

We can see that if we were to square (2.39) and (2.40) and add them together, we would get

$$x^2 + y^2 = a^2 \cos^2 \theta + a^2 \sin^2 \theta \tag{2.41}$$

This is the same as the equation of the circle given by (2.37) since

$$\cos^2 \theta + \sin^2 \theta = 1 \tag{3.42}$$

2.7.1.1 Parameters Used By Engineers

Engineers tend to use the term parameter in a different way. If we consider the equation of a circle (2.37), the value of a determines the radius of the circle and can take on any constant magnitude for each series of x and y values. If we give it a number of fixed values, we get a series of concentric circles, which have the origin of coordinates as their center, as shown in Figure 2.13(b). We can now say that we are plotting y against (or versus) x with a as a parameter. If we were to increase a in infinitesimally small steps from an initial value of zero, we would get a solid disc since the circles of increasing radii would be touching.

In EMC engineering we encounter many cases of variables that are dependent upon two or more other variables. For instance, we have the *shielding effectiveness* (SE) of a material dependent upon the frequency as well as the conductivity. We could choose the conductivity as the independent variable (x-axis) and plot the SE against conductivity for a number of fixed frequencies to get different plots on the same graph. We would then say that frequency is the parameter.

2.8 FUNDAMENTAL UNITS AND DIMENSIONS

All physical quantities can be expressed as a combination of the fundamental or basic units, also called *Systeme International* (SI) base units. The relationship between a physical quantity and the fundamental units is called its dimension. These fundamental units are shown in Table 2.3.

Table 2.3
Fundamental or SI Base Units

Physical quantity	*Name of Base Unit*	*Unit*	*Symbol*	*Dimension*
Mass	kilogram	kg	m	M
Length	meter	m	l	L
Time	second	s	t	T
Current	amperes	A	I	I
Thermodynamic temperature	degrees kelvin	K	T	K

Table 2.4
Dimensions of Commonly Used Physical Quantities

Physical Quantity	*Name of Unit*	*Unit*	*Symbol*	*Fundamental Units*
Acceleration	meters per second squared	$m\ s^{-2}$	α	LT^{-2}
Angular frequency	radians per second	rad/s	ω	T^{-1}
Capacitance	farad	F	C	$M^{-1}L^{-2}T^4I^2$
Charge	coulomb	C	Q	IT
Conductance/Susceptance	siemen	S	G	$M^{-1}L^{-2}T^3I^1$ (same as admittance)
Conductivity	siemen-meter	S−m	σ	$M^{-1}L^{-3}T^3I^1$
Current	ampere	A	I	I
Current density	amperes per square meter	$A\ m^{-1}$	J	IL^{-2}
Dielectric constant	no units	no units	ε_r	$\varepsilon/\varepsilon_o$ (same as relative permittivity)
Distance or length	meter	l	d	L
Electric field intensity	volts per meter	$V\ m^{-1}$	E	$MLT^{-3}I^{-1}$
Energy	joule	J	E or W_e	ML^2T^{-2}
Force	newton	N	F	MLT^{-2}
Frequency	hertz	Hz	f	T^{-1}
Impedance	ohm	W	R	$ML^2T^{-3}I^{-2}$ (same as resistance)
Inductance	henry	H	L	$ML^2T^{-2}I^{-2}$
Length	meter	m	l	L
Mass	kilogram	kg	m	M
Magnetic field intensity	amperes per meter	$A\ m^{-1}$	H	IL^{-1}
Magnetic flux density	tesla	T	B	$MT^{-2}I^{-1}$
Permeability	henries per meter	$H\ m^{-1}$	μ	$MLT^{-2}I^{-2}$
Permittivity	farads per meter	$F\ m^{-1}$	ε	$M^{-1}L^{-3}T^4I^1$
Power	watts	W	P	ML^2T^{-3}
Power density	watts per meter	$W\ m^{-2}$	P_d	MT^{-3}
Relative permeability	no units	no units	μ_r	μ/μ_o
Relative permittivity	no units	no units	ε_r	$\varepsilon/\varepsilon_o$ (same as dielectric constant)
Resistance/reactance	ohm	Ω	R	$ML^2T^{-3}I^{-2}$ (same as impedance)
Resistivity	ohm-meter	Ω-m	ρ	$ML^3T^{-3}I^{-1}$
Skin depth	meter	m	δ	L
Time	second	sec or s	t	T
Velocity	meters per second	$m\ s^{-1}$	v	LT^{-1}
Voltage	volt	V	V	$ML^2T^{-3}I^{-1}$
Volume	cubic meter	m^3	V_u	L^3
Wavelength	meter	m	λ	L

Consider the units of energy, for instance. Energy can be defined as the product of force and distance. Force in its turn can be defined by Newton's law as the product of mass and acceleration. Since acceleration is the rate of change of velocity, its dimensions are LT^{-2} (i.e., its SI base units are $m\ s^{-1}$); thus we get Energy = $ML^2T^{-2}$for the dimensions of energy. Since the energy is in joules, this gives us the fundamental units of the joule.

Note that joules is written with a lower-case initial letter, although it is a proper name; whereas the unit J, is denoted by an upper-case letter.

The dimensions for power in watts can be derived from the units for energy since power is the number of joules per second. Therefore, the dimensions of power are ML^2T^{-3}.

The dimensions for voltage in volts can be derived from the units for power since power is the product of volts and amps. Therefore, the dimensions of volts are $ML^2T^{-3}I^{-1}$.

The dimensions of inductance in henries can be derived in a similar manner since, according to Lenz's law, the inductance is induced *electromotive force* (EMF) divided by the rate of change of current. Thus the dimensions of henries are $ML^2T^{-2}I^{-2}$.

The dimensions of SI units most commonly used in EMC engineering are listed in Table 2.4.

2.8.1 Checking Formulas by Dimensions

Reducing a physical quantity to fundamental units is useful in verifying the correctness of formulas or in resolving ambiguities. For instance, formulas often contain the symbols μ and ε for relative permeability and permittivity, respectively, when these should have the subscripts r, that is, μ_r and ε_r.

We can also understand the physical meaning of formulas if we can derive their dimensions. For instance, if we consider the product of the electric and magnetic field intensities we know from Table 2.4 that the electric field has the dimensions of $MLT^{-3}I^{-1}$ and that the magnetic field has the dimensions of IL^{-1}; thus, the product has dimensions given by $\boldsymbol{E} \times \boldsymbol{H} = MLT^{-3}I^{-1}IL^{-1}$, which gives $\boldsymbol{E} \times \boldsymbol{H} = MT^{-3}$. Referring to Table 2.4 we can see that these are the same as the units for power density and $\boldsymbol{E} \times \boldsymbol{H}$ is known as Poynting's power flux vector. It is left to the reader to verify that the dimensions for $\sqrt{\mu/\varepsilon}$ are the same as those for resistance.

Chapter 3
Antenna Theory

This chapter deals with more advanced vector algebra that is required to understand Maxwell's equations and the theoretical aspects of antennas. It is strongly recommended that Chapter 2 be studied by engineers who are not conversant with mathematics or would like to revise the application of their knowledge to the theoretical understanding of antennas. This chapter is written in such a way as to be self-contained, and it can be entirely omitted by the reader who does not wish to study antenna theory. Although Maxwell's equations are quoted, they are explained qualitatively with a greater emphasis placed on the boundary conditions applicable to conducting surfaces. The concepts of a vector power density and radiation resistance of wire antennas are introduced. The derivation of the far-field distance for wire and aperture antennas is also included.

Vectors are denoted by bold letters, and scalars are denoted by italic characters.

3.1 UNIT VECTORS

We reviewed in Chapter 2 how to draw vectors to scale and then construct parallelograms or triangles in order to calculate the sum of two vectors. Instead of saying that we have a vector $\boldsymbol{A}$ in a particular direction we can say that we have a scalar F in the direction of a unit vector $\mathbf{r}$. The vector $\boldsymbol{A}$ can then be considered to be the dot product of the scalar F and the unit vector $\mathbf{r}$.

In rectangular coordinates unit vectors are represented by $\mathbf{i}$, $\mathbf{j}$, and $\mathbf{k}$ along the x-, y-, and z-axes, respectively. Unit vectors are sometimes denoted with a circumflex above (e.g., $\hat{\mathbf{i}}$, $\hat{\mathbf{j}}$, and $\hat{\mathbf{k}}$). In cylindrical polar coordinates unit vectors are represented by $\boldsymbol{a}_r$, $\boldsymbol{a}_\theta$, and $\boldsymbol{a}_z$ along the radial direction, the direction of angular rotation θ, and along the z-axis normal to the plane containing a_r and a_θ, respectively. In spherical polar coordinates $\boldsymbol{a}_\phi$ is used instead of $\boldsymbol{a}_z$, where ϕ is the angle of rotation in the x-y plane. The unit vectors in rectangular and polar coordinates are depicted in Figure 3.1.

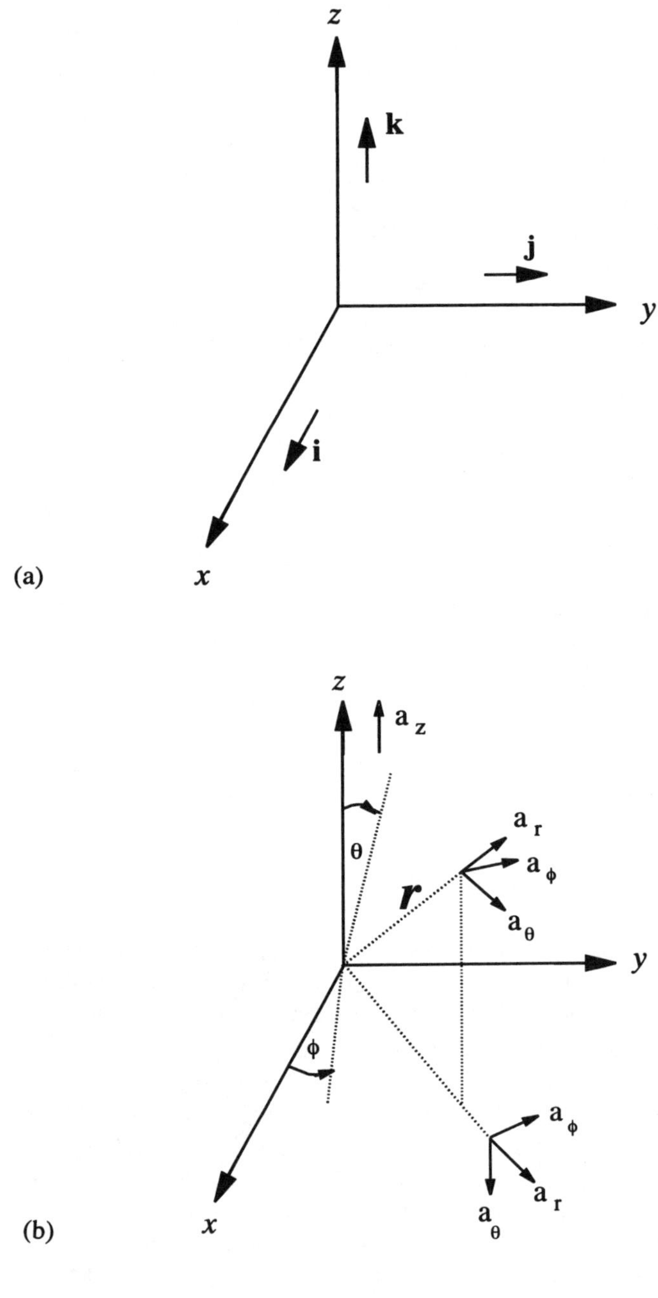

Figure 3.1 Unit vectors in (a) right-handed rectangular and (b) polar coordinates.

Unit vectors are used to express any vector in terms of three unit vectors. If we want to multiply vectors we first write them in terms of three unit vectors and then multiply them. For example if we want to calculate their cross product, we write them as

$$\boldsymbol{A} \times \boldsymbol{B} = (\mathbf{i}A_x + \mathbf{j}A_y + \mathbf{k}A_z) \times (\mathbf{i}B_x + \mathbf{j}B_y + \mathbf{k}B_z) \tag{3.1}$$

$$\boldsymbol{A} \times \boldsymbol{B} = \mathbf{i}A_x \times \mathbf{i}B_x + \mathbf{i}A_x \times \mathbf{j}B_y + \mathbf{i}A_x \times \mathbf{k}B_z + \mathbf{j}A_y \times \mathbf{i}B_x + \mathbf{j}A_y \times \mathbf{j}B_y + \mathbf{j}A_y \times \mathbf{k}B_z$$
$$+ \mathbf{k}A_z \times \mathbf{i}B_x + \mathbf{k}A_z \times \mathbf{j}B_y + \mathbf{k}A_z \times \mathbf{k}B_z$$

Since we have $\mathbf{i}A_x$ and $\mathbf{i}B_x$ in the same direction, that is, the angle between them is zero, their cross product is zero. The same holds for the cross product of $\mathbf{j}A_y$ and $\mathbf{j}B_y$ and the cross product of $\mathbf{k}A_z$ and $\mathbf{k}B_z$.We also know that the cross product of $\mathbf{i}A_x$ and $\mathbf{j}B_y$ gives us a vector $\mathbf{k}A_xB_y$ along the z-axis. On the other hand, the cross product of $\mathbf{k}A_z$ and $\mathbf{j}B_y$ gives us a vector $\mathbf{i}A_zB_y$ along the negative x-axis; that is, it is equal to $-\mathbf{i}A_zB_y$.

The same applies to the other cross products. Thus the following relationships apply.

$$\mathbf{i}A_x \times \mathbf{i}B_x = 0 \qquad \mathbf{j}A_y \times \mathbf{j}B_y = 0 \qquad \mathbf{k}A_z \times \mathbf{k}B_z = 0$$

$$\mathbf{i}A_x \times \mathbf{j}B_y = +\mathbf{k}A_xB_y \qquad \mathbf{i}A_x \times \mathbf{k}B_z = -\mathbf{j}A_xB_z \qquad \mathbf{j}A_y \times \mathbf{i}B_x = -\mathbf{k}A_yB_x$$

$$\mathbf{j}A_y \times \mathbf{k}B_z = \mathbf{i}A_yB_z \qquad \mathbf{k}A_z \times \mathbf{i}B_x = \mathbf{j}A_zB_x \qquad \mathbf{k}A_z \times \mathbf{j}B_y = -\mathbf{i}A_zB_y \tag{3.2}$$

Thus the cross product of A and B can be written as the sum of the following three vectors:

$$\boldsymbol{A} \times \boldsymbol{B} = \mathbf{i}(A_yB_z - A_zB_y) - \mathbf{j}(A_xB_z - A_zB_x) + \mathbf{k}(A_xB_y - A_yB_x) \tag{3.3}$$

which can also be written in the form of a matrix as

$$\boldsymbol{A} \times \boldsymbol{B} = \begin{vmatrix} \mathbf{i} & \mathbf{j} & \mathbf{k} \\ A_x & A_y & A_z \\ B_x & B_y & B_z \end{vmatrix} \tag{3.4}$$

3.2 SCALAR AND VECTOR FIELDS

In Chapter 2 we encountered scalars and vectors that have fixed magnitudes and directions. Now we consider scalar and vector fields. In this case we have scalars and vectors that are functions of position, that is, they are point functions. Their magnitudes and/or directions depend on position. Scalar point functions define scalar fields, and vector point functions define vector fields.

3.2.1 Spatial Rates of Change of Scalar and Vector Fields

There are three types of rates of change, namely (1) gradient or grad, (2) divergence or div, and (3) curl or rot. These rates of change can be expressed in terms of a vector operator that is called del or nabla and is represented by the symbol ∇. The operator is defined by the following relationship:

$$\nabla = \mathbf{i}\frac{\partial}{\partial x} + \mathbf{j}\frac{\partial}{\partial y} + \mathbf{k}\frac{\partial}{\partial z} \tag{3.5}$$

It should be noted that del is not a vector but an operator and operates on a scalar or vector.

The gradient, divergence, and curl of scalar and vector fields are invariant, that is, they are independent of the coordinate system.

3.2.2 Gradient of a Scalar Field

If we have a scalar point function ϕ, then at a point P surrounded by a surface S and enclosing a volume V_u, we can define the gradient of ϕ as a vector that shows the direction and magnitude of the maximum space variation at any point in space. In rectangular coordinates the gradient of the function is the sum of the individual gradients along each of the three mutually perpendicular axes and is given by

$$\text{grad } \phi = \mathbf{i}\frac{\partial \phi}{\partial x} + \mathbf{j}\frac{\partial \phi}{\partial y} + \mathbf{k}\frac{\partial \phi}{\partial z} \tag{3.6}$$

We can also write the gradient in terms of the operator del as $\nabla\phi$.

For example if we have a series of equipotentials (between two electrodes) the equipotentials will be parallel but not necessarily planar if the electrodes are not planar, as shown in Figure 3.2. The voltage potential V is a scalar quantity; but the rate of change with distance of the voltage potential is the electric field, which is a vector. The gradient is defined as the maximum rate of increase in the direction normal to the equipotentials. In the general case for rectangular coordinates the voltage gradient will have components along each of the x-, y-, and z-axes and is given by

$$\text{grad } V = \mathbf{i}\frac{\partial V}{\partial x} + \mathbf{j}\frac{\partial V}{\partial y} + \mathbf{k}\frac{\partial V}{\partial z} \tag{3.7}$$

This can also be written in terms of the operator del as ∇V.

In polar coordinates the elements of length are dr, $rd\theta$, and $r \sin \theta d\phi$, and grad V is given by

$$\text{grad } V = \mathbf{a}_r\frac{\partial V}{\partial r} + \mathbf{a}_\theta\frac{1}{r}\frac{\partial V}{\partial \theta} + \mathbf{a}_\phi\frac{1}{r\sin\theta}\frac{\partial V}{\partial \phi} \tag{3.8}$$

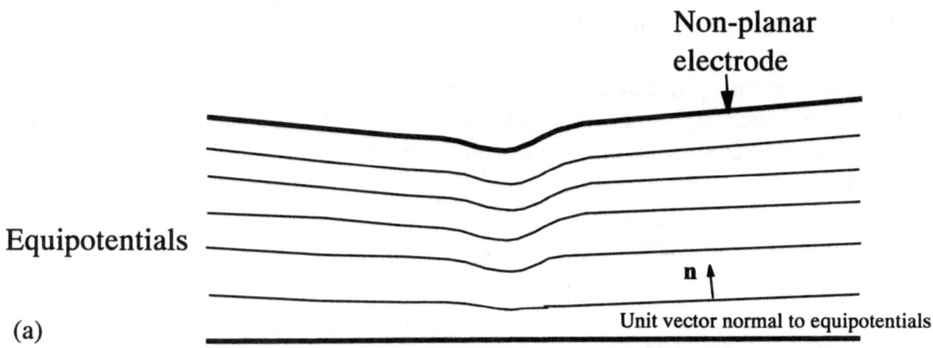

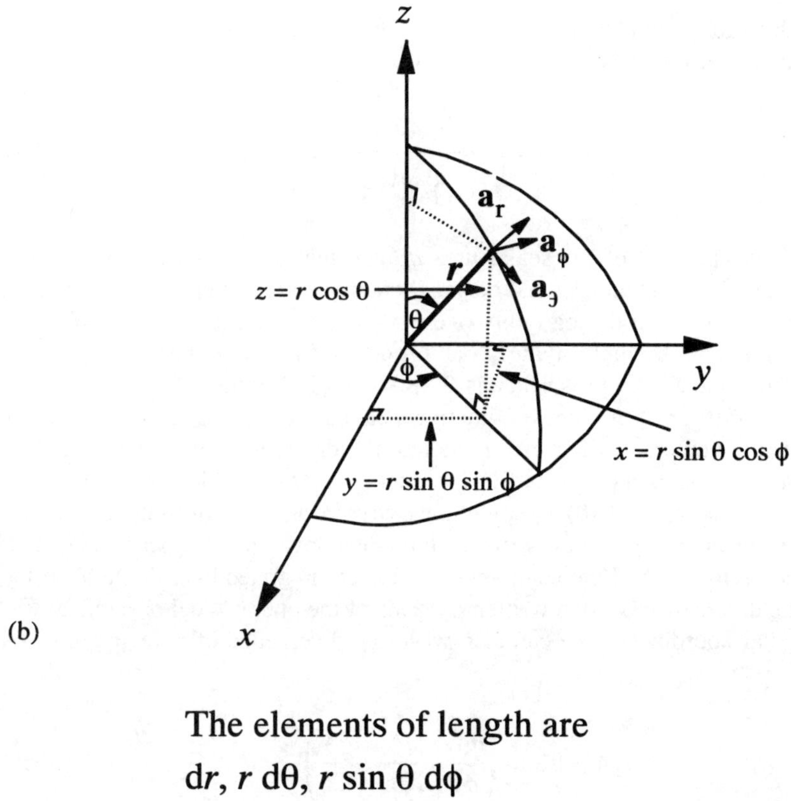

Figure 3.2 The gradient of a scalar field in (a) gradient of voltage potential and (b) spherical coordinates.

3.2.3 Divergence of a Vector Field

In the field of electromagnetism we encounter magnetic fluxes that diverge from magnetic poles. The divergence of the magnetic flux, in this case, would be defined as the total flux coming out of (diverging from) unit volume as the volume shrinks to zero. The divergence is given as

$$\operatorname{div} \boldsymbol{A} = \operatorname*{Lim}_{\text{vol}\to 0} \frac{\int \text{flux}}{\int_V dV_u} \tag{3.9}$$

In general, the divergence can be defined as the normal surface integral over a surface as the volume tends to zero, that is

$$\operatorname{div}\boldsymbol{A} = \lim_{V\to 0} \frac{\oint_s \boldsymbol{A} \cdot \boldsymbol{dS}}{\int_V dV_u} \tag{3.10}$$

The vector $\boldsymbol{dS}$ is the product of the scalar area dS and the unit vector **n** normal to the surface element dS. The element dS must be small enough so that it can be considered as a plane area and so that $\boldsymbol{A}$ varies negligibly over it. The dot product of $\boldsymbol{A}$ and $\boldsymbol{dS}$ is then taken in the usual manner by multiplying $\boldsymbol{dS}$ and $A \cos \theta$ if the angle between the vector $\boldsymbol{A}$ and the normal to the elemental surface is θ, as shown in Figure 3.3.

The integral of this dot product is sometimes called the normal surface integral of $\boldsymbol{A}$ over $\boldsymbol{dS}$ and is the numerator of (3.10). Note that the divergence of a vector field is a scalar. This bears a resemblance to the dot product of vectors, which is a scalar. The integral in the numerator of (3.10) is sometimes written as a double integral. This is because we can consider the vector $\boldsymbol{A}$ to be integrated over the element of area dS, multiplied by the vector $\boldsymbol{dS}$, and the resultant dot product integrated over the total surface S [3, p. 725]. The divergence is often written in terms of the operator del or nabla as $\nabla \cdot A$.

In rectangular coordinates, we can also write the divergence of a vector in differential form as

$$\operatorname{div}\boldsymbol{A} = \operatorname*{Limit}_{\text{vol}\to 0} \left(\frac{\partial A_x}{\partial x} + \frac{\partial A_y}{\partial y} + \frac{\partial A_z}{\partial z} \right) \tag{3.11}$$

In polar coordinates div$\boldsymbol{A}$ is given by

$$\operatorname{div}\boldsymbol{A} = \frac{1}{r^2}\frac{\partial}{\partial r}(r^2 A_r) + \frac{1}{r \sin \theta}\frac{\partial}{\partial \theta}(\sin \theta\, A_\theta) + \frac{1}{r \sin \theta}\frac{\partial A_\phi}{\partial \phi} \tag{3.12}$$

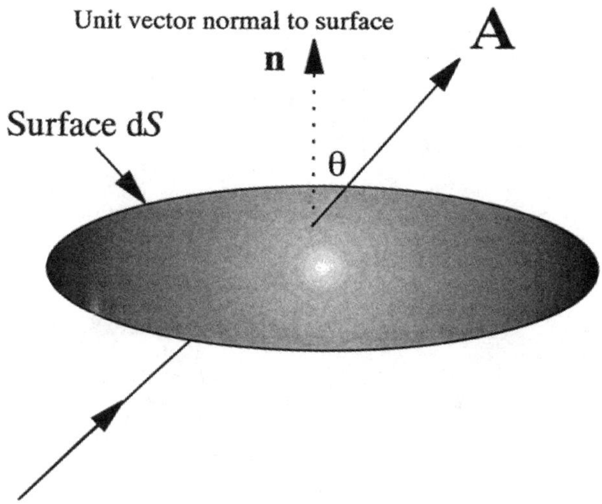

Figure 3.3 The divergence of a vector field: div $\boldsymbol{V}$ in rectangular coordinates $= \frac{\partial A_x}{\partial x} + \frac{\partial A_y}{\partial_y} + \frac{\partial A_z}{\partial z}$; div $\boldsymbol{A}$ in spherical polar coordinates is given by div $\boldsymbol{A} = \frac{1}{r^2}\frac{\partial}{\partial_r}(r^2 A_r) + \frac{1}{r \sin\theta}\frac{\partial}{\partial\theta}(\sin\theta\, A_\theta) + \frac{1}{r\sin\theta}\frac{\partial A_\theta}{\partial\phi}$.

3.2.4 Curl of a Vector

The curl was a term first used by James Clerk Maxwell [1], who took an idea that William Thomson (better known as Lord Kelvin) had of calculating a vector from another vector. The curl of a physical field defines its rotation or vorticity and is one of the most difficult concepts to interpret physically. It is best explained in terms of a physical phenomenon that we can observe in everyday life [2, pp. 2–4]. Consider an object floating downstream on the surface in the y-direction, as shown in Figure 3.4(a). If the stream is flowing in the y-direction with no eddies and has uniform velocity across its width (in the x-direction) the object would move in the y-direction alone. However, in the general case there are eddies caused by the variation in the velocities of the stream in the x- and z-directions. These variations cause the object to turn. For simplicity let us consider variations in the velocities in the x- and y-directions only for an object floating on the surface. The components of the velocities at corner P in the x- and y-directions (along PS and PQ) are v_x and v_y, respectively. The velocity gradients of these components in the y- and x-directions are $\frac{\partial v_x}{\partial y}$ and $\frac{\partial v_y}{\partial x}$, and assuming these gradients are positive, the velocities along QR and SR will be $v_x + \frac{\partial v_x}{\partial y}dy$ and $v_y + \frac{\partial v_y}{\partial x}dx$. The differences in the velocities along the opposite

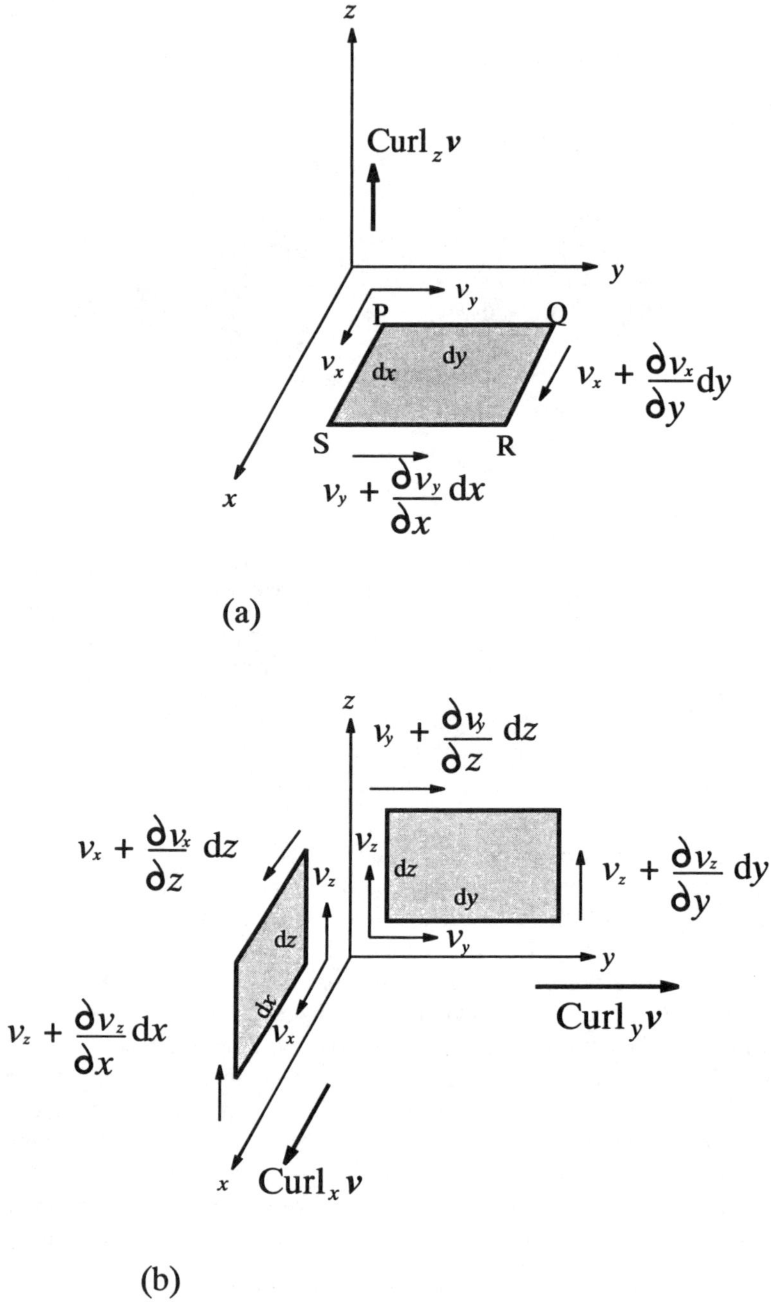

Figure 3.4 The curl of a vector field. (a) $\mathrm{Curl}_z\, \boldsymbol{v}$ component of curl $\boldsymbol{v}$. (b) $\mathrm{Curl}_x\, \boldsymbol{v}$ and $\mathrm{curl}_y\, \boldsymbol{v}$ components of curl $\boldsymbol{v}$.

sides of the object cause the object to turn. The direction of rotation will depend on which velocity gradient is the larger. If $\frac{\partial v_y}{\partial x}$ is the larger gradient the object will turn in a counterclockwise direction, looking down on the $x-y$ plane.

The curl$_z\boldsymbol{v}$ is defined [2, p. 13] as the vector in the positive z-direction that is the result of this counterclockwise rotation in the $x-y$ plane. We can see that this is the same direction that is defined by the right-hand screw rule; that is, if we turn a screw in a counterclockwise direction looking down on the $x-y$ plane, it would move in the positive z-direction. The curl$_z\boldsymbol{v}$ is given by

$$\text{curl}_z\boldsymbol{v} = \mathbf{k}\left(\frac{\partial v_y}{\partial x} - \frac{\partial v_x}{\partial y}\right) \tag{3.13}$$

where $\mathbf{k}$ is the unit vector in the positive z-direction.We can similarly define curl$_x\boldsymbol{v}$ and curl$_y\boldsymbol{v}$ along the positive x- and y-axes as shown in Figure 3.4(b). The total curl of the vector is the sum of these three individual curls and is given by

$$\text{curl } \boldsymbol{v} = \mathbf{i}\left(\frac{\partial v_z}{\partial y} - \frac{\partial v_y}{\partial z}\right) - \mathbf{j}\left(\frac{\partial v_z}{\partial x} - \frac{\partial v_x}{\partial z}\right) + \mathbf{k}\left(\frac{\partial v_y}{\partial x} - \frac{\partial v_x}{\partial y}\right) \tag{3.14}$$

Note the similarity between this vector and the cross product of two vectors that we encountered in Chapter 2 and Section 3.1. The curl can also be written in terms of the del operator as the cross product of del and the velocity vector v, and the right-hand side of (3.14) can be written as a matrix; thus, we have the curl given by

$$\nabla \times \boldsymbol{v} = \begin{vmatrix} \mathbf{i} & \mathbf{j} & \mathbf{k} \\ \frac{\partial}{\partial x} & \frac{\partial}{\partial y} & \frac{\partial}{\partial z} \\ v_x & v_y & v_z \end{vmatrix} \tag{3.15}$$

In the spherical polar coordinate system, the curl in each direction is given by

$$\text{curl}_r\boldsymbol{A} = \frac{a_r}{r \sin\theta}\left[\frac{\partial}{\partial\theta}(\sin\theta\, A_\phi) - \frac{\partial A_\theta}{\partial\phi}\right] \tag{3.16}$$

$$\text{curl}_\theta\boldsymbol{A} = a_\theta\left[\frac{1}{r \sin\theta}\frac{\partial A_r}{\partial\phi} - \frac{1}{r}\frac{\partial(rA_\phi)}{\partial r}\right] \tag{3.17}$$

$$\text{curl}_\phi \boldsymbol{A} = \frac{a_\phi}{r}\left[\frac{\partial}{\partial r}(rA_\theta) - \frac{\partial A_r}{\partial \theta}\right] \tag{3.18}$$

3.3 MAXWELL'S EQUATIONS

Maxwell's equations are based on the previous work of Gauss, Faraday, Biot, and Savart. Maxwell expressed their laws in concise terms and also introduced the idea of a displacement current, which explains the magnetic field caused by the build-up of electric charge (as in the case of the plates of a capacitor) in a similar way to the magnetic field caused by the flow of electrons in a conductor. Until Maxwell's time, it was assumed that a magnetic field could only be produced by a conduction current, that is, electrons flowing in a conductor. This ruled out the explanation of a magnetic field being produced in a region of zero conductivity. Maxwell assumed that any change in an electrostatic field produced an attendant magnetic field and vice versa. This magnetic field is related to the electrostatic field change in the same way as the magnetic field is related to the change of conduction current. Thus, if we have a static charge on an insulator such as an ebonite disk and if we rotate the disk, we will have what is known as a displacement current, and a magnetic field will be produced. It was not until 1889 (after Maxwell's death in 1879) that Rowland showed that a displacement current was the result of electrostatic build-up of charge on the electrodes of a capacitor. The electrostatic lines of force precede the charges and the attendant magnetic field.

Maxwell's equations are sometimes called the EM field equations, because only the first two are attributable to Maxwell. However, all four are normally called Maxwell's equations and are denoted as such in this book.

In the EMC field we are mainly concerned with the electric and magnetic fields in free space, that is, in source free media.

3.3.1 Maxwell's First Equation

This equation is based on Ampere's law, which states that the line integral of the magnetic field intensity is equal to the total current enclosed (i.e., the sum of the conduction and displacement currents) and is given by

$$\oint \boldsymbol{H} \cdot \boldsymbol{dl} = I_{\text{conduction}} + I_{\text{displacement}} \tag{3.19}$$

where H is the magnetic field intensity in amperes per meter (A m^{-1}) and $\boldsymbol{dl}$ is an elementary length in meters (m).

The conduction and displacement currents can be written in terms of the conduction and displacement current densities J_c and J_d, respectively, and the area. Thus (3.19) can be written as

$$\oint \boldsymbol{H} \cdot d\mathbf{l} = \int_S (\boldsymbol{J}_c + \boldsymbol{J}_d)\, dS \tag{3.20}$$

where $\boldsymbol{J}_c$ and $\boldsymbol{J}_d$ are the conduction current density and the displacement current density, respectively, in amperes per square meter (A m^{-2}) and dS is the element of area perpendicular to the current flow in square meters (m^2).

The conduction current density $\boldsymbol{J}_c$ is in turn related to the electric field intensity by the relation $J_c = \sigma \boldsymbol{E}$ where $\boldsymbol{J}_c$ is the conduction current density in amperes per square meter (A m^{-2}), σ is conductivity in siemen-meters (S-m), and $\boldsymbol{E}$ is the electric field intensity in volts per meter (V $m^{-1)}$.

The displacement current density $\boldsymbol{J}_d$ is equal to the rate of change (with time) of the electric flux (or displacement) density $\boldsymbol{D}$. Thus we can say that $\boldsymbol{J}_d$ is given by

$$\boldsymbol{J}_d = \frac{\partial \boldsymbol{D}}{\partial t} \tag{3.21}$$

where $\boldsymbol{J}_d$ is the displacement current density in amperes per square meter (A $m^{-2)}$ and $\boldsymbol{D}$ is the electric flux density in coulombs per square meter (C m^{-2}). The electric flux density can be written in terms of the electric field as

$$\boldsymbol{D} = \varepsilon \boldsymbol{E} \tag{3.22}$$

where ε is the permittivity in farads per meter (F m^{-1}) and $\boldsymbol{E}$ is the electric field in volts per meter (V m^{-1}).

We can write (3.20) as

$$\oint \boldsymbol{H} \cdot d\boldsymbol{l} = \int \left(\sigma \boldsymbol{E} + \frac{\partial}{\partial t}(\varepsilon \boldsymbol{E}) \right) dS \tag{3.23}$$

Stokes's theorem states that the curl of a vector is the maximum value of the line integral around an elementary surface as the surface shrinks to zero. Thus the curl is given by

$$\text{curl } \boldsymbol{A} = \underset{S \to 0}{\text{Limit}} \frac{\oint \mathbf{A} \cdot d\boldsymbol{l}}{\delta S} \tag{3.24}$$

Applying Stokes's theorem to (3.23) we get

$$\int_S \text{curl } \boldsymbol{H} \cdot d\mathbf{S} = \int_S \left(\sigma \boldsymbol{E} + \frac{\partial}{\partial t}(\varepsilon \boldsymbol{E}) \right) dS \tag{3.25}$$

If we assume that the surface becomes very small, that is, $dS \rightarrow \Delta S$, curl $\boldsymbol{H}$ and $\boldsymbol{E}$ can be considered to be approximately uniform over this area and almost independent of the area. Equation (3.25) can be written as

$$\text{curl}\,\boldsymbol{H} = \sigma\boldsymbol{E} + \varepsilon\frac{\partial \boldsymbol{E}}{\partial t} \tag{3.26}$$

This is Maxwell's first equation in differential vector form.

We can also write this equation in scalar differential form for each component of the magnetic field intensity H using the definition of the component of a curl given by (3.13) as

$$\frac{\partial H_z}{\partial y} - \frac{\partial H_y}{\partial z} = \varepsilon\frac{\partial E_x}{\partial t} + \sigma E_x$$

$$\frac{\partial H_x}{\partial z} - \frac{\partial H_z}{\partial x} = \varepsilon\frac{\partial E_y}{\partial t} + \sigma E_y$$

$$\frac{\partial H_y}{\partial y} - \frac{\partial H_x}{\partial x} = \varepsilon\frac{\partial E_z}{\partial t} + \sigma E_z \tag{3.27}$$

In integral form the first equation is given by (3.23), and it can be stated as: The magnetomotive force around a closed path is equal to the conduction current plus the time derivative of the electric flux density through any surface bounded by the path [2, p. 104].

3.3.2 Maxwell's Second Equation

This equation is based on Faraday's law, which states that a magnetic flux of Φ induces an *electromotive force* (EMF) in a conductor. This induced EMF has a magnitude $|V|$ that is proportional to the rate of change of the magnetic flux Φ, and it is given by

$$V = -\frac{d\Phi}{dt} \tag{3.28}$$

The negative sign indicates that the voltage opposes the flux causing it.

This voltage V is also equal to the line integral of the electric field $\boldsymbol{E}$ over the length of the conductor. This integral is found by taking the dot product of $\boldsymbol{E}$ and $\boldsymbol{dl}$ and then integrating over the whole length of the conductor. Thus we have

$$|V| = \int \boldsymbol{E} \cdot \boldsymbol{dl} \tag{3.29}$$

where $\boldsymbol{E}$ is the electric field in V m^{-1} and $\boldsymbol{dl}$ is the elemental length of the total loop. Thus we can say that

$$\oint \boldsymbol{E} \cdot \boldsymbol{dl} = -\frac{d\Phi}{dt} \tag{3.30}$$

where ϕ is the magnetic flux in webers. The flux is related to the flux density by the relation

$$\Phi = \int \boldsymbol{B} \cdot \boldsymbol{d}\mathbf{S} \tag{3.31}$$

where $\boldsymbol{B}$ is the magnetic flux density in tesla (webers per square meter) and $\boldsymbol{d}\mathbf{S}$ is the vector area, that is, the area normal to the flux density $\boldsymbol{B}$.

Combining the above two equations, we get

$$\oint_{\text{perimeter}} \boldsymbol{E} \cdot \boldsymbol{dl} = -\int_{\text{surface}} \frac{d}{dt}(\boldsymbol{B} \cdot \boldsymbol{d}\mathbf{S}) \tag{3.32}$$

which is sometimes quoted as Maxwell's second equation in integral form. It can be stated that the EMF around a closed path is equal to the time derivative of the magnetic displacement through any surface bounded by the path [2, p. 106]. By analogy with electric current, we can call the time derivative of the magnetic flux, the magnetic current. Thus we can state Maxwell's second equation as: The electric voltage around a closed path is equal to the magnetic current through the path.

To derive Maxwell's equation in differential form we use Stokes's theorem, which states that the curl of a vector is the maximum value of the line integral around an elementary surface as the surface shrinks to zero. Thus the curl is given by

$$\text{curl}\,\boldsymbol{A} = \underset{S \to 0}{\text{Limit}} \frac{\oint A \cdot \boldsymbol{dl}}{\delta S} \tag{3.33}$$

Using Stokes's theorem for a vector $\boldsymbol{E}$ we can therefore write the right-hand side of (3.32) in the form

$$\oint_{\text{perimeter}} \boldsymbol{E} \cdot \boldsymbol{d}\mathbf{l} = -\int_{\text{surface}} \text{curl}\boldsymbol{E} \cdot \boldsymbol{d}\mathbf{S} \tag{3.34}$$

Combining (3.32) and (3.34) we get

$$\int_S \operatorname{curl} \boldsymbol{E}\, d\mathbf{S} = -\frac{\partial}{\partial t}\int_S \boldsymbol{B} \cdot \boldsymbol{d}\mathbf{S} \tag{3.35}$$

If we assume that the surface becomes very small, that is $\boldsymbol{d}\mathbf{S} \rightarrow \Delta\mathbf{S}$, then curl $\boldsymbol{E}$ and $\boldsymbol{B}$ are approximately constant over this area. Equation (3.35) thus reduces to

$$\operatorname{curl} \boldsymbol{E} \cdot \Delta\mathbf{S} = -\frac{\partial}{\partial t}(\boldsymbol{B} \cdot \Delta\mathbf{S}) \tag{3.36}$$

Since $\Delta\mathbf{S}$ is independent of time, we can rewrite (3.36) in the form

$$\operatorname{curl} \boldsymbol{E} \cdot \Delta\mathbf{S} = \Delta\mathbf{S}\frac{\partial}{\partial t}(\boldsymbol{B})$$

This gives us Maxwell's second equation in differential form as

$$\operatorname{curl} \boldsymbol{E} = -\frac{\partial \boldsymbol{B}}{\partial t} \tag{3.37}$$

where $\boldsymbol{E}$ is the electric field in volts per meter (V m^{-1}) and $\boldsymbol{B}$ is the magnetic flux density in telsa (Wb m^{-2}).

3.3.3 Maxwell's Third Equation

Maxwell's third equation is based on Gauss's theorem, which states that the total outward flux is proportional to the total charge enclosed. From the definition of divergence, given by (3.9) we have

$$\operatorname{div}\boldsymbol{A} = \lim_{\text{vol}\to 0}\frac{\int \text{flux}}{\int_{V_u} dV_u}$$

In the case of Gauss's theorem we have to consider the divergence of the electric flux density $\boldsymbol{D}$. Writing the total outward flux in terms of this divergence we have

$$\text{total flux} = \operatorname{div} \boldsymbol{D}\int_{V_u} dV_u = \int_{V_u} \rho \, dV_u \tag{3.38}$$

If we consider a volume that has an infinitesimal size of ΔV_u we can assume that the charge distribution ρ is constant over it, and (3.38) can be written as

$$\text{div}\,\boldsymbol{D}\int_{\Delta V_u} dV_u = \rho \int_{\Delta V_u} dV_u$$

which reduces to Maxwell's third equation, namely

$$\text{div}\,\boldsymbol{D} = \rho \tag{3.39}$$

where $\boldsymbol{D}$ is the electric flux density or charge per unit area ($= \varepsilon\boldsymbol{E}$) in coulombs per square meter (C m^{-2}), ε is the permittivity in farads per meter (F m^{-1}), $\boldsymbol{E}$ is the electric field intensity in volts per meter (V m^{-1}), and ρ is the charge distribution per unit volume in coulombs per cubic meter (C m^{-3}).

In integral form this equation can be written as

$$\int_S \boldsymbol{D}\cdot d\boldsymbol{S} = \int_{V_u} \rho\, dV_u \tag{3.40}$$

where S is a closed surface. It can be stated that the total electric displacement through the surface enclosing a volume is equal to the total charge within the volume [2, p. 104].

3.3.4 Maxwell's Fourth Equation

Maxwell's fourth equation is the magnetic flux equivalent of the electric flux defined by the third equation. The divergence of the magnetic flux density $\boldsymbol{B}$ is equal to the distribution of the magnetic poles enclosed. Maxwell's fourth equation states that the divergence of $\boldsymbol{B}$ is zero. We would expect this to be the case since there are no free north or south magnetic poles. A north pole would travel to the south pole of a magnet and a south pole would travel to the north pole. Thus the poles would always appear in pairs and the resultant magnetic pole strength would be zero. Maxwell's fourth equation can be written in differential vector form as

$$\text{div}\,\boldsymbol{B} = 0 \tag{3.41}$$

where $\boldsymbol{B}$ is the magnetic flux density in tesla (Wb m^{-2}). In integral form this equation can be written as

$$\int_S \boldsymbol{B}\cdot d\mathbf{S} = 0 \tag{3.42}$$

where S is a closed surface. It can be stated that the net magnetic flux emerging through any closed surface is zero [2, p. 104].

3.4 BOUNDARY CONDITIONS

We must consider what happens to the fields at the interfaces between two media. The conditions that govern these fields are known as boundary conditions. In general, the media are dielectrics; but in the EMC field, the media are usually free space and a conductor. Let us first define a conductor. A conducting medium is defined as one in which an electric field or voltage potential is always accompanied by a movement of electric charges carried by free electrons. There can be no static charge in a perfect conductor since any electric field causes the electric charges to redistribute themselves until the electric field is zero [2, p. 2]. However, all conductors have a finite conductivity, and at high frequencies the electric field penetrates the conductor to a depth that is inversely proportional to the conductivity. The electric field decays exponentially and asymptotically approaches zero as it penetrates the conductor. As the conductivity approaches infinity (i.e., that of a perfect conductor) the depth of penetration gets smaller and the current forms a thinner "skin" on the surface of the conductor. The skin depth is the distance from the surface of the conductor at which the electric field is equal to $1/e$ (1/2.718) of its magnitude at the surface.

We shall consider the tangential and normal components of the E and H fields in each medium on either side of the interface, as shown in Figure 3.5.

3.4.1 Tangential Component of the Electric Field

The tangential component of the electric field is continuous if both the media are dielectrics, as shown in Figure 3.5. This means that the tangential component of the E field is the same inside each media near the interface; that is, we can say that

$$E_{t1} = E_{t2} \tag{3.43}$$

where E_{t1} and E_{t2} are the tangential electric fields in mediums 1 and 2, respectively, in volts per meter (V m^{-1}).

3.4.1.1 Tangential Component of the Electric Field for Conductors

If the second medium is a perfect conductor, the tangential component of the electric field E_{t2} is zero. We know that this must be the case since the electric field is the voltage per unit length, and the voltage applied across a perfect conductor will immediately fall to zero. This is to be expected from Ohm's law ($V = IR$), since the resistance is zero for a

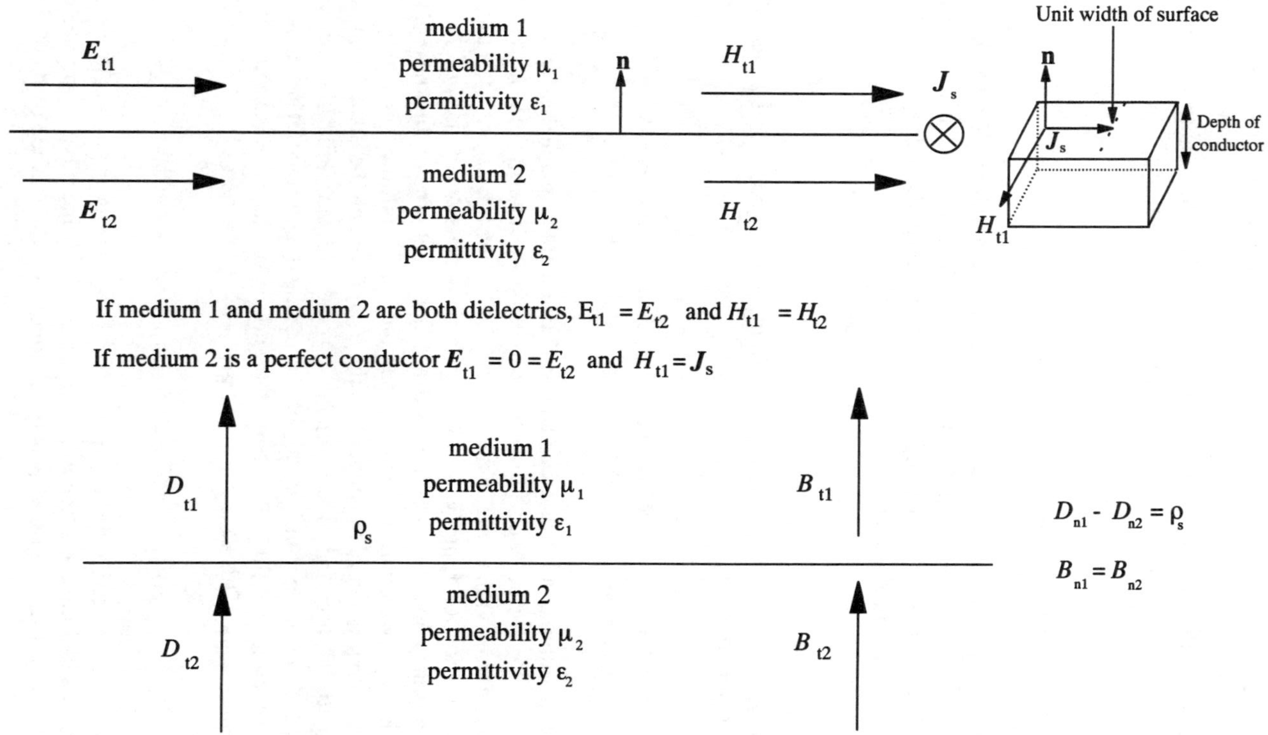

Figure 3.5 Boundary conditions at the interface of two media: (a) for the tangential E and H fields, and (b) for the normal D and B fields.

perfect conductor and thus the voltage will also be zero. Since the tangential electric field is zero just inside the conductor, it must also be zero just outside the conductor in order to satisfy (3.43).

For conductors with finite conductivity the electric field falls to zero at a distance inside the conductor that depends on the frequency in addition to the permeability and conductivity of the conductor.

3.4.2 Tangential Component of the Magnetic Field

The tangential component of the magnetic field $\boldsymbol{H}$ is continuous if both the media are dielectrics and as long as the current density and the electric flux density are finite. This means that the tangential component of the H field is the same inside each media near the interface; that is, we can say that

$$H_{t1} = H_{t2} \tag{3.44}$$

where H_{t1} and H_{t2} are the tangential magnetic fields in mediums 1 and 2, respectively, in amperes per meter (A m^{-1}).

3.4.2.1 Tangential Component of the Magnetic Field at the Interface With a Conductor

If the second medium is a perfect conductor, the tangential magnetic field is zero just inside the conductor surface but not zero just outside the interface in the first medium. This would appear to violate the condition of (3.44), but this is not the case since the equation is only true for finite values of the current density. In the case of a perfect conductor, the finite current flows on the surface as a sheet of zero thickness. This makes the current density (which is the current per unit area) infinite since the area under consideration is the product of the depth into the conductor and the distance on the surface perpendicular to the direction of flow of the current. We can define the surface current density $\boldsymbol{J}_s$ as the current per unit width of the conducting surface. The tangential component of the magnetic field just outside the surface of the conductor is equal to this surface current density. This relationship is given by

$$\boldsymbol{J}_s = \mathbf{n} \times H_{t1} \tag{3.45}$$

where $\boldsymbol{J}_s$ is the surface current density in amperes per meter (A m^{-1}), H_{t1} is the tangential magnetic field in medium 1 in amperes per meter (A m^{-1}), and $\mathbf{n}$ is the unit vector normal to the surface. The surface current density is induced on the surface of the conductor by the EM field.

In general, the tangential magnetic field at the interface between a dielectric and a conductor can be said to be discontinuous by an amount equal to the surface current density (or current per unit width) of the conductor surface; that is, we can say that

$$J_s = H_{t1} - H_{t2} \tag{3.46}$$

3.4.3 Normal Component of the Electric Field

Let us consider the normal component of the electric flux density $\boldsymbol{D}$. The difference between the normal components of the electric flux density $\boldsymbol{D}$ is equal to the surface charge density; that is, we can say that the normal component of the electric flux density is discontinuous by an amount equal to the surface charge density. This is given by the relation

$$D_{n1} - D_{n2} = \rho_s \tag{3.47}$$

where D_{n1} and D_{n2} are the normal electric flux densities in mediums 1 and 2, respectively, in coulombs per square meter (C m^{-2}) and ρ_s is the surface charge density in coulombs per square meter (C m^{-2}).

Since $\boldsymbol{D} = \varepsilon \boldsymbol{E}$, we can write (3.47) as

$$\varepsilon_1 E_{n1} - \varepsilon_2 E_{n2} = \rho \tag{3.48}$$

where ε_1 and ε_2 are the permittivities in mediums 1 and 2, respectively, in farads per meter (F m^{-1}) and E_{n1} and E_{n2} are the electric field intensities in mediums 1 and 2, respectively, in volts per meter (V m^{-1}).

3.4.3.1 Normal Component of the Electric Field for a Chargefree Boundary

In a chargefree boundary the surface charge density ρ_s is zero, so the electric flux density is continuous and (3.47) can be written as

$$D_{n1}=D_{n2} \text{ and (3.48) reduces to } \varepsilon_1 E_{n1}=\varepsilon_2 E_{n2}$$

where ε_1 and ε_2 are the permittivities in mediums 1 and 2, respectively, in farads per meter (F m^{-1}) and E_{n1} and E_{n2} are the electric field intensities normal to the boundary in mediums 1 and 2, respectively, in volts per meter (V m^{-1}).

3.4.3.2 Normal Component of the Electric Field at the Interface With a Conductor

If the second medium is a perfect conductor, D_{n2} is zero, so $D_{n1} = \rho_s$ where D_{n1} is the normal electric flux density in coulombs per square meter (C m^{-2}) in medium 2 and ρ_s is the surface charge density also in coulombs per square meter (C m^{-2}). The surface charge density is induced on the surface of the conductor by the EM field.

Since the tangential component of the electric field in medium 2 is zero, it follows that at the interface of a dielectric and a perfect conductor, the resultant electric field in the dielectric is normal to the interface.

3.4.4 Normal Component of the Magnetic Field

The normal component of the magnetic flux density is continuous at the boundary of two dielectric media. This means that $B_{n1} = B_{n2}$, where B_{n1} and B_{n2} are the magnetic flux densities in telsa in mediums 1 and 2, respectively, applies. Since the magnetic flux density is equal to the product of the magnetic field and the magnetic permeability, we can say that

$$\mu_1 H_{n1} = \mu_2 H_{n2} \tag{3.49}$$

where H_{n1} and H_{n2} are the magnetic fields normal to the boundary in mediums 1 and 2, respectively, in amperes per meter (A m^{-1}) and μ_1 and μ_2 are the magnetic permeabilities in mediums 1 and 2, respectively, in henries per meter (H m^{-1}).

3.4.4.1 Normal Component of the Magnetic Field at the Interface With a Conductor

In the case of the boundary between a conductor and a dielectric, the magnetic permeability of the conductor is much larger than that of a dielectric. This means that $\mu_2 >> \mu_1$. Thus, it follows from (3.49) that the normal component of the magnetic field inside the conductor H_{n2} will be much smaller than that in the dielectric, H_{n1}.

3.5 FIELDS DUE TO A RADIATING DIPOLE

In order to derive the EM fields of a radiating dipole, we can imagine it to be made up of elementary dipoles attached end to end. Each elementary dipole can be considered to be fed by a current that radiates and results in electric and magnetic fields in the surrounding free space.

3.5.1 Fields Due to a Current Element

Let us consider a current element, as shown in Figure 3.6, that is short enough for the current I to be uniform across it. The element is also assumed to be a thin wire so that the current flows longitudinally through it and there are no components of the current in its transverse cross section, that is, there are no radial or azimuthal currents. We usually want to calculate the fields at a radial distance from the element, so it is better to use spherical

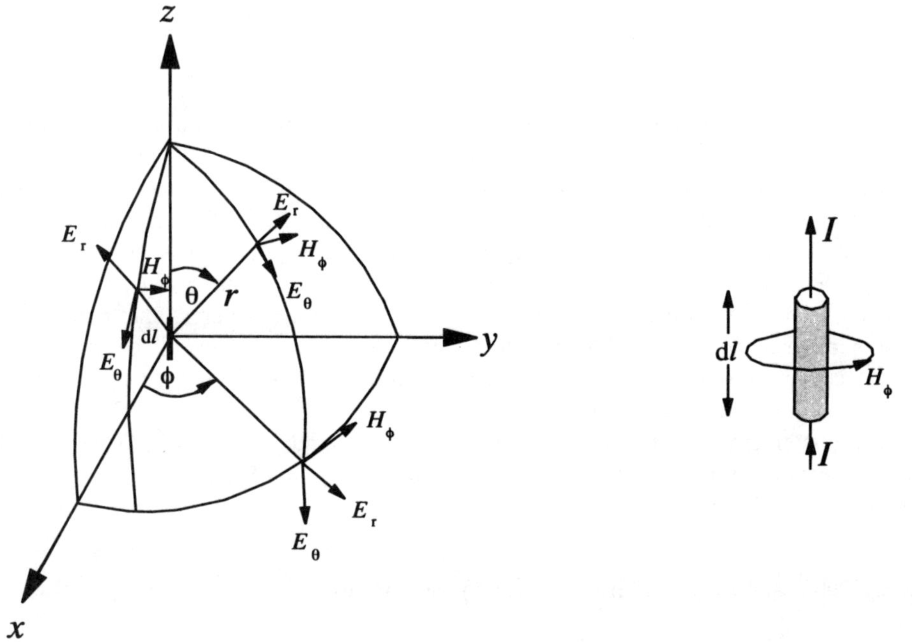

Figure 3.6 Fields due to a radiating current element.

polar coordinates rather than rectangular coordinates. If the current through the element is alternating, then the current has a magnitude of $I \cos \omega t$. There is a finite time for the EM wave to travel from the element to a point P at a distance r. If the velocity of the EM wave is v meters per second, the time for the wave to reach P will be r/v sec. This means that there will be a phase delay of $\omega r/v$ rad if the angular frequency is ω rad/s.

The only component of the magnetic field is that given by right-hand screw rule. This is the component in the ϕ-direction in the plane transverse to the current element.

In the case of the electric field we would not expect to have a component of the electric field in the transverse plane perpendicular to the vertical element (since this would be at right angles to the electric polarization vector). Thus the components of the ***E*** field would be in the θ- and the radial r-directions only.

The instantaneous values of the component fields can be derived by applying Maxwell's equations [2, p. 304] and are given by the expressions

$$H_\phi = \frac{Idl \sin\theta}{4\pi}\left[-\frac{\omega \sin \omega(t - r/v)}{rv} + \frac{\cos \omega(t - r/v)}{r^2}\right] \tag{3.50}$$

$$E_\theta = \frac{Idl \sin\theta}{4\pi\varepsilon}\left[-\frac{\omega \sin \omega(t - r/v)}{rv^2} + \frac{\cos \omega(t - r/v)}{r^2 v} + \frac{\sin \omega(t - r/v)}{\omega r^3}\right] \tag{3.51}$$

$$E_r = \frac{2Idl\cos\theta}{4\pi\varepsilon}\left[\frac{\cos\omega(t - r/v)}{r^2 v} + \frac{\sin\omega(t - r/v)}{\omega r^3}\right] \tag{3.52}$$

In (3.50) the first term on the right-hand side is the radiation field or distant field term and the second term is the induction field term. The radiation field term is not present for steady (dc) currents and takes account of the finite time of propagation, of a EM wave, which does not apply in the dc case. It contributes to the flow of energy away from the element, whereas the induction field term contributes to energy that is stored in the field during one-quarter of the cycle and returned during the next quarter. We can see that the radiated field term is inversely proportional to the distance r, whereas the induction field term is inversely proportional to the square of the distance. This means that the induction field decays rapidly as the distance increases and becomes negligible compared with the radiation field at large values of r.

In the electrostatic case, $\omega \to 0$, and thus the $1/r$ and $1/r^3$ terms in (3.51) as well as the $1/r^3$ term in (3.52) will be zero (since $\sin\omega(t-r/v) = 0$). This leaves us with the $1/r^2$ terms only for E_θ and E_r.

3.5.2 Fields at Large Distances From Wire Antennas

At large distances the terms that are inversely proportional to the square and cube of the distance will be reduced to very small magnitudes and, thus, can be ignored. This means that the radial component of the electric field E_r can be ignored. The second and third terms of E_θ and the second term of H_ϕ can also be ignored. We can therefore write the electric and magnetic fields at large distances from the dipole in terms of the E_θ and H_ϕ, which are given by

$$E_\theta = \frac{Idl\sin\theta}{4\pi\varepsilon}\left[\frac{\omega\sin\omega(t - r/v)}{rv^2}\right] \tag{3.53}$$

$$H_\phi = \frac{Idl\sin\theta}{4\pi}\left[\frac{\omega\sin\omega(t - r/v)}{rv}\right] \tag{3.54}$$

3.6 POWER FLUX DENSITY FOR A PLANE WAVE

The power flux density is a vector known as Poynting's power flux vector, and its direction is in the direction of propagation of the EM wave. In the case of a plane wave the electric and magnetic fields are perpendicular to each other and the plane containing these vectors is perpendicular to the direction of propagation of the wave as shown in Figure 3.7(a).

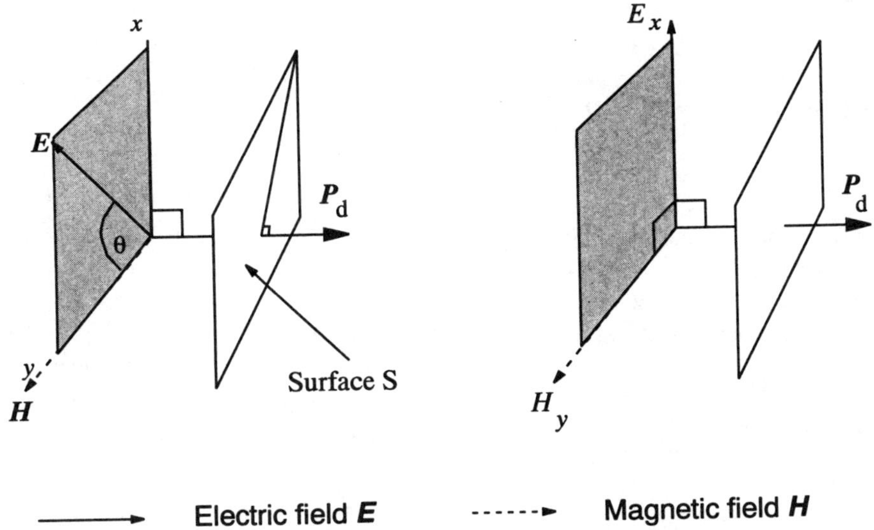

(a)

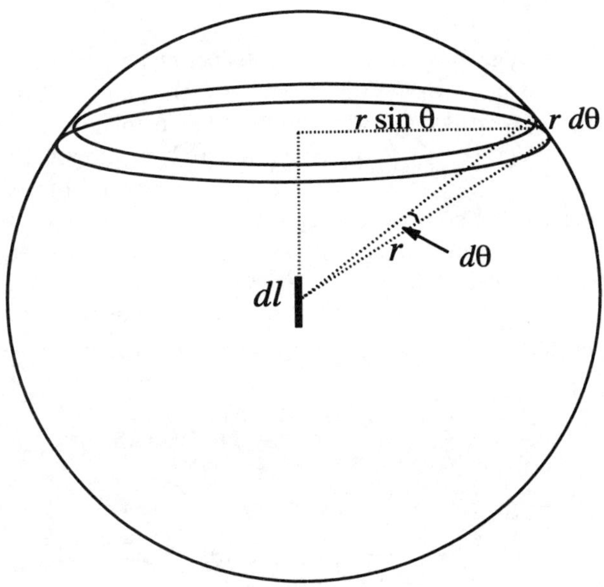

(b)

Figure 3.7 Power flux density: (a) Poynting's power flux vector ***P***—general and plane wave cases, and (b) calculation of total power radiated over a sphere.

The power density $\boldsymbol{P}_d$ is defined as the vector or cross product of the electric and magnetic field intensities and is given by $\boldsymbol{P}_d = \boldsymbol{E} \times \boldsymbol{H}$. The magnitude of the power flux density is given by the real part of this product, since this is the only part that contributes to the power. The magnitude is given by

$$P_d = Re(\boldsymbol{E} \times \boldsymbol{H}) \tag{3.55}$$

where *Re* stands for the real part of the expression in the brackets, $\boldsymbol{E}$ is the electric field in volts per meter (V m^{-1}), and $\boldsymbol{H}$ is the magnetic field in amperes per meter (A m^{-1}). Only the mutually perpendicular components of $\boldsymbol{E}$ and $\boldsymbol{H}$ contribute to the power flow.

For sinusoidally alternating fields, the magnitude of the average (in time) Poynting's power flux vector is half the magnitude of that given by (3.55). This is because the *root mean square* (RMS) of the electric and magnetic fields is $1/\sqrt{2}$ of their amplitudes or peak values. The average power density $P_{d(ave)}$ is given by

$$P_d(\text{ave}) = \frac{1}{2}\text{Re}(\boldsymbol{E} \times \boldsymbol{H}^*)$$

where $\mathbf{P}_{d(ave)}$ is the average power density in watts per meter (W m^{-2}), $\boldsymbol{E}$ is the amplitude of electric field in volts per meter (V m^{-1}), and $\boldsymbol{H}$ is the amplitude of the magnetic field in amperes per meter (A m^{-1}). To calculate the total average power radiated in the far field, we take the average power density (i.e., the product of the expressions for H_ϕ and E_θ given by (3.53) and (3.54)) and integrate it over a sphere. In order to perform this integration we consider an elemental thin annular shell on a sphere as shown in Figure 3.7. Its width is $rd\theta$ and its circumference is $2\pi r \sin \theta$, and thus its area is $2\pi r^2 \sin \theta \, d\theta$. The total average power radiated is found by integrating this annular shell with respect to θ from 0 to π in order to get the area over the whole sphere. Note that we only need to consider the amplitudes of H_ϕ and E_θ, and thus the total average power P_{ave} is given by

$$P_{ave} = \frac{1}{2}\int_0^\pi \left(\frac{\omega^2 I^2 dl^2 \sin^2\theta}{16\pi^2 r^2 v^3 \varepsilon}\right) 2\pi r^2 \sin \theta \, d\theta \tag{3.56}$$

$$= \frac{\omega^2 I^2 dl^2}{16\pi v^3 \varepsilon} \int_0^\pi (\sin^3 \theta \, d\theta)$$

since $\sin^3\theta$ can be written as $\sin \theta(1 - \cos^2\theta)$, we can integrate this expression more readily to give

$$P_{ave} = \frac{\omega^2 I^2 dl^2}{16\pi v^3 \varepsilon}\left[-\cos \theta + \frac{\cos^3\theta}{3}\right]_0^\pi$$

Evaluating the expression in brackets yields 4/3, and so average power is given by

$$P_{\text{ave}} = \frac{I^2 dl^2 \omega^2}{12\pi v^3 \varepsilon} \tag{3.57}$$

For free-space propagation the velocity v is equal to the speed of light c, which is in turn equal to $1/\sqrt{\mu_0 \varepsilon_0}$. Additionally, $\varepsilon = \varepsilon_0$, $\omega = 2\pi f$, and $c = f/\lambda$. Thus the total average power is given by

$$P_{\text{ave}} = \frac{\pi I^2 dl^2}{3\lambda^2} \sqrt{\frac{\mu_0}{\varepsilon_0}} \tag{3.58}$$

This is the total average power radiated by a current element carrying a uniform current along its length. This current element is called a doublet or a Hertzian dipole. In practice, a Hertzian dipole cannot be realized; although a very short dipole $<\lambda/10$ is a fairly good approximation. The current amplitude along the length of a radiating wire decreases uniformly from a maximum at the center to zero at the ends and a triangular distribution of current is a good approximation for a short dipole. This would give an average current equal to half the maximum, and thus $I^2/4$ should be substituted for I^2 in (3.58) to give the total average power as

$$P_{\text{ave}} = \frac{\pi I^2 dl^2}{12\lambda^2} \sqrt{\frac{\mu_0}{\varepsilon_0}} \tag{3.59}$$

3.7 WAVE IMPEDANCE FOR A PLANE WAVE

We can think of an EM wave as having an impedance in the same way as a transmission line has a characteristic impedance. The wave impedance, ξ, is usually called the intrinsic impedance and is defined as the ratio of the electric and magnetic field intensities. We can see that dimensionally this would be the case since the units of the electric and magnetic fields are volts per meter (V m^{-1}) and amperes per meter (A m^{-1}), so the units of their ratio gives us impedance. The electric and magnetic fields are very complex near the antenna, but in the far field are given by the expressions of (3.53) and (3.54).

$$\xi = \frac{E_\theta}{H_\phi} = \frac{1}{\varepsilon v} \tag{3.60}$$

For free space, $\varepsilon = \varepsilon_0$ and $v = c = 1/\sqrt{\mu_0 \varepsilon_0}$.

Thus the free-space wave impedance ξ_0 given by (3.60) can be written as

$$\xi_0 = \sqrt{\frac{\mu_0}{\varepsilon_0}} \tag{3.61}$$

3.8 RADIATION RESISTANCE

The total average power given by (3.58) is equivalent to the familiar expression of $I^2R/2$ for average power in the case of a sinusoidally alternating current, since the rms current is $1/\sqrt{2}$. In this case since the element is radiating, we call its resistance the radiation resistance. Thus we get an expression for the radiation resistance of the current element given by

$$R = \frac{2\pi}{3}\sqrt{\frac{\mu_0}{\varepsilon_0}}\left(\frac{dl}{\lambda}\right)^2 \tag{3.62}$$

Inserting the values of $4\pi \times 10^{-7}$ and $1/36\pi \times 10^9$ for μ_0 and ε_0, respectively, into (3.62) we get the expression $R = 80\pi^2\left(\frac{dl}{\lambda}\right)^2$ for the radiation resistance of a current element dl, through which a constant current I flows.

If we assume that the current distribution is triangular, as a more realistic situation we get the expression for the power density given by (3.59) and, hence, the radiation resistance is given by

$$R = 20\pi^2\left(\frac{l}{\lambda}\right)^2 \tag{3.63}$$

which is a fairly good approximation for dipoles of less than $\lambda/4$ [2, p. 312]. The corresponding expression for short monopoles of length $<\lambda/8$ is given by

$$R = 10\pi^2\left(\frac{l}{\lambda}\right)^2 \tag{3.64}$$

Note that in this case the length l is twice the physical length of the monopole, since if we have a monopole on an infinite ground plane, it has an image and therefore its length appears to be doubled.

3.9 FAR FIELD OF ANTENNAS

The distance from an antenna at which a plane wave can be assumed to exist depends on the whether we have a wire or an aperture antenna.

3.9.1 Far Field for Wire Antennas

We can define the far-field distance as the distance from the dipole at which the magnitude of the induction field is equal to the radiated field. Comparing the amplitudes of these two terms given by (3.50) we get

$$\frac{\omega}{rv} = \frac{1}{r^2} \tag{3.65}$$

where the velocity v is the speed of light c and ω is the angular velocity in radians per second and, therefore, is equal to $2\pi f$. Also since $f = c/\lambda$, (3.65) reduces to $r = \lambda/2\pi$, which is the distance from wire antennas at which the intrinsic impedance of the wave has a purely resistive value of 120π. This is the impedance of a plane wave; and thus, this is also the distance at which a plane wave can be assumed to exist. In practice, the far-field distance is taken as double this value, that is, λ/π is usually taken as the far-field distance.

3.9.2 Far Field for Aperture Antennas

Aperture antennas such as horns and reflectors tend to operate at higher frequencies mainly in the microwave region above 300 MHz. In this region, the approach is similar to that used in optics and the far field is known as the Fraunhofer region. Let us consider a point along the axis as the center of a family of spheres with radii that increase in steps of $\lambda/2$ from R, to $R + n\lambda/2$, where n is an integer. The spheres divide the aperture into a series of annular regions, three of which are shown in Figures 3(a, b). In the perspective view, for reasons of clarity, the plane of the aperture is shown transparent and the outer spherical sections are shown in outline.

We can see that as we move away from the aperture the size of the Fresnel zones increase and the number of the zones decrease. The limit of the Fresnel region is taken as the distance at which the whole aperture (of diameter D) is just one Fresnel zone. This means that if the distance P along the axis is R, the distance from P to the edge of the aperture is $(R + \lambda/2)$. Using Pythagoras's theorem we get

$$\left(R + \frac{\lambda}{2}\right)^2 = \left(\frac{D}{2}\right)^2 + R^2$$

$$\lambda R = \frac{D^2}{4} - \frac{\lambda^2}{4}$$

For $D > \lambda$, λ^2 is negligible compared with D^2; hence we can say that

$$R = \frac{D^2}{4\lambda} \tag{3.66}$$

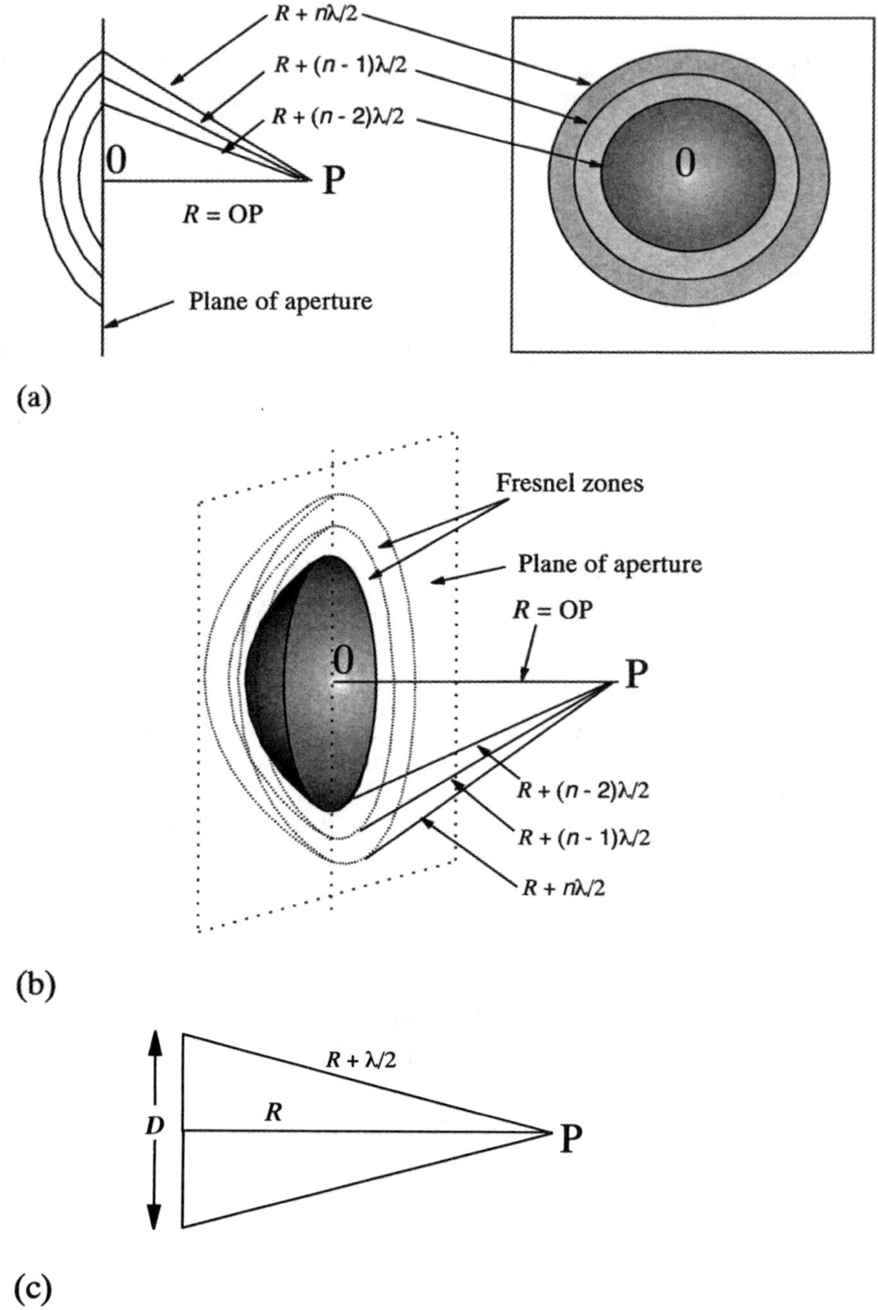

Figure 3.8 Fresnel zones and Fraunhofer region: (a) cross sectional and front views of spheres with center P, (b) perspective view of Fresnel zones, and (c) calculation of the Fraunhofer distance.

Distances less than $D^2/4\lambda$ are considered to be in the Fresnel region; however $D^2/4\lambda$ is not the point at which the Fraunhofer region starts since the path difference between the center and the edge of the aperture is $\lambda/2$. The Fraunhofer or far field is considered as the point at which the path differences between points on the aperture are negligible. The Fraunhofer region is usually taken as $2D^2/\lambda$ or $4D^2/\lambda$. If we compare the ratio of the gain at each of these two distances with that at infinity G_0 [4, p. 188] we get that the ratio of G to G_0 is 0.94 at $2D^2/\lambda$ and 0.99 at $4D^2/\lambda$. Thus the distance $4D^2/\lambda$ more accurately represents the far-field distance for an aperture antenna, although the $2D^2/\lambda$ is often the distance used for EMC measurements.

REFERENCES

[1] Mahin, P. J., ''Maxwell's Grand Unification,'' *IEEE Spectrum*, Vol. 29, No. 3, March 1992, p. 46.

[2] Jordan, E. C., *Electromagnetic Waves And Radiating Systems*, London: Constable and Company Ltd., 1953.

[3] Fewkes, J. H., and J. Yarwood, *Electricity, Magnetism And Atomic Physics, Volume 1*, second edition, London: University Tutorial Press Ltd., 1962.

[4] Silver, S., *Microwave Antenna Theory And Design*, first edition, New York: McGraw-Hill Book Company Inc., 1949.

Chapter 4
Antennas for Frequencies Below 1 MHz

This chapter describes the antennas used at frequencies from 50 Hz to 1 MHz. The mechanism of radiation is explained qualitatively, and effective height/length is defined. Electric and magnetic wave impedances are introduced and the difference between receiving and transmitting antennas is explained by considering the dipole as an example. These antennas are commonly called *receive and transmit antennas*, and often abbreviated to Rx and Tx. At frequencies below 1 MHz the wavelength is greater than 300 m, and thus the antennas used have to be small fractions of a wavelength in order to have antennas of manageable proportions. The physical distances involved (between the antenna and the *device under test* (DUT)) are also a few meters, which means that the electrical distances (in terms of wavelength) are small fractions. This leads to measurements being performed in the near field of antennas. Under these conditions the radiation pattern (which is the radiated power in the far field) is not relevant. The antennas used at these frequencies tend to be simple wire antennas such as dipoles, monopoles, and loops. The dipoles are generally used for receiving, whereas monopoles and loops are used for receiving as well as transmitting. The antennas also tend to have preamplifiers used to overcome the loss experienced between the antenna and the receiver or transmitter. This loss is partially a result of connecting a balanced antenna to an unbalanced line (such as a coaxial line) as well as a result of the impedance mismatch between the antenna and transmission line to which it is connected. This chapter also explains the concept of radiation power factors and describes the circuits used to balance and match the antenna to the transmission line.

4.1 MECHANISM OF RADIATION

When an electron is suddenly accelerated or decelerated the effect of its field takes a finite time to reach a distant point. This time depends on the distance and the velocity of the light. If the electron is subjected to a sinusoidally alternating acceleration, the field also

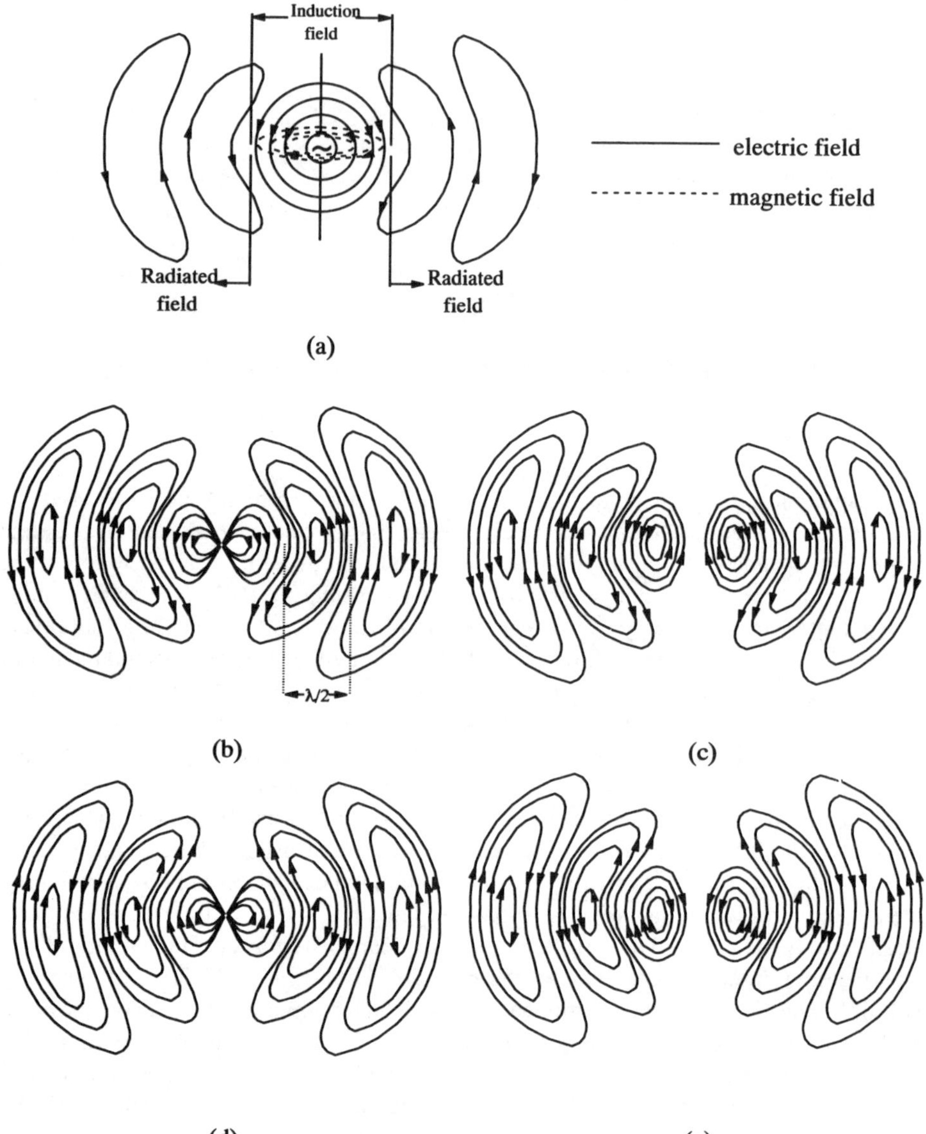

Figure 4.1 Mechanism of radiation: (a) induction field near a radiating dipole, (b) radiated field at $\omega t = 0$, (c) radiated field at $\omega t = \pi/2$, (d) radiated field at $\omega t = \pi$, (e) radiated field at $\omega t = 3\pi/2$.

changes in a similar manner. Since the magnitude of the field depends on the acceleration experienced by the electron, it depends on the frequency f of the alternating current [1]. At higher frequencies (i.e., in the kilohertz range and above) the charge distribution on the antenna reverses sign more rapidly than the collapsing field lines near the antenna. This reversing field repels the collapsing field and pushes it away so that the latter forms closed loops that expand as they travel away from the antenna at the speed of light. In the induction field near a vertical dipole, the electric field lines are in the vertical plane, as shown in Figure 4.1(a), and the magnetic field is in the horizontal plane. Beyond this region the electric radiated field is shown in Figures 4.1(b–e) at every quarter of a cycle [2].

4.2 NEAR AND FAR FIELDS OF RADIATION

Near the antenna where interference effects predominate, the resultant wave is of a complex nature and the power density (the power per unit area) does not decrease monotonically (i.e., it does not decrease consistently) as the distance from the antenna increases but follows an oscillatory pattern. Because of this dependence of the field on distance, standards specify the distance at which measurements have to be performed. This also explains why the measurements of power at 3 m and 10 m do not follow the inverse square law. The near field is reactive as explained in Chapter 3, and the space is occupied mainly by stored energy in the electric and magnetic fields. In the far field where a plane wave exists, the electric and magnetic fields are perpendicular to each other and to the direction of propagation of the wave. Figure 4.2 shows a plane wave propagating in the z-direction with the electric field in the y-direction, hence showing the spatial distribution (the variation with distance) of the electric and magnetic fields at a fixed instant in time. It is important to remember that these fields also vary with time in a similar sinusoidal way, which results in the propagation of the wave as in the case of shallow water waves in a ripple tank. The electric and magnetic fields are in time phase with each other, unlike the situation in circuits containing inductors and capacitors where the phase between the fields is 90 deg.

4.3 WAVE IMPEDANCE

In the near field of an antenna, the intrinsic impedance of the wave is larger or smaller than it is beyond the Fresnel region, in the far field. The intrinsic wave impedance in the far field is that of a plane wave, that is, 120π or 377Ω. However, in the near field (distances $<\lambda/2\pi$ for wire antennas) the intrinsic wave impedance ξ depends on the type of radiator. In the near field of magnetic sources such as loops, the intrinsic wave impedance ξ_H at a distance r is approximately equal to $240\pi^2 r/\lambda$, whereas in the near field of electric sources such as monopoles and dipoles, the wave impedance ξ_E is approximately equal to $60\lambda/r$. We can see that at $r = \lambda/2\pi$ the wave impedances for both fields are equal to 120π. The

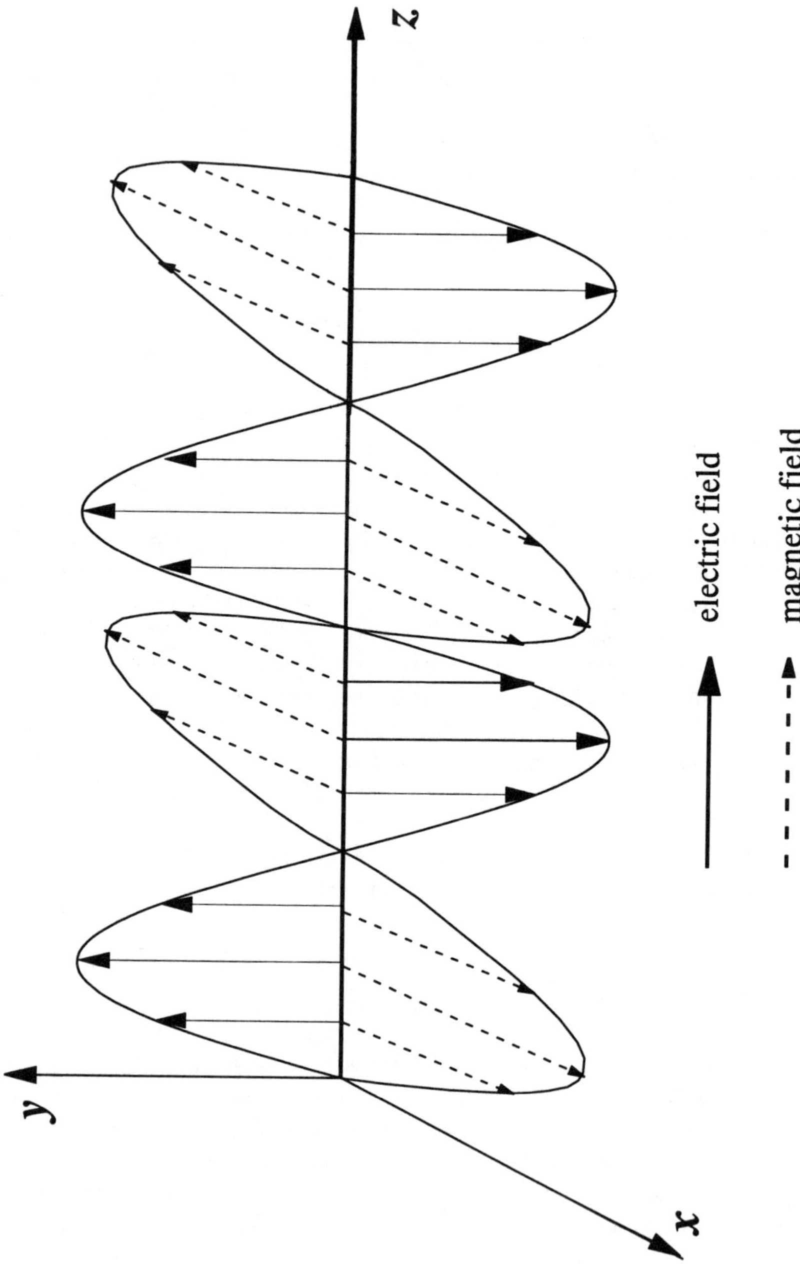

Figure 4.2 Plane wave in free space.

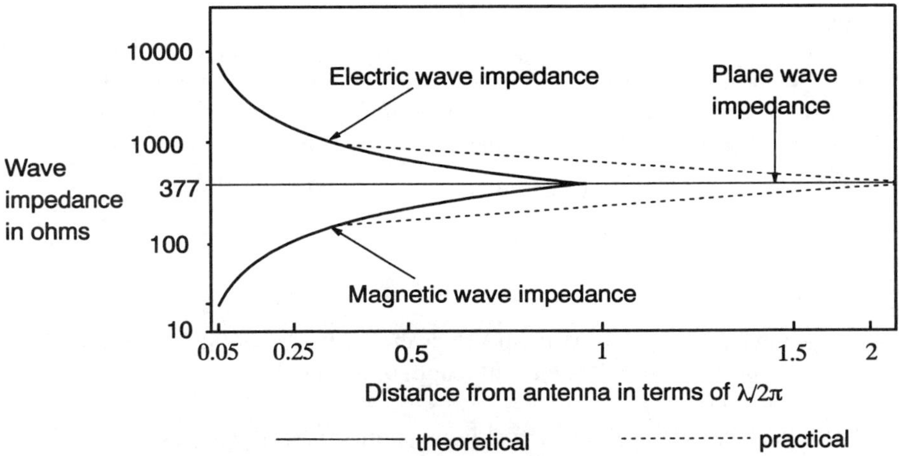

Figure 4.3 Difference between receptor and transmitter antennas.

intrinsic wave impedances for electric and magnetic sources are shown in Figure 4.3 for distances in terms $\lambda/2\pi$. Thus 1 on the x-axis represents a distance of $\lambda/2\pi$. The solid curves show the theoretical values of the electric and magnetic wave impedances ξ_E and ξ_H whereas the dotted curves show the actual impedances. It can be seen that in the case of the dotted curves the impedances asymptotically approach 377 Ω at a point beyond the value of 1, at a value approaching 2, that is, a physical distance of λ/π. It will also be seen that in the near field of magnetic sources the wave has a low impedance, whereas in the near field of electric sources the wave has a high impedance. Low wave impedances induce high currents, whereas high wave impedances induce low currents. EMC standards specify the type of field required at different frequencies.

4.4 DIFFERENCE BETWEEN RECEIVE AND TRANSMIT ANTENNAS

The reciprocity theorem states that an antenna has the same characteristics whether it is radiating or receiving. These characteristics include the impedance and radiation pattern. However, one of the main differences between a radiating and receiving antenna is the orientation of the antenna's radiation pattern for optimum transmission and reception [5, p. 546]. The optimum pattern is used during transmission to direct the signal in a given direction(s). If there is noise in a particular direction, then increasing the directivity of the antenna in that direction will increase the signal-to-noise ratio in that direction. However, in reception, the optimum condition is not the maximum received power but the highest signal-to-noise ratio. If the noise is coming from the same direction as the signal, then increasing the directivity in that particular direction will not increase the signal-to-noise ratio, since both the signal and the noise will be increased.

Another important difference between a transmit and receive antenna is the distribution of current along the length. In the case of a thin transmitting dipole that is fed centrally, the current distribution is approximately sinusoidal, as shown in Figure 4.4(a). The thinner the antenna, the closer is the approximation to a sinusoidal distribution. The instantaneous currents I_1 and I_2 in the upper and lower arms of the dipole, respectively, are given by

$$I_1 = I_{max1} \sin \beta(H - z) \tag{4.1}$$

$$I_2 = I_{max2} \sin \beta(H + z) \tag{4.2}$$

where I_{max1} and I_{max2} are the maximum values or amplitudes of the currents in the upper and lower arms of the dipole, H is the half length of the dipole, and β is the phase constant or wave number $2\pi/\lambda$.

The input impedance Z_0 of the dipole is approximately given by

$$Z_a = -jZ_0 \cot(\beta H) \tag{4.3}$$

where Z_0 is the characteristic impedance given by

$$Z_0 = 120\left[\log_e\left(\frac{2H}{a}\right) - 1\right] \tag{4.4}$$

where a is the radius of the conductor of the dipole.

In the case of a receiving dipole, the current distribution is a function of the length,

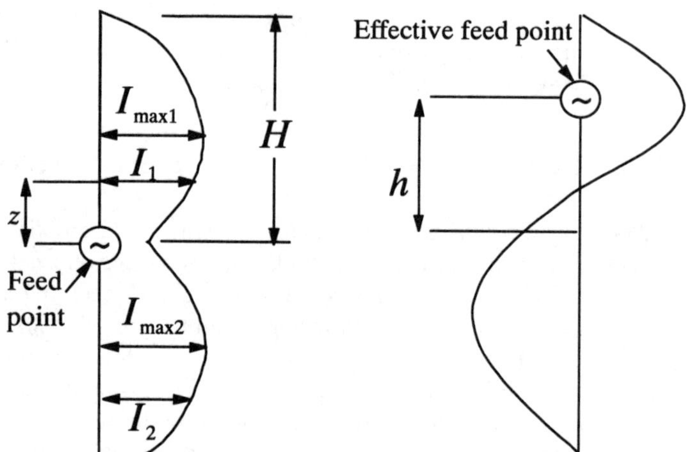

Figure 4.4 The current distribution for (a) transmit and (b) receive dipole.

the direction of arrival of the incident wave, and the impedance of the load connected to the dipole. In most cases the load is the receiver. The current does not approximate a sinusoidal current distribution, except in the case of resonant half-wave dipoles. We can think of the receiving dipole as an unsymmetrically fed dipole. Consider the dipole of Figure 4.4(b), which is fed at a point at a distance h from the center. If we ignore the radiation resistance and assume that the total input impedance is made up of the sum of the impedances of the two sections of lengths $(H - h)$ and $(H + h)$, then the total impedance is given by

$$Z_{\text{total}} = Z(H + h) + Z(H - h) \tag{4.5}$$

where

$$Z_{(H - h)} = \frac{1}{2} jZ_{01}\cot \beta(H - h) \tag{4.6}$$

and

$$Z_{(H + h)} = \frac{1}{2} jZ_{02}\cot \beta(H + h) \tag{4.7}$$

where Z_{01} and Z_{02} are the characteristic impedances of the two sections. As Z_{01} gets larger, Z_{02} gets smaller. If we assume that these characteristic impedances are approximately equal, they will each be equal to Z_0 given by (4.4).

4.5 SMALL ANTENNAS

At frequencies below 1 MHz where the wavelength is greater than 300 m, antennas have electrical sizes (the size in terms of wavelength) that are fractions, that is, magnitudes that are <<1. These fractions are usually taken as $<\lambda/8$ for monopoles and $<\lambda/4$ for dipoles. The antennas are called small antennas, although their physical size may be several meters. Another definition of a small antenna by Wheeler [3, p. 6–1] is one that occupies a small fraction of a radiansphere, which is defined as a sphere of radius $\lambda/2\pi$. It will be noted that this is the far-field distance for wire antennas that we encountered in Chapter 3, where the magnitude of the induction and radiation fields are equal. Within this sphere, the stored field predominates, whereas outside this sphere the radiation field predominates.

4.6 BALUNS

The transmission line connecting an antenna to the receiver or transmitter is usually a coaxial line. When a dipole is connected to a coaxial line, one arm is connected to the inner conductor and the other arm is connected to the outer conductor. Since the arms are not coupled to the line in the same way, an imbalance occurs. If the dipole is fed by a two-wire line, there would be no imbalance since the two-wire parallel line is connected to each arm of the dipole in the same way. However, the loss experienced by two-wire lines makes them unsuitable except at very low frequencies. Let us consider a coaxial line feeding a dipole as shown in Figure 4.5(a). The currents I_1 and I_2 that flow along the inner and outer conductors are equal and opposite. At a fixed point in time, I_1 is flowing down the inner conductor (and the right arm of the dipole) and I_2 is flowing up the inside of the cylindrical outer conductor [5, p. 539]. The current I_2 flows to the top of the outer conductor and at point A divides into two, with I_3 flowing down the outside of the conductor and I_2-I_3 flowing along the left arm of the dipole. The equivalent circuit is shown in Figure 4.5(b). The magnitude of the current I_3 depends on the impedance to the ground Z_g provided by this path along the outer conductor. In order to make I_3 as small as possible, we use a balun, which ideally balances the line by canceling this outside current [9, p. 365]. In practice, I_3 is made as small as possible by making Z_g as large as practicable. Balun is an abbreviation of balanced–unbalanced. The balun can consist of a quarter wave line in the form of a skirt of effective length equal to a quarter of a wavelength, as shown in Figure 4.5(c). The skirt is a concentric cylinder whose lower end is connected to the outer conductor at point C. The impedance between points A and B depends on the Q of the shorted quarter wave section. In theory, if the section is exactly an effective quarter wave, this impedance would be infinite; in practice, it can be set to a very high value for a band of frequencies. This type of balun is sometimes called a Bazooka balun [9, p. 366].

In another arrangement for a balun shown in Figure 4.5(d), the outer conductor feeds the left arm of the dipole and the inner conductor is connected to the right arm of the dipole. A transmission line of approximately a quarter of a wavelength is attached between the right arm and the outer conductor at C. These baluns are relatively narrow-band. Ferrite devices can be used to increase the bandwidth as in the case of the combined balun-transformer described in Section 4.8.

4.7 RADIATION POWER FACTOR

A term called the radiation power factor was introduced by Wheeler [3, Chapter 6] in 1947. It is defined as the reciprocal of the Q factor of the combination of the reactance of the antenna X_a and its effective radiation resistance R_e. The radiation power factor PF_{rad} of the antenna is the ratio of R_e to X_a and is given by

$$PF_{rad} = \frac{R_e}{X_a} \tag{4.8}$$

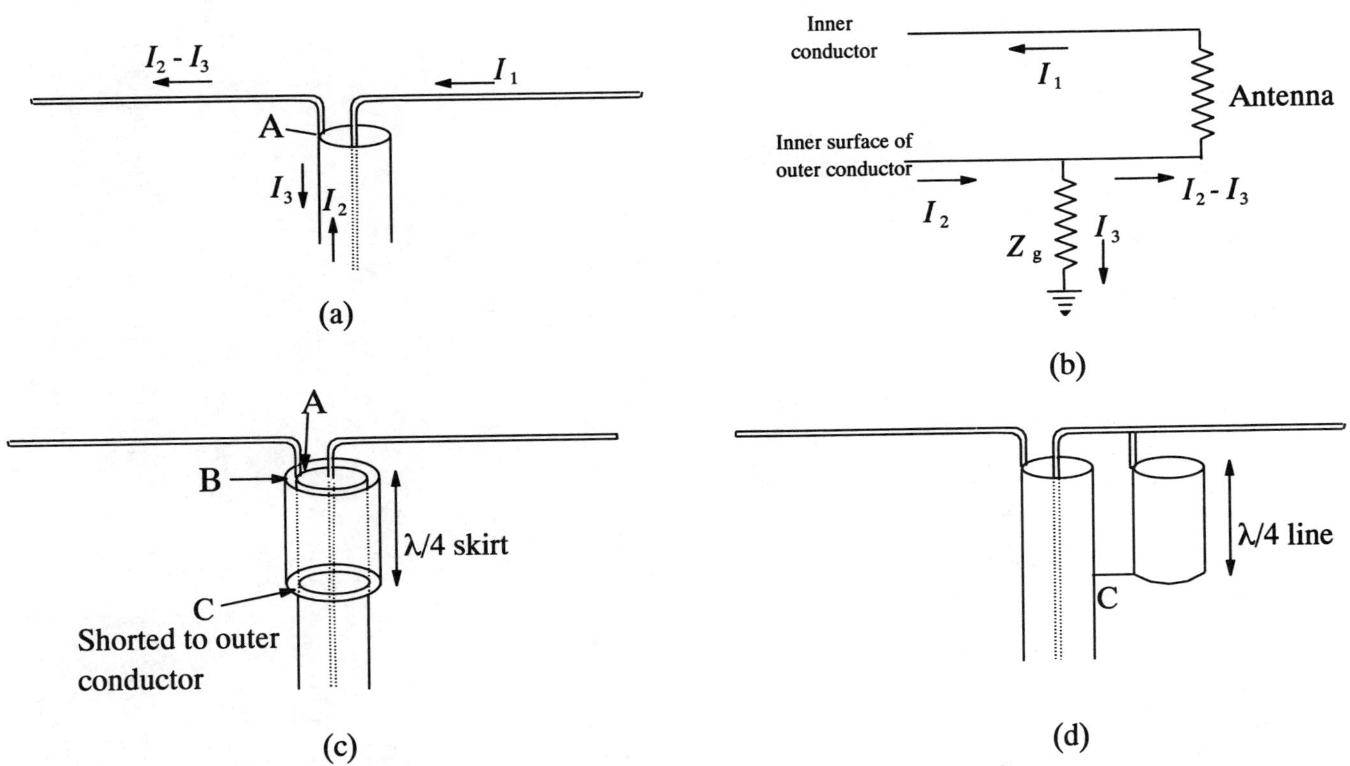

Figure 4.5 Types of baluns: (a) coaxial lien feeding a dipole without a balun, (b) equivalent circuit of the dipole and coaxial line without a balun (c) Bazooka balun consisting of an outer concentric skirt, (d) coaxial balun.

This radiation power factor is <<1 and is proportional to the size of the antenna. It limits the operating efficiency of the antenna. The radiation efficiency of a transmitting antenna represents the percentage of available power from a transmitter that is radiated into free space. In the case of receiving antennas it is the percentage of power from free space that is delivered to the receiver and is a measure of the ability of the received signal to overcome the noise level in the circuitry. A small dipole behaves like a capacitor, whereas a small loop behaves like an inductor. The equivalent circuits of a small dipole and a small loop are shown in Figure 4.6.

In the case of a small dipole the antenna has a capacitive reactance 1/ωC, and the power it radiates can be assumed to be coming from a small inductance L_a in parallel with a much larger resistance R_a. The inductance L_a has a reactance that is <<1/ωC. The combination of the resistance R_a and the inductance L_a is an impedance Z_e that is given by

$$\frac{1}{Z_e} = \frac{1}{j\omega L_a} + \frac{1}{R_a} \tag{4.9}$$

$$Z_e = \frac{j\omega L_a R_a}{R_a + j\omega L_a}$$

$$Z_e = \frac{j\omega L_a}{1 + j\omega L_a\ R_a}$$

Using the binomial expansion $(1 + x)^{-1} \simeq (1 - x)$, we get

$$Z_e = j\omega L_a\left(1 - \frac{j\omega L_a}{R_a}\right) \tag{4.10}$$

$$Z_e = j\omega L_a - \frac{j^2\omega^2 L_a^2}{R_a} \tag{4.11}$$

Since $j^2 = -1$, (4.11) can be written as

$$Z_e = j\omega L_a + \frac{\omega^2 L_a^2}{R_a}$$

The total antenna impedance Z_a is this impedance Z_e in series with the capacitance C and is given by

$$Z_e = -\frac{j}{\omega C} + j\omega L_a + \frac{\omega^2 L_a^2}{R_a}$$

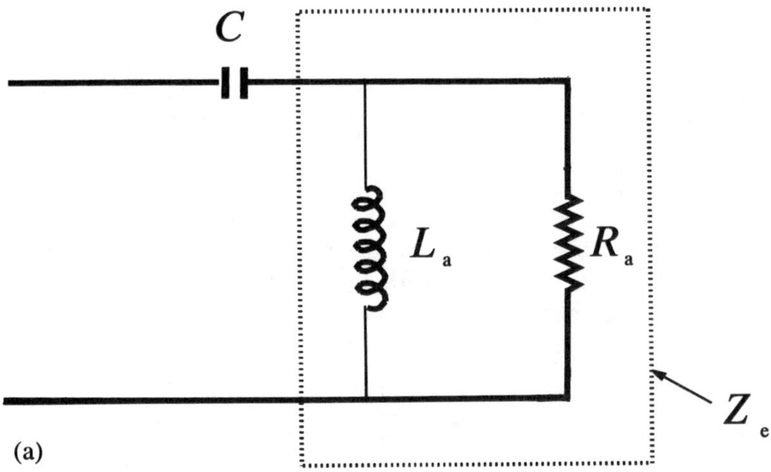

(a)

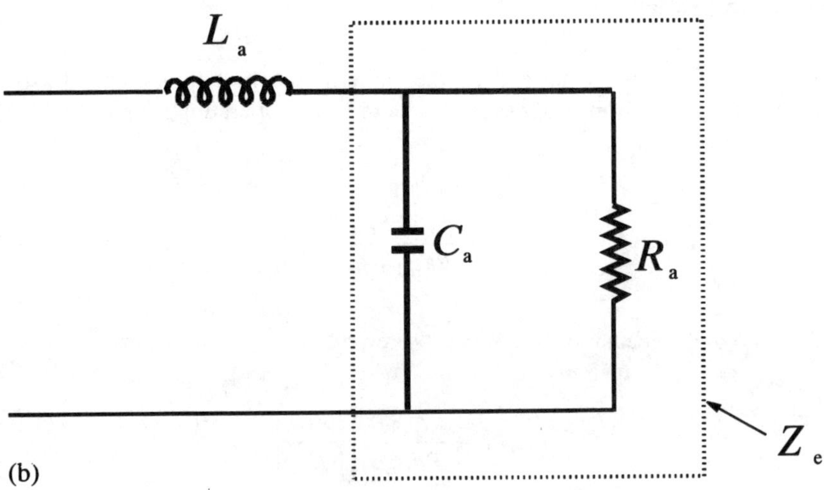

(b)

Figure 4.6 Equivalent circuits for (a) a small dipole and (b) a loop.

Since $j\omega L_a << j/\omega C$, it can be ignored. The total antenna impedance Z_a is therefore given by

$$Z_a = -\frac{j}{\omega C} + \frac{\omega^2 L_a^2}{R_a} \tag{4.12}$$

The first term on the right-hand side of (4.12) is the effective reactance X_e and the second term is the effective resistance R_e. The radiation power factor PF_{rad} is the ratio of the resistive to the reactive parts and is given by

$$PF_{rad} = \frac{\omega^3 L^2 C}{R_a} \tag{4.13}$$

Thus we can see that the radiation power factor is proportional to the cube of the frequency.

4.7.1 Operating Efficiency

The operating efficiency is defined as the ratio of the radiation power factor to the combination of the radiation and loss power factors. The operating efficiency η_{op} is given by

$$\eta_{op} = \frac{PF_{rad}}{PF_{rad} + PF_{loss}} \tag{4.14}$$

Using the binomial expansion, (4.14) can be rewritten as

$$\eta_{op} = 1 - \frac{PF_{loss}}{PF_{rad}} + \left[\frac{PF_{loss}}{PF_{rad}}\right]^2 \tag{4.15}$$

To a first approximation, the third term of (4.15) can be ignored. Thus we can see that the smaller the ratio of PF_{loss} and PF_{rad}, the greater is the operating efficiency. Small antennas have comparable loss to their radiation power factors, and so their operating efficiency is small. As the size of antennas increase, their operating efficiencies also increase. Tuning units with reactances opposite to the reactance of the antenna are used, so the tuning units used for dipoles and loops consist mainly of inductors and capacitors, respectively.

For passive antennas operating over narrow frequency bands, adjustable tuning units (step or continuous) are used and the relative bandwidth is less than the radiation power factor PF_{rad}. The efficiency is limited by the PF_{loss}, which is the dissipative loss in the antenna and the tuning circuit. In broadband operation fixed tuning units are used. The relative bandwidth is much greater than PF_{rad}, and the efficiency is limited by the ability of the passive network to match the antenna to the receiver or transmitter.

4.8 MATCHING ANTENNAS

The antenna can be regarded as a transformer that matches the impedance of a transmission line (usually 50Ω for a coaxial line) to 377Ω for a plane wave in free space. The antenna must be matched to (have the same impedance as) the characteristic impedance of the transmission line that feeds it or connects it to the receiver. If the antenna is not matched, maximum power transfer will not take place. The antenna can be matched to the transmission line by inserting a transformer or tuning circuit between them. In the case of an electric dipole, which has a capacitive impedance, a transformer with double inductive tuning is a common device used for tuning a dipole [3, Figure 6–1], as shown in Figure 4.7(a). A combined balun-transformer incorporating a ferrite core such as that shown in Figure 4.7(b) can be used to obtain bandwidths of 8:1. The ferrite core maintains a high impedance over a large bandwidth [9, p. 368].

4.9 EFFECTIVE LENGTH AND EFFECTIVE HEIGHT

There are many definitions of effective length and effective height depending upon the application. There are the general definitions used in electric engineering and specific ones used in EMC engineering.

4.9.1 Effective Length

The IEEE dictionary [10] defines effective length in two ways, depending on whether the antenna is radiating or receiving.

1. For an antenna radiating linearly polarized waves, the effective length is defined as the length of a straight conductor that is oriented perpendicular to the direction of maximum radiation having a uniform current equal to that at the antenna terminals and producing the same far-field strength as the antenna. Thus if an antenna has a triangular current distribution, its effective length would be the length it would have to be if its maximum current I_{max} (usually at the feed point) were to be distributed uniformly across its entire length.

2. In the case of the same antenna receiving linearly polarized waves from the same direction, the effective length is defined as the ratio of the open-circuit voltage developed

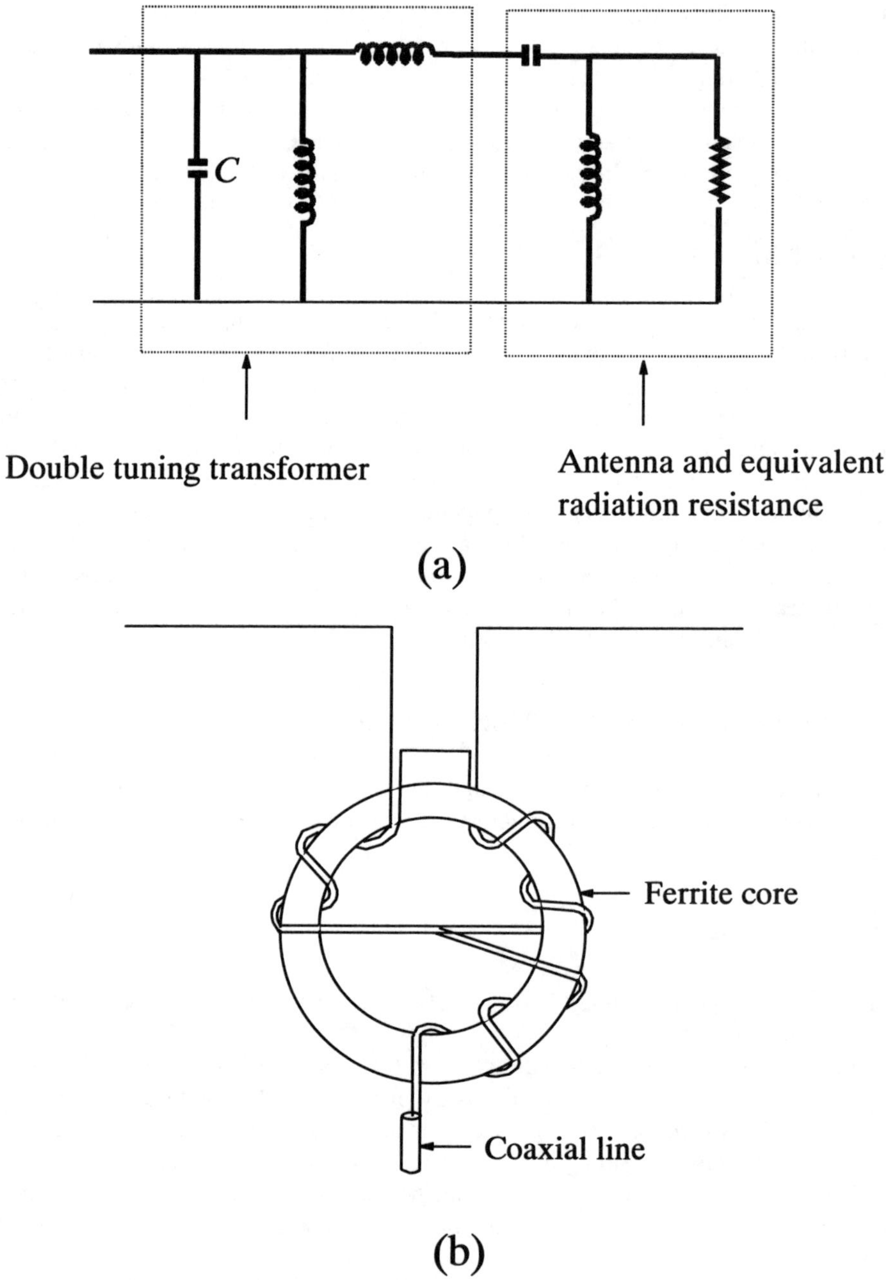

Figure 4.7 (a) Double inductive transformer and (b) combined balun-transformer for a dipole.

at the terminals of the antenna to the component of the electric field strength in the direction of antenna polarization.

4.9.2 Effective Height

The IEEE dictionary [10] defines the term effective height in a number of ways.

1. The unqualified definition is the height of the center of radiation above the effective ground level.

2. For low-frequency applications the term is applied to loaded or unloaded vertical antennas and defined as the moment of the current distribution in the vertical section, divided by the input current.

In EMC low-frequency applications the term effective height is generally used in the two ways.

1. Firstly it is defined in one of the ways that the IEEE dictionary defines the effective height, that is, as the moment of the current distribution in the vertical section, divided by the input current. The moment of current distribution is the current at a particular point multiplied by its distance from the position of maximum current. The position of maximum current occurs at the input terminals in the case of radiating antennas.

For a short thick monopole of length $< 0.1\lambda$ the effective height h_{eff} is given by

$$h_{\text{eff}} = \frac{\int_0^l I(z)\,dz}{I_{\max}} \tag{4.16}$$

where h_{eff} is the effective height in meters, l is the physical length of the monopole in meters, and $I(z)$ is the current at a distance z along the monopole in amperes. The current $I(z)$ is given by

$$I(z) = \frac{I_{\max}}{l} z \tag{4.17}$$

Thus (4.16) can be written as

$$h_{\text{eff}} = \int_0^l \frac{I_{\max}}{l} z\,dz \Big/ I_{\max} \tag{4.18}$$

Integrating with respect to z gives

$$h_{\text{eff}} = \left[\frac{I_{\max}}{l}\frac{z^2}{2}\right]_0^l \Big/ I_{\max} \tag{4.19}$$

which reduces to

$$h_{\text{eff}} = \frac{l}{2} \tag{4.20}$$

Thus we can see that the effective height is only equal to half its physical height/length.

2. The other way in which the effective height is defined is similar to the IEEE definition of the effective length for a receive antenna but broadened to cover magnetic as well as electric fields. It is defined as the ratio of the antenna open-circuit voltage to the strength of the field component being measured. The effective height [8] of a magnetic radiator such as a loop antenna can be considered as the number of volts measured per telsa (T) of incident magnetic flux density, whereas in the case of an electric radiator such as a dipole it is the number of volts per meter (V m^{-1}) of incident electric field. In other words, it acts as a conversion between telsas (or volts per meter) and volts.

$$h_{\text{eff}} = \frac{V}{\boldsymbol{B}} \text{ or } \frac{V}{\boldsymbol{E}} \tag{4.21}$$

where V is the voltage in volts at the output terminals of the antenna, $\boldsymbol{B}$ is the incident magnetic flux density in telsas, and $\boldsymbol{E}$ is electric field in volts per meter.

4.10 *E* FIELD ANTENNAS

Electric field wire antennas are in general linear wires carrying current with the electric field in the space surrounding the antenna having the same polarization as the direction of the current in the wire. At low frequencies, below 1 MHz *E*-field antennas are mainly dipoles and monopoles. The parallel element is another antenna that is used to provide an electric field for susceptibility measurements. Dipoles are usually used as receiving probes, and monopoles are used for transmitting as well as receiving.

4.10.1 Small Dipole

At low frequencies the dipole is usually used as a receiving probe to give relative values rather than absolute values of electric field strengths. For example, it may be used to check the integrity of a shielded room or equipment casing by identifying leakage from rogue seams and apertures. It can be used for absolute measurements if it has been calibrated or if it is used in a calibrated test site. A small radiating dipole behaves like a current element, and it is sometimes called a Hertzian dipole or a doublet. We have seen (Chapter 3) that the instantaneous values of the component fields shown in Figure 4.8(a) can be derived by applying Maxwell's equations [5, p. 304] and are given by the expressions

$$H_\phi = \frac{Idl \sin\theta}{4\pi}\left[\frac{\omega \sin \omega\,(t - r/v)}{rv} + \frac{\cos \omega(t - r/v)}{r^2}\right] \tag{4.22}$$

$$E_\theta = \frac{Idl \sin \theta}{4\pi\varepsilon}\left[-\frac{\omega \sin \omega\,(t - r/v)}{rv^2} + \frac{\cos \omega\,(t - r/v)}{r^2 v} + \frac{\sin \omega\,(t - r/v)}{\omega r^3}\right] \tag{4.23}$$

$$E_r = \frac{2Idl \cos \theta}{4\pi\varepsilon}\left[\frac{\cos \omega(t - r/v)}{r^2 v} + \frac{\sin \omega\,(t - r/v)}{\omega r^3}\right] \tag{4.24}$$

where I is the current in amperes, dl is the length of the element in meters, ω is the angular frequency in radians per meter, r is the radial distance from the element in meters, ε is the permittivity of the medium in farads per meter, and v is the velocity of the wave in the medium in meters per second.

If we consider the RMS values of a time-varying field of the form $e^{-j\omega t}$ and use the cosine term as the reference phasor, as shown in Figure 4.8(b), then the sine term lags behind the cosine phasor and can be represented by a phasor of $-j$, as shown in Figure 4.8(c). Similarly, the negative sine term can be represented by a phasor of $+j$, as shown in Figure 4.8(d). The phase constant β can be used to replace the ratio ω/v. Equations (4.22) to (4.24) can therefore be rewritten in the forms

$$H_\phi = \frac{j\beta^2 Idl}{4\pi}\left[\frac{1}{\beta r} - \frac{j}{(\beta r)^2}\right]\sin \theta\; e^{-j\beta r} \tag{4.25}$$

$$E_\theta = 30j\beta^2 Idl\left[\frac{1}{\beta r} - \frac{j}{(\beta r)^2} - \frac{1}{(\beta r)^3}\right]\sin \theta\; e^{-j\beta r} \tag{4.26}$$

$$E_r = 60\beta^2 Idl\left[\frac{1}{(\beta r)^2} - \frac{j}{(\beta r)^3}\right]\cos \theta\; e^{-j\beta r} \tag{4.27}$$

Near the antenna, that is, at distances $<0.01\lambda$ the $1/r$ and $1/r^2$ terms can be ignored in comparison with the $1/r^3$ term. Thus, only the last terms in E_r and E_θ need to be considered. The H_ϕ component can be ignored since it does not have a $1/r^3$ term. The E_r and E_θ components are given by

$$E_r = -\frac{j60Idl}{r^3}\cos \theta e^{-j\beta r} \tag{4.28}$$

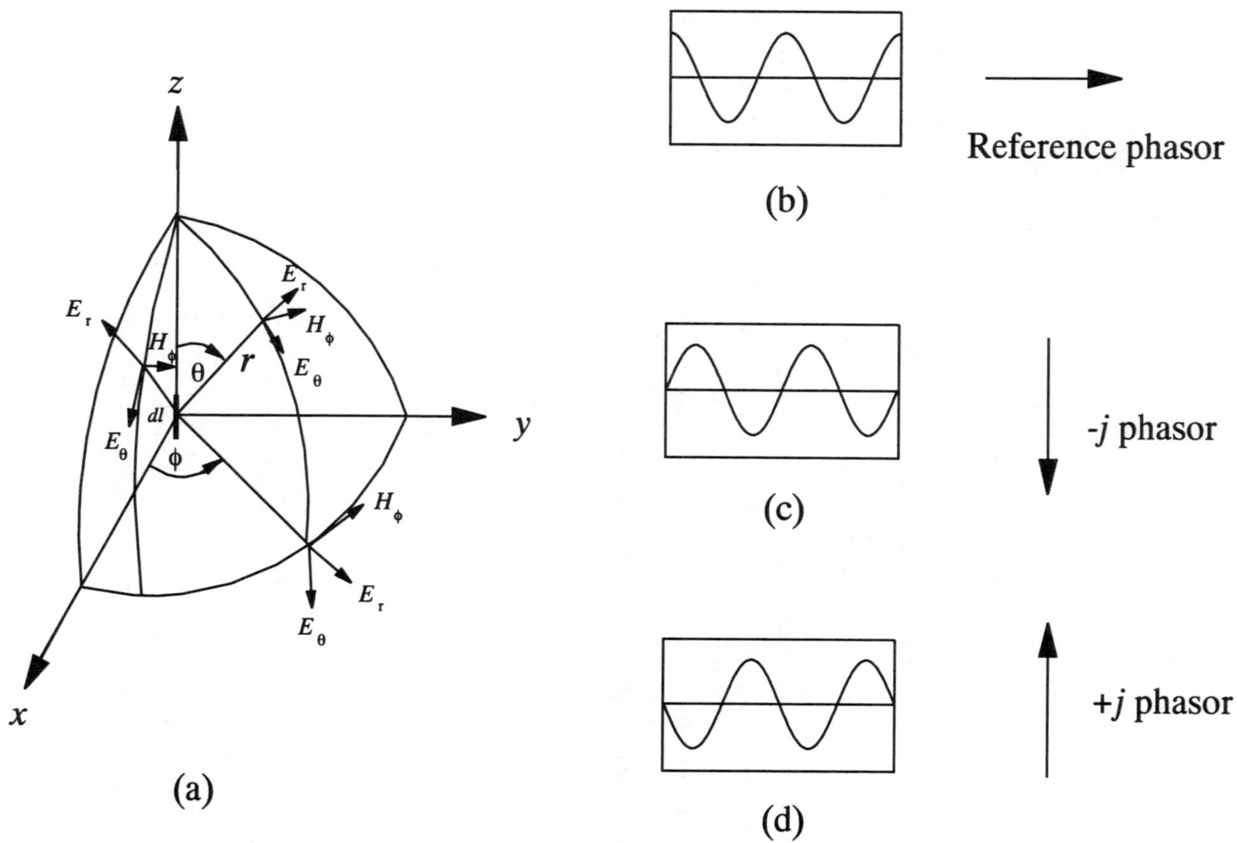

Figure 4.8 The electric and magnetic fields due to a small dipole. (a) Orientation of the electric and magnetic field vectors form a radiating dipole; (b) cosine term; (c) postive sine term; and (d) negative sine term.

$$E_\theta = -\frac{j30Idl}{r^3} \sin \theta e^{-j\beta r} \tag{4.29}$$

It can be seen that the E_r vector has twice the magnitude of the E_θ vector, and the phase between the two vectors is 90 deg. The ratio of E_r to E_θ in the near field is given by

$$\frac{E_r}{E_\theta} = 2 \cot \theta \tag{4.30}$$

4.10.2 Short Monopoles

Monopoles are used at low frequencies for transmitting as well as receiving. Monopoles can be made more broadband by making the diameter of the conductor larger. Length-to-diameter ratios *(l/d)* of 10 to 10^4 are common [4]. A linear dipole with an *l/d* ratio of 5,000 and a bandwidth of 3% can have its bandwidth extended to 30% by keeping its length constant but having its *l/d* ratio decreased to 260 [9, p. 333]. The monopole could be fed by a coaxial line, with the center conductor connected to the conductor of the monopole and the outer conductor connected to the ground plane.

For monopoles of height $H < 0.2\lambda$ the approximate reactance X_a is given by

$$X_a = -Z_a' \cot\left(\frac{2\pi H}{\lambda}\right) \tag{4.31}$$

where the value of Z_a' [5, p. 512] has been found empirically (by comparison with experimental results) to be given by

$$Z'_a = 60\left[\log_e\left(\frac{H}{a}\right) - 1\right] \tag{4.32}$$

where a is the radius of the conductor of the monopole.

For very thin monopoles, that is, where the radius is much smaller than the length, the current distribution is approximately sinusoidal and independent of the size of the ground plane [4]. However, for thick monopoles or rods, the current distribution is no longer sinusoidal and the current distribution on the monopole as well on the ground plane are functions of the radius of the ground plane.

4.10.3 Ground Plane Dependence

In cases where the ground plane is small in terms of wavelengths, the outer edge of the ground plane diffracts the incident radiation in all directions. The currents on the top and bottom surfaces of the ground plane are equal in magnitude, but they must be opposite in direction since the net current at the edge is zero. This outer edge diffraction becomes more significant as the size of the ground plane gets smaller since the currents are larger near the monopole. Edge diffraction can alter the input impedance by more than 100% and the gain by more than 6 dB compared with a monopole on an infinite ground plane.

For monopoles with infinite and zero ground planes, the peak directivity occurs at the horizon, that is, at $\theta = 90$ deg as shown in Figure 4.9(a, c). However, in general, it can be said that as the size of the infinite ground plane is decreased, the peak directivity occurs at smaller angles ($\theta < 90$ deg), although this variation is not monotonic. The directivity at $\theta = 90$ deg for a short monopole with no ground plane is 1.76 dBi, whereas with a large but finite ground plane the directivity at $\theta = 90$ deg is -1.249 dBi, and with an infinite ground plane the directivity is 4.77 dBi [4]. The magnitude of the peak directivity for a short monopole with no ground plane is also smaller, 1.76 dBi, as compared with 4.77 dBi for a large or infinite ground plane. The radiation resistance R_{rad} of a monopole with no ground plane is $20\pi^2(h/\lambda)^2$, whereas with a large or infinite ground plane the radiation resistance is $40\pi^2(h/\lambda)^2$ as shown in Figure 4.9.

At frequencies up to 1 MHz where the wavelength is greater than 300 m it is possible to have a ground plane consisting of wires instead of a solid one. A ground plane of eight or more radial wires is often used as shown in Figure 4.9(d).

4.10.4 Top-Loaded Monopoles

A monopole can be loaded by attaching a disk at the top or by bending it into the shape of an L or T as shown in Figure 4.10. Top loading can increase the effective length by a factor of two; this increases the gain of the antenna. Top loading can also decrease the capacitive reactance of a short antenna [5, p. 511]. In the case of a receive antenna, a large shunt capacitance requires a high inductive tuning circuit; this lowers the efficiency of the total circuit and, hence, the signal received.

In the case of transmitting antennas the current I depends on the voltage V and the capacitive reactance X_{a} of the antenna; that is, $I = V/X_{\text{a}}$. The power radiated is given by

$$P_{\text{rad}} = I^2 R_{\text{rad}} \tag{4.33}$$

where R_{rad} is the radiation resistance of the antenna.

Since the current $I = V/X_{\text{a}}$, the power radiated can be written as

$$P_{\text{rad}} = \frac{V^2 R_{\text{rad}}}{X_{\text{a}}^2} \tag{4.34}$$

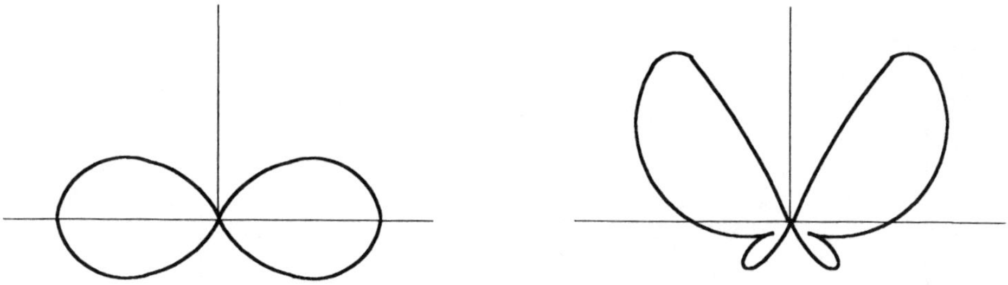

(a) (b)

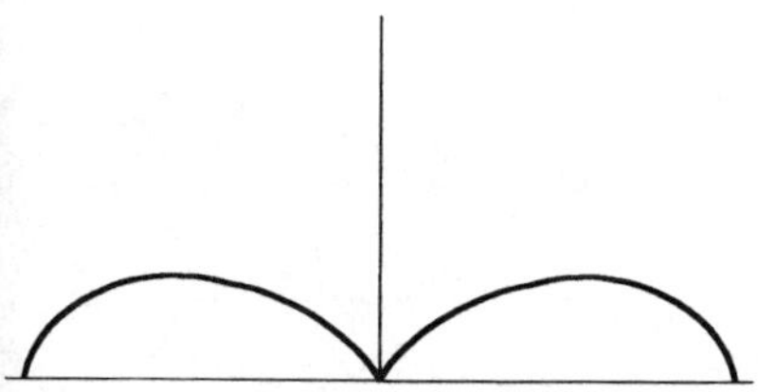

(c) Infinite ground plane

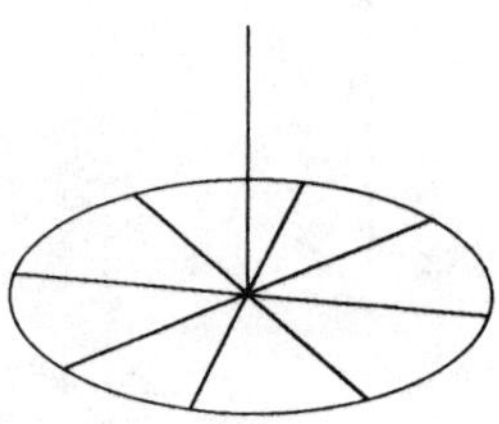

(d)

Size of ground plane	zero	large but finite	infinite
Peak directivity in dBi	1.76	4.77	4.77
Directivity at horizon in dBi	1.76	-1.249	4.77
Radiation Resistance in Ω	$20\pi^2\left(\frac{h}{\lambda}\right)^2$	$40\pi^2\left(\frac{h}{\lambda}\right)^2$	$40\pi^2\left(\frac{h}{\lambda}\right)^2$

Figure 4.9 Variation of the radiation pattern for a monopole with the size of ground plane: (a) zero ground plane, (b) finite ground plane of diameter $10\lambda/\pi$, (c) infinite ground plane, and (d) monopole with radial wire ground plane.

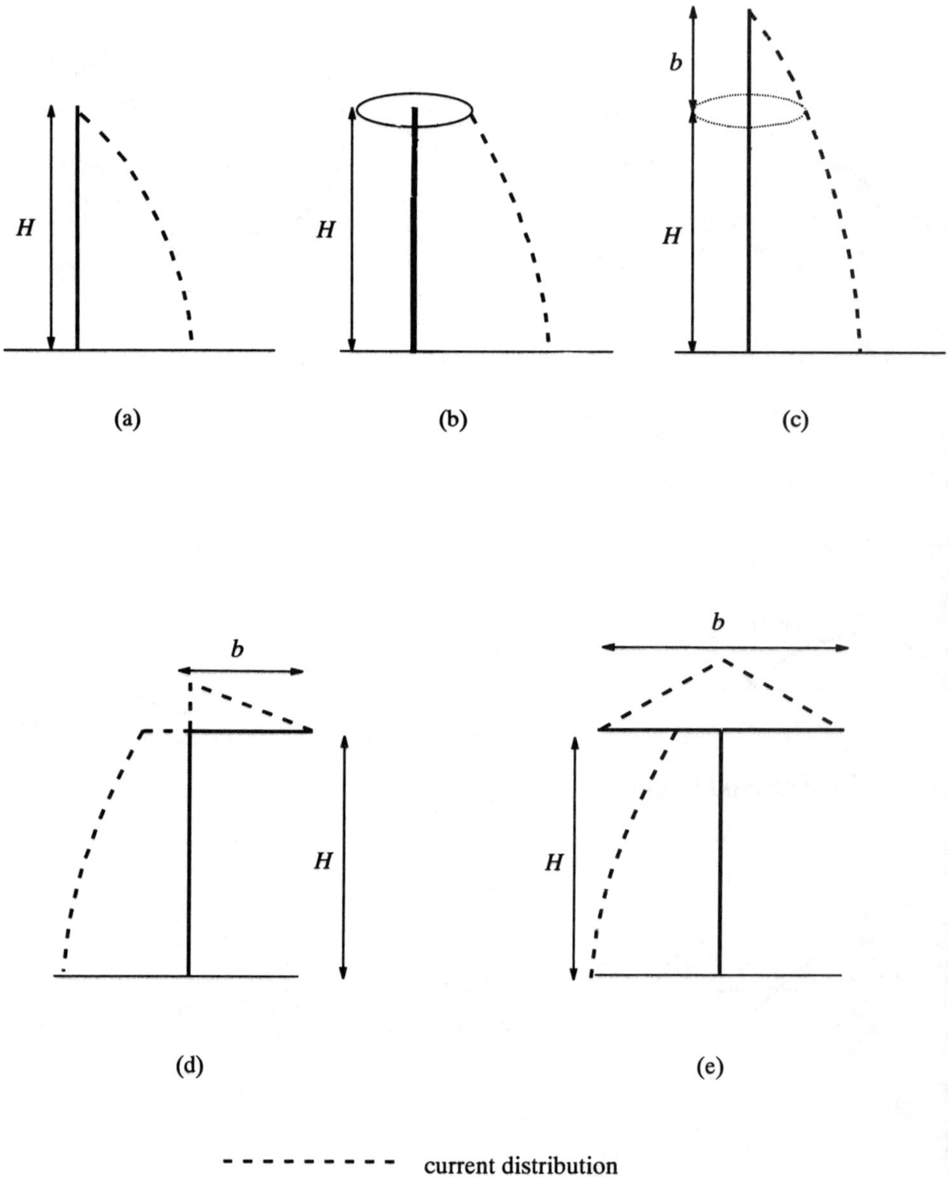

Figure 4.10 Top-loaded monopoles: (a) unloaded monopole, (b) top hat monopole, (c) current distribution on a monopole of length $H + b$, (d) L-shaped antenna, and (e) T-shaped antenna.

If X_a is large, the voltage has to be increased to very large values to radiate even reasonable amounts of power. Thus, by reducing the magnitude of X_a we can increase the power radiated. Top loading by use of a top hat changes the current distribution of the unloaded monopole as shown in Figure 4.10(a) to that shown in Figure 4.10(c). This shows the current distribution that would be present if the monopole was extended by a length b equivalent to the effective increased length of the top-loaded monopole. In the case of the L or T antennas shown in Figure 4.10(d, e) the bent sections draw the same amount of current that would be drawn by an additional length b connected to the antenna. This additional length b is approximately equal to the physical length of the horizontal portion of the L and T antennas. Without top loading the radiation resistance would be $40\pi^2(h/\lambda)^2$. However, for a loaded monopole of total effective length $(H + b) < 0.1\lambda$ the radiation resistance R_{rad} is given by

$$R_{rad} = 160\pi^2\left(\frac{H}{\lambda}\right)^2\left[1 + \frac{H}{H+b} + \frac{1}{4}\left(\frac{H}{H+b}\right)^2\right] \quad (4.35)$$

We have seen from (4.20) that the effective height of a short, thick monopole is half its physical length. By top loading, the effective height can be increased to 0.8 of its physical length in the frequency range 10 kHz to 40 MHz. In the lower frequency range, 20 Hz to 150 kHz, an effective height of around 0.63 of the physical length can be obtained. A preamplifier can be used to provide a match between the high impedance of the rod antenna and the receiver so that no signal loss occurs between them. This preamplifier effectively gives an antenna correction factor of around 12 dB, as a loss of around this figure would occur in its absence.

4.10.5 Parallel Element *E*-Field Generator

The parallel element *E*-field generator provides an electric field for susceptibility testing to military standards such as MIL-STD 462. It consists of two parallel horizontal wires about 2-m long that are fed by a voltage of the order of 50 V to 75 V [7]. The operating bandwidth is from 10 kHz to 30 MHz in two bands, and the electric field produced between the elements is approximately 200 V m^{-1}. At 1 m from the antenna and along the center line, the electric field strength is approximately 20 V m^{-1}. In a typical arrangement, the parallel wires are connected across the secondary of an impedance-matching transformer as shown in Figure 4.11(a). The primary of the transformer forms the load of a transmission line as shown in Figure 4.11(c).

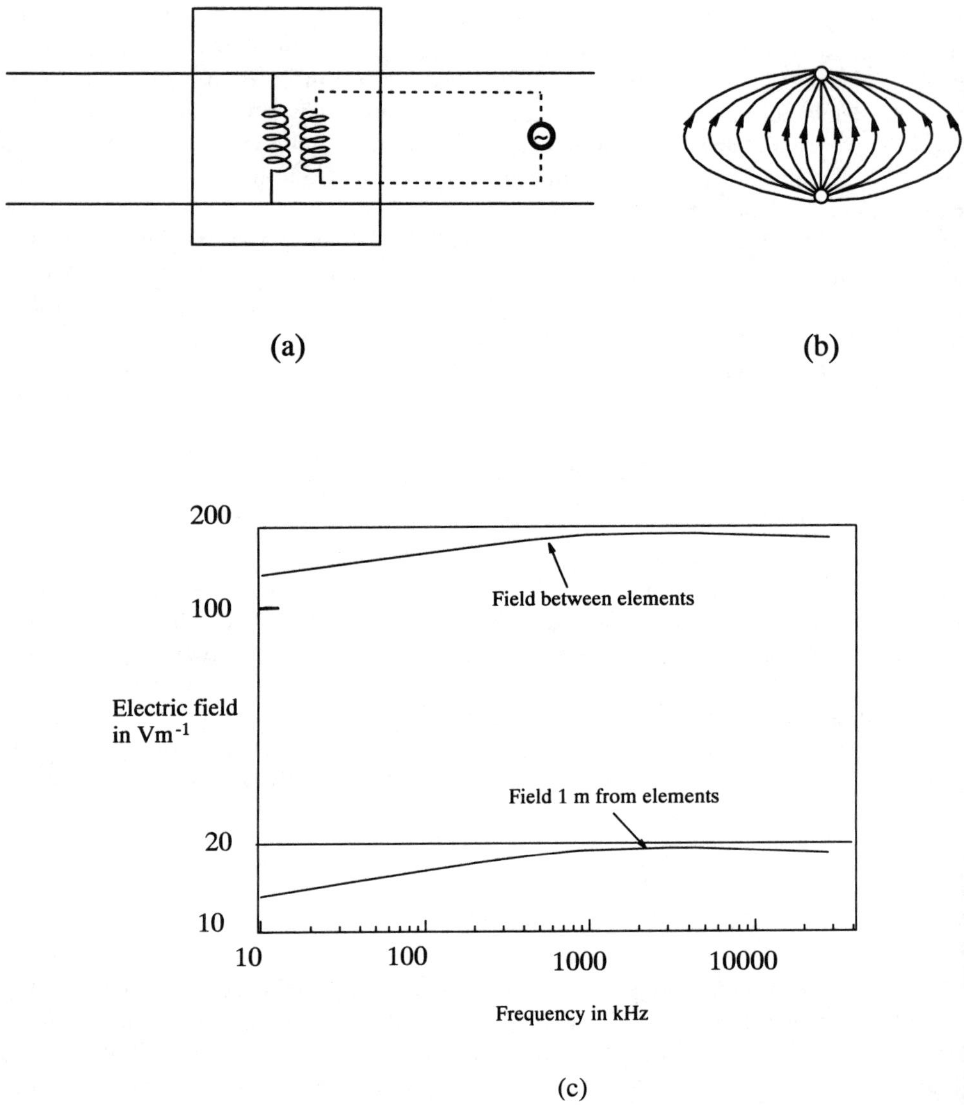

Figure 4.11 Parallel element E field generator: (a) typical feed arrangement for the parallel element, (b) side view showing electric field lines, and (c) variation of the electric field with frequency.

4.11 *H* FIELD ANTENNAS

For magnetic field measurements, loops or coils of wire are used for receiving as well as transmitting. At frequencies below 1 MHz the loops are small in terms of wavelength. The loops can consist of a single turn of wire or multiturn coils. At very low frequencies (e.g., at mains frequencies) a pair of coils, known as Helmholtz coils, are used to provide a uniform magnetic field.

4.11.1 Helmholtz Coils

Helmholtz coils are used to provide uniform magnetic fields for immunity and susceptibility measurements. They consist of a pair of coils of equal diameter, wound as flat rings as shown in Figure 4.12. The coils are placed in a parallel configuration and at a distance apart equal to the radius of either coil. The coils are wound in the same sense, that is, both clockwise or counterclockwise, and the same current flows in the both coils. This arrangement produces a uniform magnetic field in the area between the coils. The device under test (DUT) should be placed midway between the pair of coils and also near the common axis of the pair.

If a single coil had been used, the magnetic lines of force would have been like that shown in Figure 4.12(b).

Consider an element *dl* at a point A on a coil with radius *a,* as shown in Figure 4.12. The magnetic field $\delta \boldsymbol{H}$ at point P on the axis due to the element *dl* carrying a current *I* is given by

$$\delta \boldsymbol{H} = \frac{I dl \sin \theta}{4\pi x^2} \tag{4.36}$$

where x is the distance between P and the element and θ is the angle between the element *dl* and the line joining it to P.

The direction of the magnetic field vector $\delta\mathbf{H}$ is at an angle α to the plane containing the coil. This is the same as the magnitude of the vector H_ϕ in the near field given by (4.25) since the first term can be ignored for very small electrical distances (i.e., fractions of λ) of $r = x$. This vector can be resolved into vertical and horizontal components of $\delta H \cos \alpha$ and $\delta H \sin \alpha$. If we consider another element diametrically opposite at point B, the magnetic field at P due to it will have a vertical component of $\delta H \cos \alpha$ that is in the opposite direction (vertically downwards) to that due to the element at A and will cancel out the vertical component of the field due to A. This is true for all diametrically opposite points, so for a circular coil there will be no resultant vertical magnetic field. Thus there will only be a horizontal magnetic field (parallel to the axis of the coil), found by taking the line integral of $\delta H \sin \alpha$ around the coil circumference ($2\pi a$). The resultant magnetic field is given by

$$H_r = \oint_0^{2\pi a} \delta H \sin \alpha \tag{4.37}$$

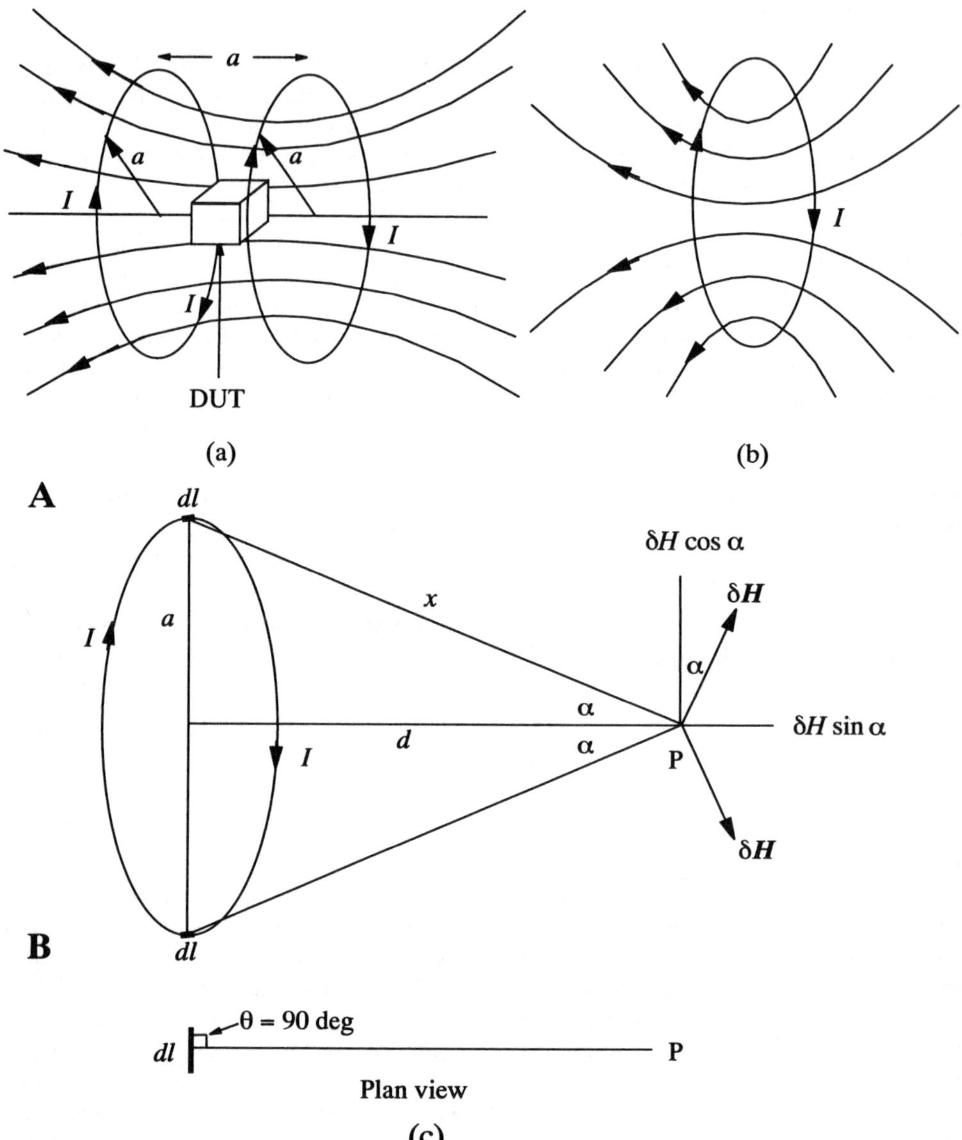

Figure 4.12 Magnetic fields due to Helmholtz coils: (a) Helmholtz arrangement of coils; (b) magnetic field lines of force for a single coil; (c) calculation of magnetic field intensity at a point P, due to a current element on a coil.

The magnitude of $\delta \boldsymbol{H}$ is given by (4.36), but in this case the angle θ between the element and the line joining it to P is 90 deg; and thus $\sin\theta$ is equal to 1. The resultant H_r can therefore be written as

$$H_r = \oint_0^{2\pi a} \frac{Idl \sin\alpha}{4\pi x^2} \tag{4.38}$$

$$H_r = \frac{Ia \sin\alpha}{2x^2} \tag{4.39}$$

If d is the distance along the axis, then by using Pythagoras's theorem, we can see that $\sin\alpha$ can be written as

$$\sin\alpha = \frac{a}{\sqrt{a^2 + d^2}} \tag{4.40}$$

And since $x^2 = a^2 + d^2$, (4.39) can be rewritten in the form

$$H_r = \frac{Ia^2}{2(a^2 + d^2)^{3/2}} \tag{4.41}$$

In the case of a pair of coils each having N turns and radius a, separated by a distance a, the resultant magnetic field H_t along the common axis and midway between the coils (at $d = a/2$) is given by

$$H_t = \frac{2NIa^2}{2(a^2 + a^2/4)^{3/2}} \tag{4.42}$$

This reduces to

$$H_t = \frac{8NI}{5a\sqrt{5}} \tag{4.43}$$

It can be seen that the resultant magnetic field is proportional to the current I and the number of turns on the coils. The diameter of the conductor used to wind the coils determines the current handling capacity. Maximum currents of 20A are common. The magnetic field is also inversely proportional to the radius a of the coils, but the smaller radius will also restrict the size of the DUT. The diameters of commercially available Helmholtz coils vary between 0.5m and 3m.

Sets of two (biaxial) or three (triaxial) coils are used to provide uniform magnetic fields along two or three mutually perpendicular axes. It is important to remember that the field is only uniform near the common axis of each pair of coils; thus the larger the DUT, the larger the coil diameter required.

4.11.2 Small Magnetic Loops

Single-turn loops are used to receive magnetic fields, whereas multiturn loops are used for receiving as well as transmitting. The loop is sometimes called a magnetic dipole. The small loop has fields similar to that from a small dipole but with the *E* and *H* fields interchanged. The polarization of the electric field in the horizontal plane is tangential to the loop as shown in Figure 4.13, unlike the situation of the dipole where the electric field is in the same direction as the axis of the dipole. The electric and magnetic fields radiated by a small single-turn loop are given by

$$E_\phi = 30\beta^3 IdA\left[\frac{1}{\beta r} - \frac{j}{(\beta r)^2}\right]\sin\theta\, e^{-j\beta r} \tag{4.44}$$

$$H_\theta = \frac{\beta^3 IdA}{4\pi}\left[\frac{1}{\beta r} - \frac{j}{(\beta r)^2} - \frac{1}{(\beta r)^3}\right]\sin\theta\, e^{-j\beta r} \tag{4.45}$$

$$H_r = \frac{j\beta^3 IdA}{2\pi}\left[\frac{1}{(\beta r)^2} - \frac{j}{(\beta r)^3}\right]\cos\theta\, e^{-j\beta r} \tag{4.46}$$

where I is the current in the loop in amperes, dA is the area of the loop in square meters, and β is the phase constant and equals $2\pi/\lambda$ m^{-1}.

If the loop has N turns, each of the expressions in (4.44) to (4.46) must be multiplied by N.

Near the antenna, that is, at distances $<0.01\lambda$, the $1/r$ and $1/r^2$ terms can be ignored in comparison with the $1/r^3$ term. Thus, only the last terms in H_r and H_θ need to be considered. The E_ϕ component can be ignored since it does not have a $1/r^3$ term. The H_r and H_θ terms in the near field are given by

$$H_r = \frac{IdA}{2\pi r^3}\cos\theta e^{j\beta r} \tag{4.47}$$

$$H_\theta = -\frac{IdA}{4\pi r^3}\sin\theta e^{j\beta r} \tag{4.48}$$

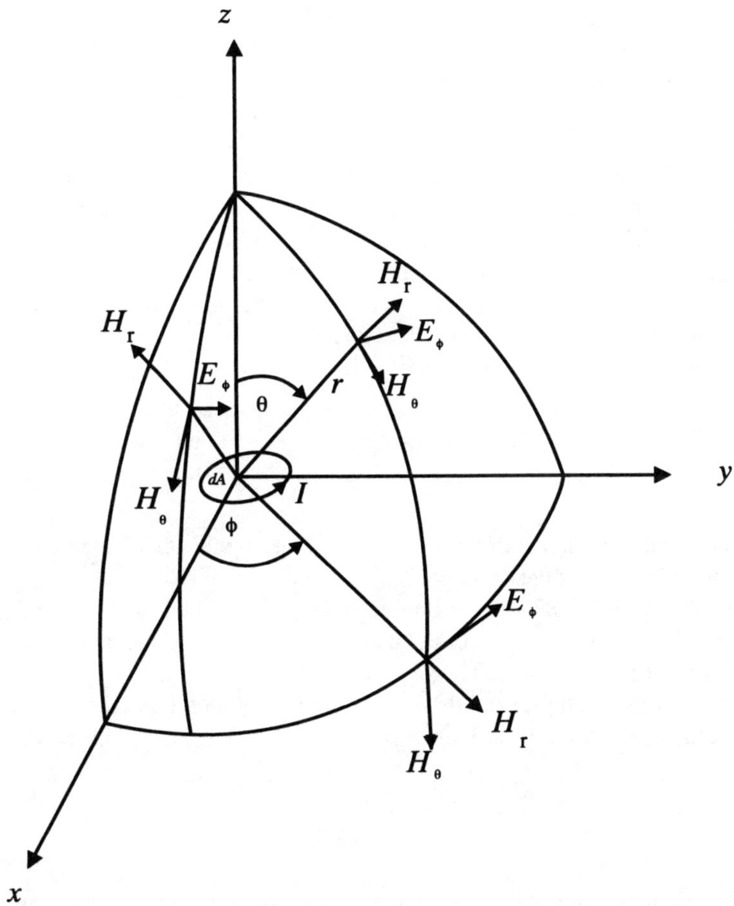

Figure 4.13 Fields due to a small radiating loop.

It can be seen that the H_r vector has twice the magnitude of the H_θ vector, and the phase between the two vectors is 90 deg. The ratio of H_r to H_θ in the near field is given by

$$\frac{H_r}{H_\theta} = -2 \cot \theta \tag{4.49}$$

4.11.3 Single-Turn Shielded Loops

Shielded loops reject the electric field and sample the magnetic field over a small area. Only a short length of the loop conductor is exposed to the incident magnetic field. The shield is sometimes called a Faraday shield. There are two types of shielded loops, namely, the balanced and unbalanced types. The shield is made of metal whose thickness is several skin depths to prevent any direct interaction between the currents on the internal and external surfaces of the shield. The effective terminals CD are at the small gap formed between the two halves of the loop.

The loop can be assumed to consist of a series combination of an internal impedance Z_{int}, an external inductance L_{ext}, and a radiation resistance R_{rad}. In practice, there is also a distributed capacitance (between the sides of the turn) in parallel with these impedances, but this is often omitted since it can be accommodated by having a larger variable capacitance in parallel. The internal impedance Z_{int} consists of an inductance L_{int} in series with a resistance R_{int}. Thevenin's equivalent circuit for a receiving loop is shown in Figure 4.14(a). The voltage V_l developed at the load is given by

$$V_l = \frac{V_{oc} Z_l}{(Z + Z_l)} \tag{4.50}$$

where Z_l is the load impedance in ohms, Z is the input impedance of the loop in ohms, and V_{oc} is the open-circuit voltage in volts.

In the case of the unbalanced loop, the coaxial line is bent into a semicircle and the inner conductor is extended beyond the outer conductor and connected to the solid semi-circle as shown in Figure 4.14(b). The semicircular loop of length πb and the transmission line of length h connect the terminals CD to the load, which has an impedance Z_l.

The incident magnetic flux density produces a voltage of V_{oc} at the gap CD that is given by

$$V_{oc} = j\omega A\boldsymbol{B} \tag{4.51}$$

where V_{oc} is the open-circuit voltage in volts, ω is the angular frequency in radians per second, A is the area of the loop in square meters, and $\boldsymbol{B}$ is the flux density in telsas. It is assumed that the magnetic flux density is uniform over the area of the loop. This would be the case for a small loop.

It should be noted that only the component of the magnetic field density that is perpendicular to the plane of the loop is instrumental in inducing the voltage produced. In the general case, the relative orientation between the loop and the incident wave is shown in Figure 4.14(c), and the induced voltage is given by

$$V_{oc} = j\omega A\boldsymbol{B}\cos\theta \sin\phi \tag{4.52}$$

where V_{oc} is the open-circuit voltage in volts, ω is the angular frequency in radians per second, A is the area of the loop in square meters, $\boldsymbol{B}$ is the flux density in telsas, and ϕ and θ are the angles following the normal polar notation.

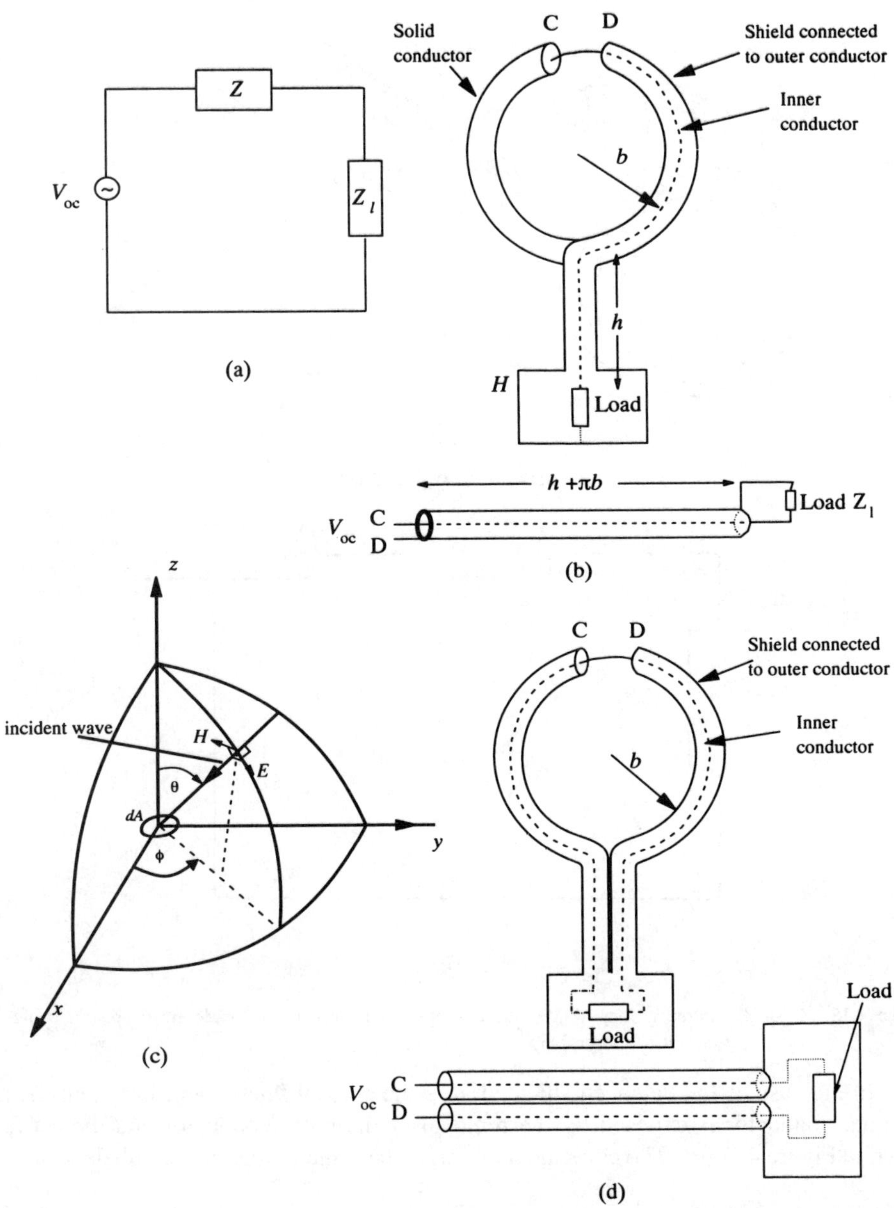

Figure 4.14 Shielded single-turn loops: (a) equivalent circuit for a receiving loop, (b) shielded unbalanced single-turn loop, (c) plane wave incident on a loop, and (d) shielded balanced single-turn loop.

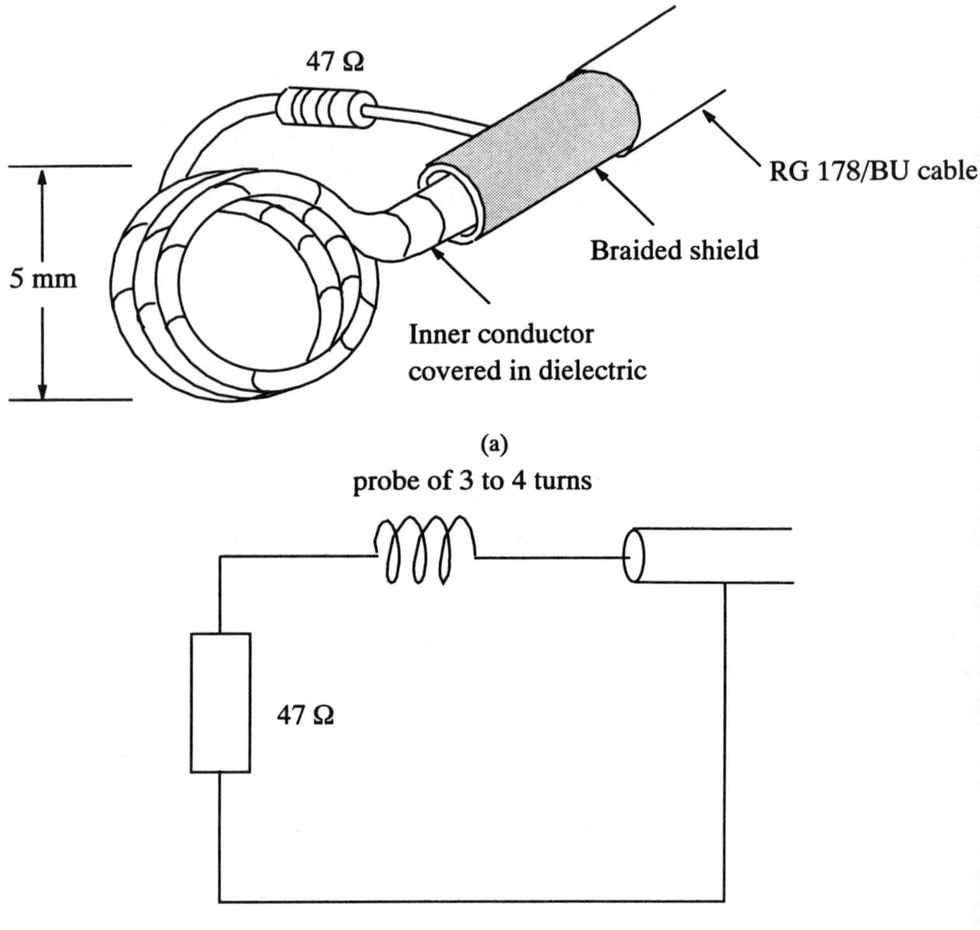

Figure 4.15 A small multiturn loop probe: (a) a simple multi-turn probe made from coaxial cable; (b) equivalent circuit of the probe.

In the case of the balanced shielded loop the coaxial line is bent into a circle, and the inner conductor is exposed by the removal of the outer conductor and dielectric as shown in Figure 4.14(d). The equivalent circuit is the same as that of the unbalanced loop.

4.11.4 Simple Multiturn Loop Probe

A simple probe can be constructed by forming the inner conductor of a coaxial cable into a three- or four-turn coil as shown in Figure 4.15(a). The coil is then connected to the outer conductor via a 47-Ω resistor [11]. The equivalent circuit is shown in Figure 4.15(b).

The signal generator reading is in volts, although the magnetic flux density (in telsas) is being measured. The effective height of a loop antenna as defined in Section 4.9 is the number of volts measured per telsa of incident magnetic flux density. The area of the loop is usually given in centimeters, and incident magnetic flux density is of the order of picotelsas (pT). Thus, the effective height can be defined in volts per picotelsas, as in MIL-STD 461A [8], and is given by the relation

$$\frac{V}{\boldsymbol{B}} = 2\pi NAf \times 10^{-16} \tag{4.53}$$

where $\boldsymbol{B}$ is the magnetic flux density in picotelsas (10^{-12} telsa), N is the number of turns, A is the area of the loop in square centimeters (cm^2), and f is the frequency in hertz (Hz, cycles per second).

REFERENCES

[1] Lucas, J. G., "Antenna Systems," *Student Quart. J.*, June 1967, pp. 186–195

[2] Connor, F. R., *Introductory Topics in Electronics and Telecommunications—Antennas*, second edition, London: Edward Arnold, 1992.

[3] Johnson, R. C., and H. Jasik, *Antenna Engineering Handbook*, New York: McGraw-Hill Book Company, 1984.

[4] Weiner, M. M., "Monopole Element at the Centre of a Circular Ground Plane Whose Radius is Small or Comparable to a Wavelength," *IEEE Trans. on Antennas and Propagation*, Vol. 35, No. 5, May 1987, pp. 488–495.

[5] Jordan, E. C., *Electromagnetic Waves and Radiating Systems*, London: Constable and Company Ltd., 1953.

[6] Fewkes, J. H., and J. Yarwood, *Electricity, Magnetism and Atomic Physics, Volume* 1, second edition, London: University Tutorial Press Ltd., 1962.

[7] Shepherd, D., "A Five-Decade System For Testing RFI Susceptibility" *RF Design*, September 1988, pp. 20–23.

[8] MIL-STD 461A, *Electromagnetic Emission Interference Characteristics Requirements for Equipment*, Washington, DC: U.S. Government Printing Office, August 1968.

[9] Balanis, C. A., Antenna Theory—Analysis And Design, New York: Harper & Row, 1982.

[10] *IEEE Standard Dictionary of Electrical and Electronics Terms*, fourth edition (F. Jay, ed.), 1988.

[11] Torselli, J. V., "Electromagnetic Compatibility In Pocketable Equipment," *IEEE Trans. on EMC*, Vol. 16, November 1974, pp. 337–344.

Chapter 5
Antennas for Frequencies Between 1 MHz and 1 GHz

At frequencies between 1 MHz and 1 GHz wire as well as aperture antennas are used. The wire antennas can be realized as free-standing wires or as printed circuits on a dielectric support. Since the wavelength at 1 GHz is 30 cm, resonant dipoles and monopoles are both manageable sizes. This chapter describes monopoles, dipoles and loop antennas. It also discusses folded and triangular dipoles and the biconical, which can all be considered to be variants of the dipole. Other antennas described are the log-periodic and Yagi antenna, which are really arrays of dipoles and the discone antenna, which is a variant of the monopole. The helical antenna can be considered as a type of loop. The bilog antenna is a combination of the triangular (bowtie) antenna and the log periodic. The broadband antennas such as the biconical, log spiral (which is described in Chapter 6), and log periodic can be considered as variants of *frequency-independent* (FI) antennas, although only the log spiral is a truly FI antenna. Double-ridged horns are also used at these frequencies, but these are described in Chapter 6.

5.1 RESONANT MONOPOLES

A monopole is connected to the inner conductor of the coaxial line and the ground plane is connected to the outer conductor as shown in Figure 5.1(a). At frequencies in the megahertz range, monopoles can be an appreciable fraction of or larger than a wavelength; therefore, these monopoles can be considered to be large monopoles. When a monopole is approximately a quarter-of-a-wavelength long it is resonant and radiates the maximum power at this frequency. It has a purely resistive impedance of 36.8Ω at resonance. This resistance varies slightly with the gap between the inner and outer conductors of the coaxial line feeding the antenna. The ratio of the radii of the outer and inner diameters (*b/d*) should be 1.868, for a line of characteristic impedance 37Ω [14, p. 605]. Experi-

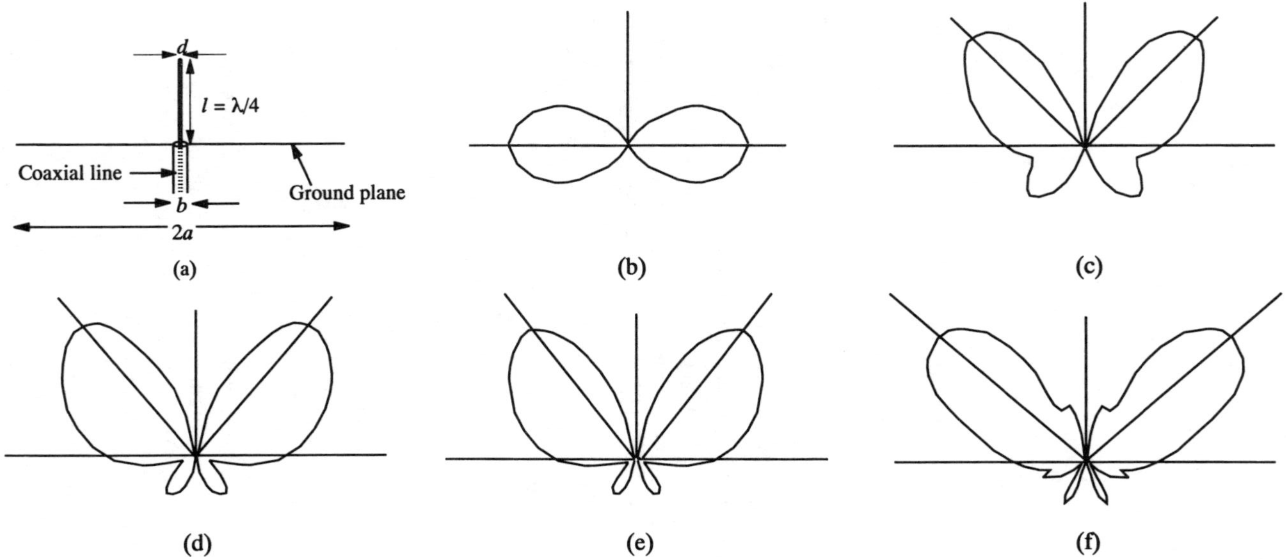

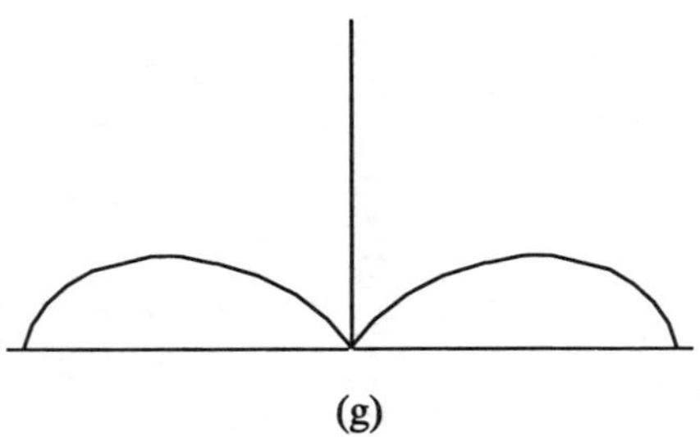

Size of ground plane	Zero	Large but finite	Infinite
Peak directivity in dBi	1.88	5.16	5.16
Directivity at horizon in dBi	1.88	- 0.859	5.16
Radiation resistance in Ω	19.43	36.54	36.54

Figure 5.1 Variation of the radiation pattern of a resonant monopole with the size of the ground plane: (a) monopole fed by coaxial line, (b) zero ground plane, (c) ground plane diameter = $3\lambda/\pi$, (d) ground plane diameter = $4\lambda/\pi$, (e) ground plane diameter = $5\lambda/\pi$, (f) ground plane diameter = $6.5\lambda/\pi$, and (g) infinite ground plane.

mental data show that the resonant length of a monopole occurs when the length is 0.236λ and the ratio of the diameter to length *(l/d)* of the conductor is 0.00318λ. At 100 MHz the monopole would have a diameter of 9.5 mm and its length would be 0.708 m. When the monopole is >λ/4 long, it exhibits an inductive impedance. The radiation pattern varies with the size of ground plane [10], as in the case of the short monopoles described in Chapter 4. Figures 5.1(b–g) show the variation of the radiation pattern of the resonant monopole with the electrical diameter (the diameter in terms of wavelength) of the ground plane. As the ground plane diameter is increased from zero to 6.5λ/π the main lobe (position of peak directivity) moves from the horizon (θ = 90 deg) to smaller angles and then increases again, but the variation is not monotonic (it does not increase or decrease consistently). The peak directivity varies from 1.88 dBi for zero ground plane to 5.16 dBi in the case of an infinite ground plane. The radiation resistance varies between 19.43Ω and 36.54Ω, as shown in the table given in Figure 5.1.

5.2 DISCONE

The discone, which is sometimes called a *Kandoian* (after its inventor) antenna, consists of a disk mounted on a truncated cone. The disk and cone could be solid or made of wires. The discone antenna can be considered to be a stub antenna (a thick monopole) that has been top loaded by connecting a disk to its free end and in which the ground plane has been modified to form a cone [11, p. 723]. Alternatively it can be considered as a variant of the biconical antenna where one of the cones has been flattened to a disk. The antenna is simple to construct and also easy to feed. The center conductor of the coaxial cable feeding the antenna is connected to the disk and the outer conductor of the coaxial cable is connected to the top (or narrow) end of the cone as shown in Figure 5.2(a). The azimuth radiation pattern of the discone is omnidirectional, as in the case of a vertical dipole, and the radiated electric field vector is vertically polarized. The diameter of the narrow end of the cone, C_{min}, is inversely proportional to the bandwidth of the antenna, and the slant height L is a function of the frequency [1]. The slant height must be greater than a quarter of a wavelength at the lowest operating frequency (i.e., $L > \lambda_c/4$).

Nail [1] has found that the disk-to-cone spacing affects the impedance match, and that the optimum impedance match to a 50Ω transmission line is independent of the cone flare angle ϕ for angles between 25 deg and 90 deg. The optimum impedance match is obtained when the disk-to-cone spacing is $0.3C_{min}$ (where C_{min} is the diameter of the top of the cone) and the disk diameter D is $0.7C_{max}$ (where C_{max} is the diameter of the base of the cone). This applies when C_{min} is electrically small (<λ/75) at the high-frequency end and the diameter w of the feed pin is not taken into account. It is also assumed that the disk-to-cone spacing s is << the diameter of the disk D. At large angles the discone acts like a high pass filter, in that once the cone length L exceeds λ/4, the match to a 50-Ω line remains good over an extremely large-frequency range. The radiation pattern of the discone antenna is omnidirectional in the azimuth plane as in the case of a monopole, and in the elevation plane the pattern is tilted downward as shown in Figure 5.2(b).

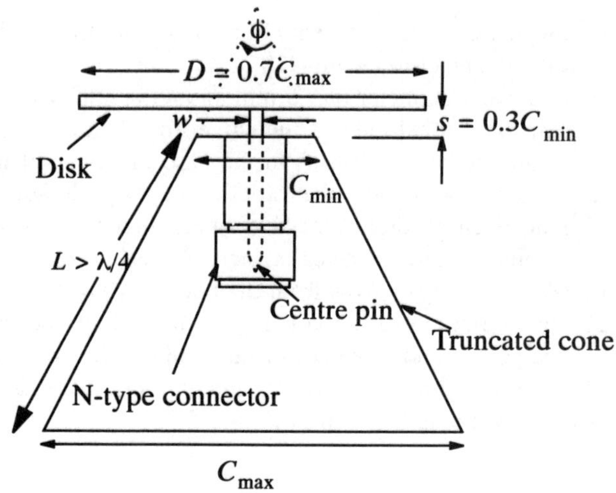

(a)

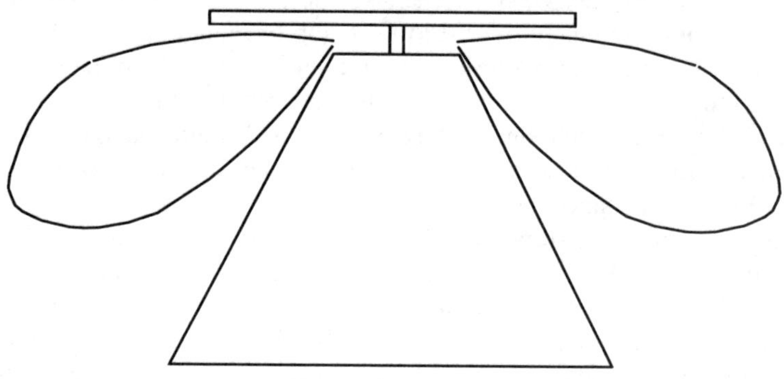

(b)

Figure 5.2 The discone or Kandoian antenna: (a) physical characteristics of a discone antenna; (b) radiation pattern of a discone in the elevation plane.

5.3 CAVITENNA

The Cavitenna is used in the frequency range 20 MHz to 300 MHz. It is a top-loaded monopole that is mounted by the use of magnetic clamps to the side or ceiling of a shielded room. The wall (or ceiling) forms the ground plane for the monopole, and the room is excited like a cavity. At certain frequencies the room resonates, and these resonances account for the spurious readings obtained when measurements are made at certain positions and for some frequencies. They also account for the lack of repeatable measurements.

5.4 RESONANT AND LARGE DIPOLES

When the length of a dipole approaches half a wavelength, the dipole resonates and maximum power is radiated. The radiation resistance is found to be about 73Ω, and the condition for zero reactance occurs when the wavelength is about 0.49λ for thin wires but greater than 0.5λ for thicker wires, which are used for wider bandwidths [15, p. 547]. The peak linear directivity at the resonant frequency is 1.64, which is 2.15 dBi. The radiation pattern in the azimuth plane (assuming sinusoidal current distribution) is omnidirectional, as shown in Figure 5.3(a). In the elevation plane we have the familiar figure eight for a vertical center driven dipole as shown in Figure 5.3(b) with the main lobes on the horizon at $\theta = \pm 90$ deg. As the length of the dipole increases above half a wavelength, the pattern in the elevation plane gets narrower and the peak directivity increases, as shown in Figure 5.3(c, d); and when the length is 1.25λ, minor lobes appear as shown in Figure 5.3(e). As the length is increased further the lobes at the horizon become narrower and the sidelobes get larger. For longer electrical lengths the main lobes are no longer on the horizon. The pattern breaks up so that there are four main lobes, as shown in Figure 5.3(f–j).

The effective height h of a half-wave dipole depends on the angle θ between the axis of the dipole and the point of observation [6, p. 4–24] and is given by

$$h = \frac{\lambda}{\pi} \frac{\cos\left(\frac{\pi}{2}\cos\theta\right)}{\sin\theta} \tag{5.1}$$

We can see that on the axis where angle θ is 90 deg the effective height is λ/π.

The impedance Z_d, of a half-wave resonant dipole is given by

$$Z_d = R(kl) - j\left[120\left(\log_e\frac{l}{a} - 1\right)\cot(kl) - X(kl)\right] \tag{5.2}$$

where k is the phase constant $2\pi/\lambda$, a is the radius of the conductor, and $R(kl)$ and $X(kl)$ are functions shown in Figure 5.4.

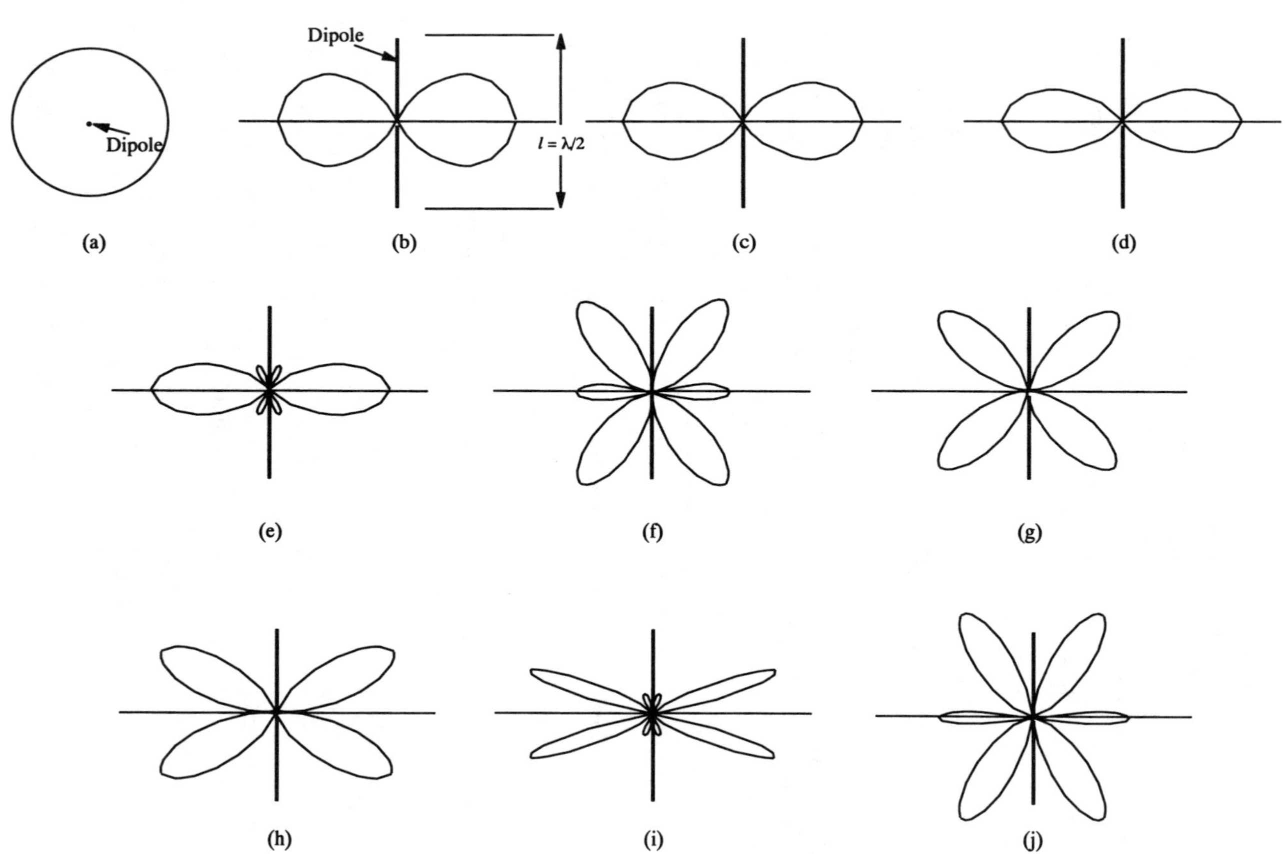

Figure 5.3 Radiation patterns for large and resonant dipoles: (a) azimuth plane pattern; (b) elevation pattern for λ/2 dipole; (c) elevation pattern for 3λ/4 dipole; (d) elevation pattern for 1λ dipole; (e) elevation pattern for 1.25λ dipole; (f) elevation pattern for 1.5λ dipole; (g) elevation pattern for 1.75λ dipole; (h) elevation pattern for 2λ dipole; (i) elevation pattern for 2.5λ dipole; and (j) elevation pattern for 2.75λ dipole.

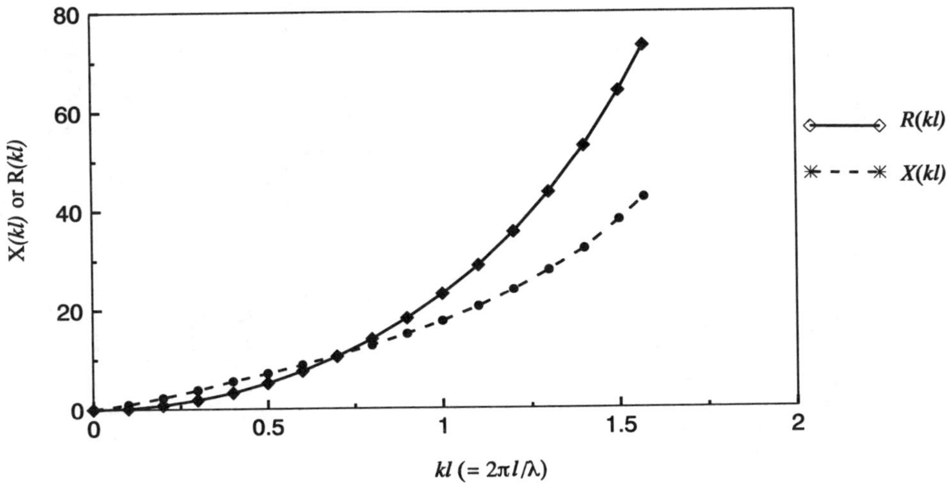

Figure 5.4 Functions $R(kl)$ and $X(kl)$.

5.5 FOLDED DIPOLES

The folded dipole is a very popular antenna because of its ease of construction and its impedance. The dipole is formed by joining two cylindrical dipole elements of equal length $2l$ at their ends and driving one of the elements at its center, as shown in Figure 5.5(a). The dipole connected to the signal source is known as the driven element, and the dipole connected to the ends of the driven element is known as the parasitic element. The dipoles can be identical or of different radii r_1 and r_2. The folded dipole is often constructed from commercially available twin wire transmission lines [12, p. 10]. The solid dielectric will reduce the phase velocity of the EM wave so that its wavelength is no longer equal to the free-space wavelength but reduced to a value of $\lambda/\sqrt{\varepsilon_r}$, where ε_r is the dielectric constant of the material. The wavelength of the transmission line currents are reduced by a proportionate amount. However, the dielectric material has a negligible effect on the antenna currents, and thus the resonant length remains unchanged. One method of construction that satisfies both requirements is to use shorting elements so that the distance between the shorted portions is $2l/\sqrt{\varepsilon_r}$, as shown in Figure 5.5(b); this ensures that the lengths of the radiating elements are correct for air, but the transmission line modes are satisfied in the dielectric.

The excitation of the folded dipole can be considered as the superposition of common and differential modes, as shown in Figure 5.5(c). In the case of the common mode, the driving voltages in the two elements are both V and they have the same polarity. The

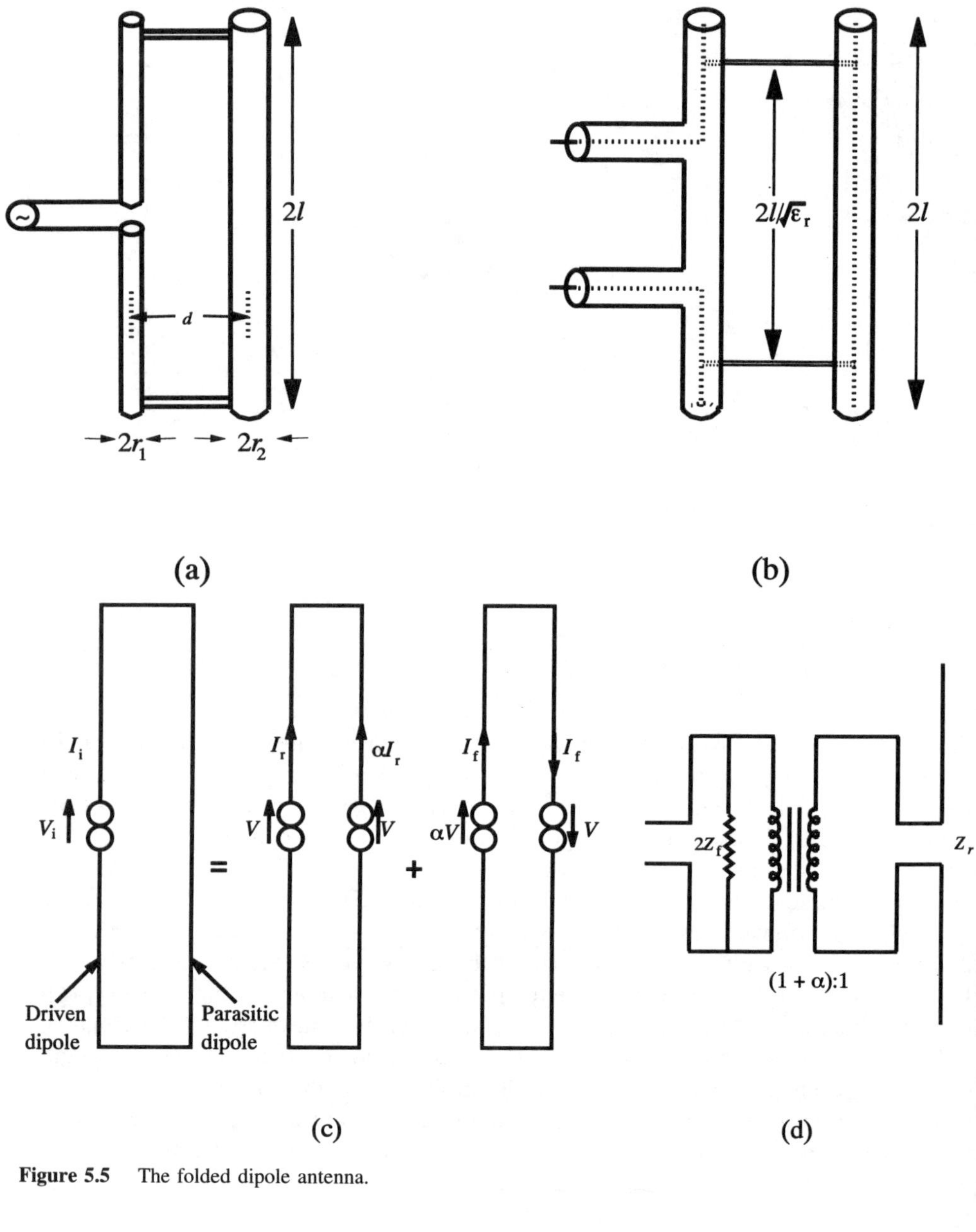

Figure 5.5 The folded dipole antenna.

currents in the driven and parasitic arms are I_r and αI_r, respectively, where α is the current division factor and is given by

$$\alpha = \frac{\cosh^{-1}[(v^2 - u^2 + 1)/2v]}{\cosh^{-1}[(v^2 + u^2 - 1)/2uv]} \tag{5.3}$$

where v (= d/r_1) is the ratio of the dipole spacing to the radius of the driven dipole and u (= r_2/r_1) is the ratio of the radii of the parasitic and driven elements.

The differential mode driving voltages in the two elements are V and αV, and they have opposite polarity. The currents in the driven and parasitic arms are both I_f.

For a half-wave folded dipole whose elements are of equal length $2l$, the input impedance is close to that of a 300-Ω twin wire transmission line.

The input impedance of a folded dipole of total length $2l$ less than or equal to $\lambda/2$ is given by

$$Z_{in} = \frac{2(1+\alpha)^2 Z_r Z_f}{2Z_f + (1+\alpha)^2 Z_r} \tag{5.4}$$

where α is the current division factor given by (5.3), Z_r is the impedance of the common mode and is given by

$$Z_r = \frac{V}{(1+\alpha)I_r} \tag{5.5}$$

and Z_f is impedance of the differential mode and is given by

$$Z_f = \frac{(1+\alpha)V}{2I_f} \tag{5.6}$$

These equations assume that the element spacing d and the radii r_2 and r_1 are all much smaller than the length $2l$ of the elements.

When the element length $2l$ is equal to $\lambda/2$, the impedance Z_f is much larger than $Z_r(1 + \alpha)^2$ and (5.4) reduces to

$$Z_{\lambda/4} = Z_r(1+\alpha)^2 \tag{5.7}$$

The effective height h of a half-wave folded dipole [6, p. 4–24] depends on the angle θ between the axis of the dipole and the point of observation and is given by

$$h = \frac{2\lambda}{\pi} \frac{\cos\left(\frac{\pi}{2}\cos\theta\right)}{\sin\theta} \tag{5.8}$$

The equivalent circuit of the folded dipole is shown in Figure 5.5(d).

5.6 TRIANGULAR DIPOLES

A dipole can be made more broadband by using triangular sheets of metal instead of straight wires, as shown in Figure 5.6(a). This type of antenna is sometimes called a bowtie antenna. The bowtie has radiation patterns and input impedance similar to the biconical, but it is much lighter in weight and simpler to construct. The antenna is characterized by the flare angle α of the apices and the height A (from the apex to the opposite side) of the triangles. Simplifying the construction to a wire outline results in significant degradation of broadband performance, but at low frequencies mesh can be used instead of solid sheets as long as the mesh spacing is smaller than one-tenth of a wavelength at the lowest operating frequency.

The larger the values of A and α the more the radiation pattern in the elevation (z-y) plane varies from the azimuth (x-y) plane and the higher are the sidelobes [6, p. 29–5]. Optimum design (a compromise between pattern and impedance characteristics) is obtained by selecting values of α between 60 deg and 80 deg and values of A up to 0.58λ.

When a bowtie antenna of typical dimensions, 15 in by 10.5 in, is mounted $\lambda/4$ in front of a flat reflecting screen, as shown in Figure 5.6(b), its gain is increased and its back radiation is reduced. Placing the antenna in front of a screen of size 56 cm by 48 cm results in a 5-dB improvement (approximately) in gain over the 470 MHz to 900 MHz frequency range, as shown in Figure 5.6(c).

5.7 BICONICAL

The biconical antenna was first used by Oliver Lodge in 1892 [11, p.340]. Although broadband, this antenna is not frequency-independent when the cones are truncated, as would be the case for a practical antenna. The infinite biconical shown in Figure 5.7(a) acts as a guide for spherical waves in the same way as an infinite transmission line acts as a guide for plane waves. The feed at the terminals produces a voltage V between the cones and a total current I flows on the surface of the cones at a distance r from the terminals. The characteristic impedance is defined as the ratio of V to I in the same way as it is defined for transmission lines. If we consider only the first or fundamental mode, that is, the *transverse electromagnetic* (TEM) mode, the electric and magnetic fields are both transverse to the direction of propagation of the wave and perpendicular to each

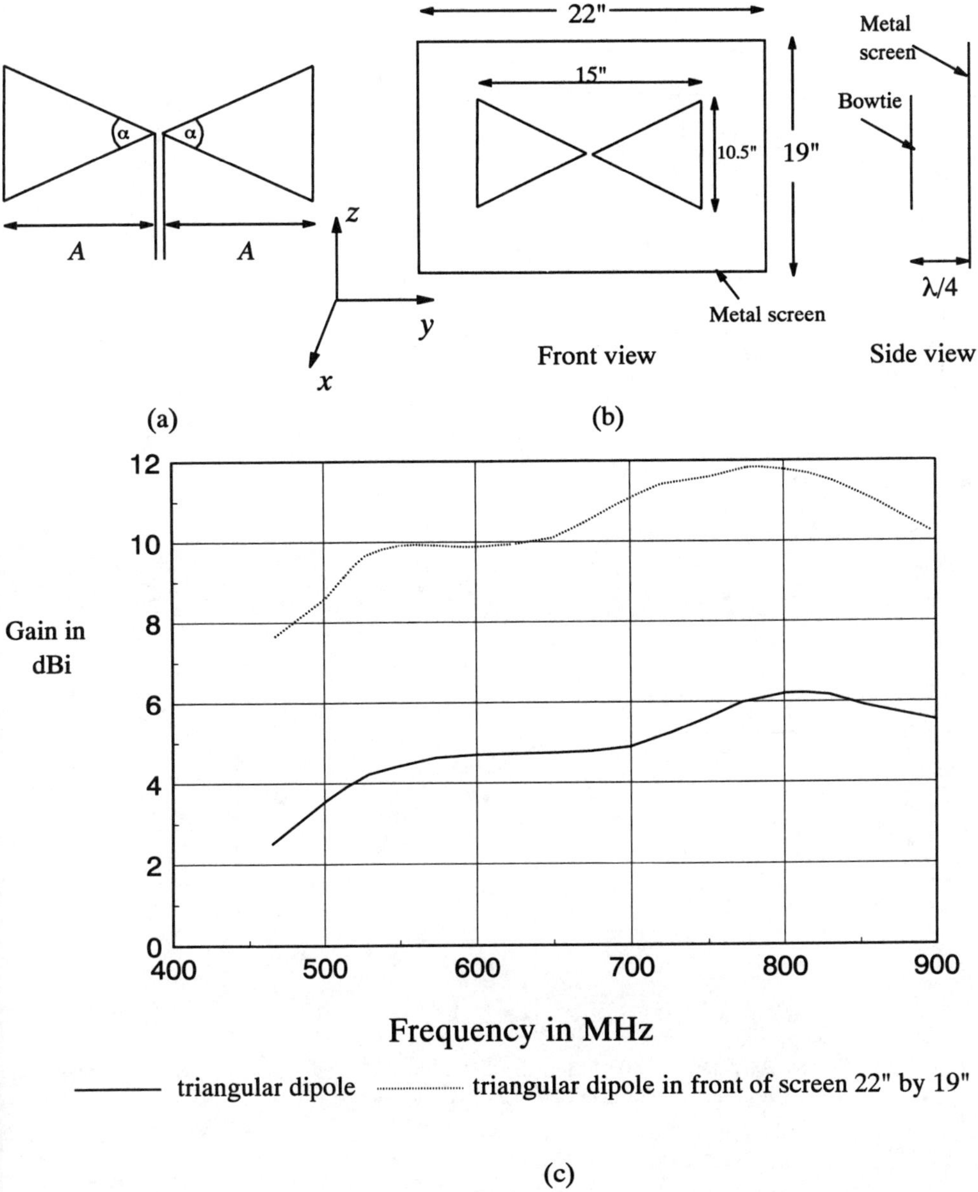

Figure 5.6 The triangular dipole or bowtie antenna: (a) physical characteristics of the bowtie; (b) the bowtie in front of a metal screen; and (c) variation of gain with frequency for a bowtie placed in front of a screen.

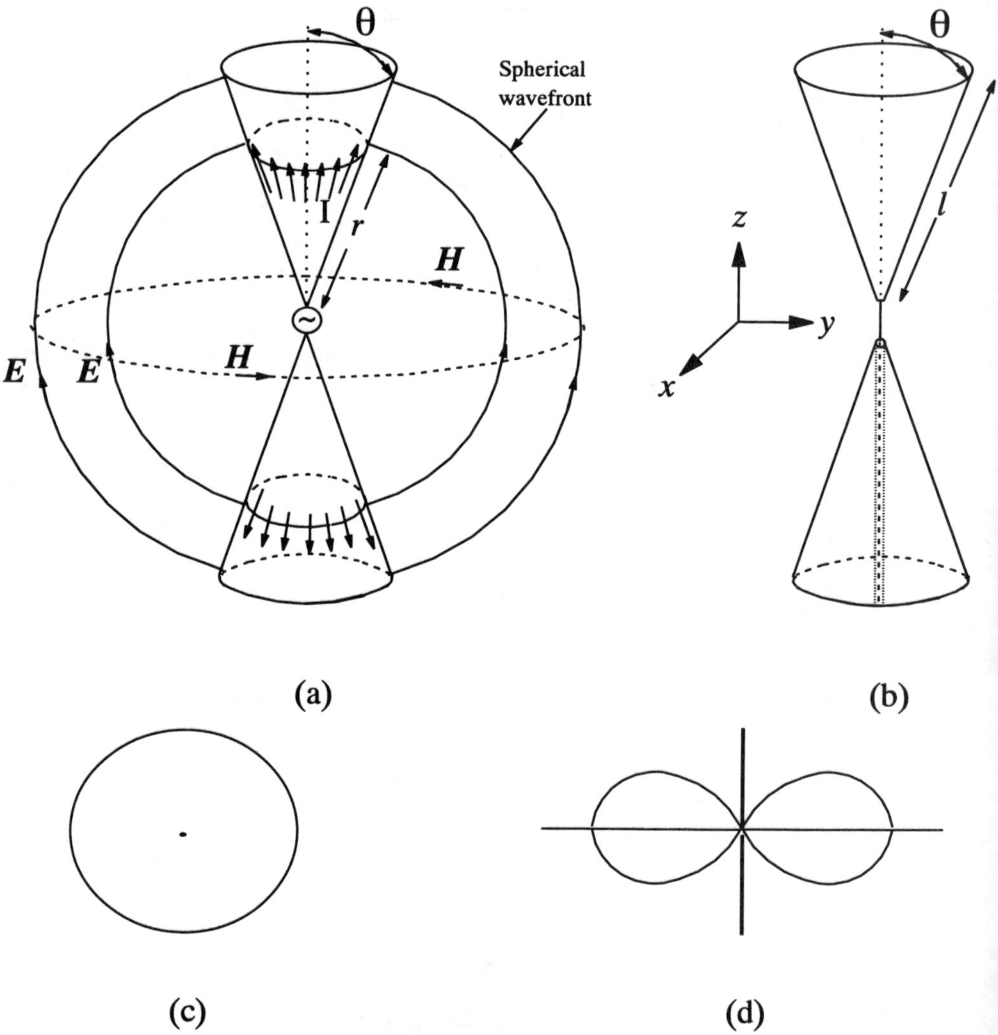

Figure 5.7 The biconical antenna: (a) TEM mode; (b) feeding details; (c) radiation pattern in x-y or azimuth plane; and (d) radiation pattern in y-z or elevation plane.

other. The electric field lines are great circles through the polar axis, and the magnetic lines are circles perpendicular to the axis.

Using Maxwell's second equation we have in this case

$$\nabla \times \boldsymbol{E} = -j\omega\mu \boldsymbol{H} \tag{5.9}$$

Expanding the left-hand side of (5.9) in spherical coordinates we get

$$\nabla \times \boldsymbol{E} = \frac{\mathbf{a}_r}{r^2 \sin\theta}\left[\frac{\partial(r\sin\theta E_\phi)}{\partial\theta} - \frac{\partial(rE_\theta)}{\partial\phi}\right] + \frac{\mathbf{a}_\theta}{r\sin\theta}\left[\frac{\partial E_r}{\partial\phi} - \frac{\partial(r\sin\theta E_\phi)}{\partial r}\right] + \frac{\mathbf{a}_\phi}{r}\left[\frac{\partial(rE_\theta)}{\partial r} - \frac{\partial E_r}{\partial\theta}\right] \tag{5.10}$$

The electric field does not have a component in the ϕ or radial (r) directions, and the E_θ component is independent of ϕ. Thus, the right-hand side of (5.10) reduces to

$$\nabla \times \boldsymbol{E} = \frac{\mathbf{a}_\phi}{r}\frac{\partial(rE_\theta)}{\partial r} \tag{5.11}$$

In the case of the right-hand side of (5.9) we only have to consider the H_ϕ component, and this gives

$$-j\omega\mu\boldsymbol{H} = -\mathbf{a}_\phi j\omega\mu H_\phi \tag{5.12}$$

Combining (5.11) and (5.12) gives us

$$\frac{1}{r}\frac{\partial(rE_\theta)}{\partial r} = -j\omega\mu H_\phi \tag{5.13}$$

If we consider Maxwell's first equation for a nonconducting medium $\sigma = 0$, we have

$$\nabla \times \boldsymbol{H} = j\omega\varepsilon\boldsymbol{E} \tag{5.14}$$

For the TEM mode we again only have the H_ϕ and E_θ components; thus, Maxwell's first equation is reduced to

$$\frac{\partial(rH_\phi)}{\partial r} = -j\omega\varepsilon(rE_\theta) \tag{5.15}$$

It can be shown that the H_ϕ component [11, p. 345] is given by

$$H_\phi = \frac{1}{r \sin\theta} H_0 e^{-j\beta r} \tag{5.16}$$

where H_0 is the amplitude of the magnetic field.

The intrinsic impedance Z_0 is defined as the ratio of E_θ to H_ϕ and is given by

$$Z_0 = \frac{E_\theta r \sin\theta}{H_0} e^{j\beta r} \tag{5.17}$$

from which we can write E_θ as

$$E_\theta = \frac{Z_0 H_0}{r \sin\theta} e^{-j\beta r} \tag{5.18}$$

We can calculate the voltage between the points at the radial distance r by taking the line integral of the electrical field E_θ (with respect to θ) along a great circle between the two points. The angle θ varies between $-\theta_1$ and $+\theta_1$ over this path, where θ_1 is the cone semiangle. Thus the voltage is given by

$$V(r) = Z_0 H_\theta e^{-j\beta r} \int_{\theta_1}^{\pi - \theta_1} \frac{d\theta}{\sin\theta} \tag{5.19}$$

The biconical is usually fed by a coaxial line as shown in Figure 5.7(b). The radiation pattern of the biconical is similar to that of a dipole. In the azimuth (x-y) plane and in the elevation (y-z) plane it is omnidirectional and a figure eight, respectively, as shown in Figures 5.7(c, d). Typical gains of biconical vary between 0 dBi and 4 dBi, and beamwidths in the elevation plane are between 20 deg and 100 deg.

The broadband impedance characteristics of the biconical antenna occur when the cone half-angle θ_{hc} lies between 30 deg and 60 deg [6, p. 4–12]. This angle is not critical and is usually selected to match as closely as possible the characteristic impedance of the transmission line feeding the antenna.

The characteristic impedance of a biconical antenna is given by

$$Z = 120 \log_e \left[\cot\left(\frac{\theta_{hc}}{2} \right) \right] \tag{5.20}$$

where θ_{hc} is the cone half-angle.

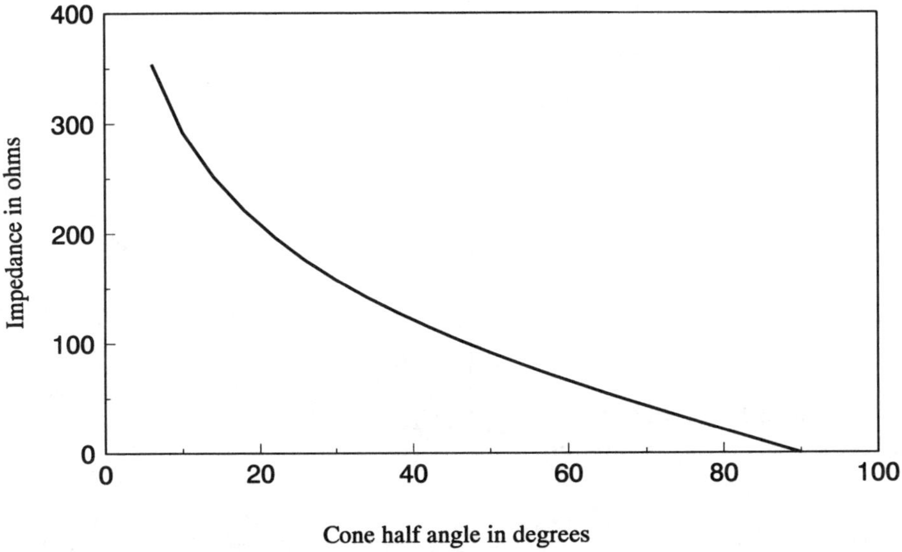

Figure 5.8 Variation of impedance with cone half-angle.

Using (5.20) for a biconical antenna of half-angle 60 deg, the characteristic impedance is calculated as 66Ω. The variation of the impedance with cone half-angle is shown in Figure 5.8. Measured values of the characteristic impedance for a 6:1 bandwidth biconical of half-angle 60 deg and cone diameter of one wavelength (at the lowest frequency) has been found to be a constant 50Ω over the bandwidth with a VSWR of <1.2 [11, p. 64].

For small cone angles $\cot(\theta_{hc}/2)$ is approximately equal to $2/\theta_{hc}$ and, thus, the characteristic impedance of a small-angled biconical antenna is given by

$$Z = 120 \log_e\left[\frac{2}{\theta_{hc}}\right] \tag{5.21}$$

5.8 YAGI–UDA

This antenna is often called a Yagi, after the student of Professor Uda of Japan who first presented the paper in English and performed many of the measurements; however, it was designed by Uda and should therefore be called a Uda–Yagi. It consists of a single driven dipole and a number of slightly shorter parasitic dipoles in front of it, as shown in Figure 5.9(a). The driven element is a resonant or tuned dipole, or it could be a folded dipole. Parasitic dipoles are those that are not directly fed but have currents induced in them in the same way as the current from the primary of a transformer induces a current in the

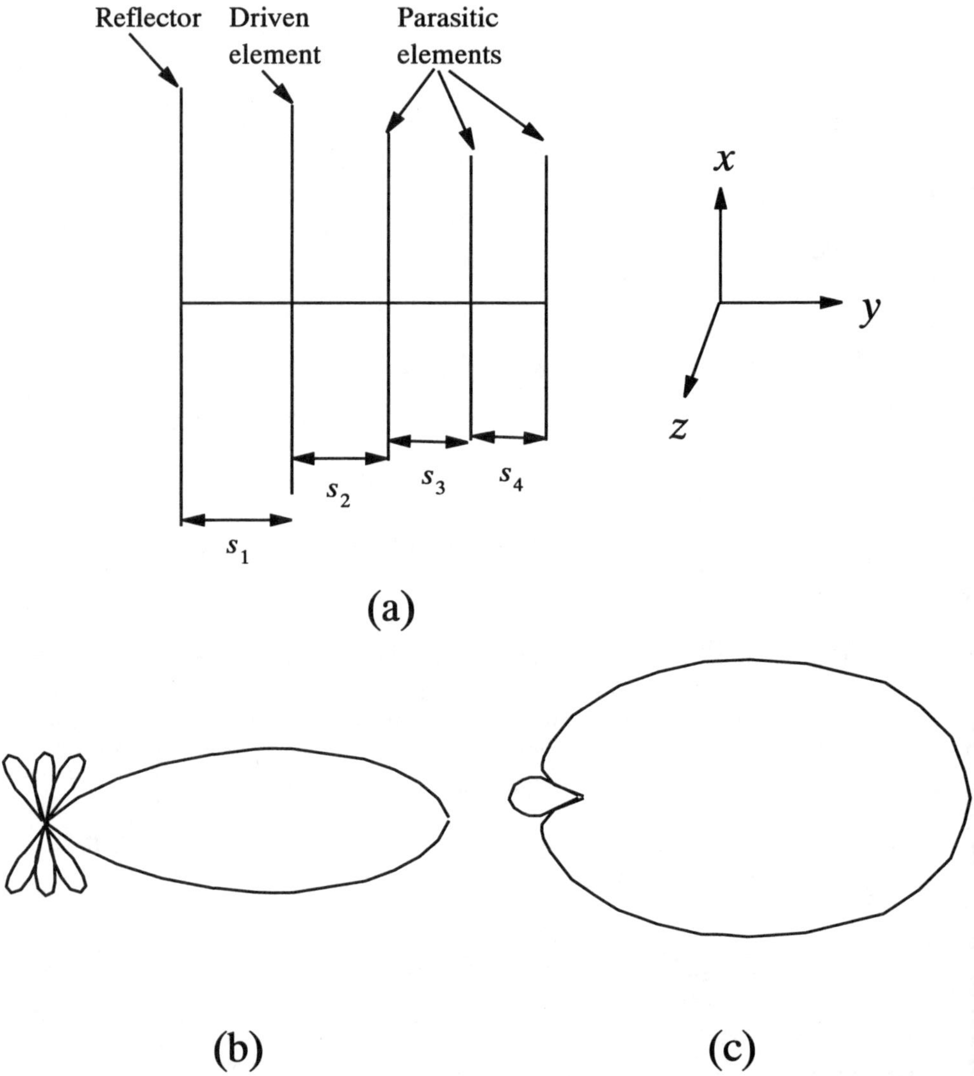

Figure 5.9 (a) The Yagi–Uda dipole array; (b) radiation pattern in the *x-y* plane; (c) radiation pattern in the *z-y* plane; and (d) uniform Yagi-Uda array.

secondary. The parasitic dipole can then reradiate like directly fed radiating dipoles. These parasitic elements are called directors. A reflector that is slightly longer than the driven element is placed behind the driven element. All elements are parallel to each other and in the same plane. The antenna radiates in the end fire direction from the reflector toward the shorter directors. Increasing the number of directors, increases the gain, but with diminishing returns; so, in practice, there is not a large increase in gain when the number

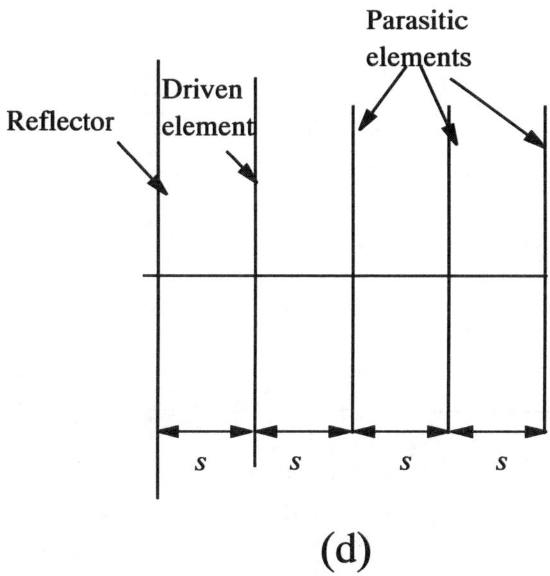

Figure 5.9 (continued).

of parasitic elements is above about 12 [5, Vol. 2, p. 765]. The radiation pattern in the *x-y* plane is shown in Figure 5.9(b), and that in the *y-z* plane is shown in Figure 5.9(c). In the case of uniform Yagi arrays, all the elements have the same circular cross section, the spacings between the elements are the same, and all directors have the same physical length, as shown in Figure 5.9(d). The directors are made the same length for ease of construction, but they do not carry equal currents.

The gain obtainable for different numbers of elements has been computed for uniform Yagis made of wires that have diameters from 0.0025λ to 0.02λ and spacings between 0.15λ and 0.3λ [7]. The curve of gain versus the number of elements for a conductor diameter of 0.0025λ shows an average gain of 0.9 dB for each additional element above three, but the gain flattens off after about five elements, as shown in Figure 5.10.

Comparisons of measured results and computed values suggest that the elements behave as though they are slightly longer than their physical lengths. This is thought to be due to an "end effect," and when the elements are slightly shortened, empirically, optimum performance can be obtained at a particular design frequency.

Increasing the number of parasitic elements reduces the radiation resistance of the antenna so that it does not match the transmission line feeding it. A folded dipole is required to match the resistance of the antenna to the transmission line in these cases. When four elements are used and spaced 0.2λ apart, the input resistance is 45Ω. The antenna can be made more broadband by designing the directors for the upper end of the frequency band and the reflector for the lower end. An antenna with five elements (one reflector and three directors) can have a bandwidth of 60%, but its gain is reduced to

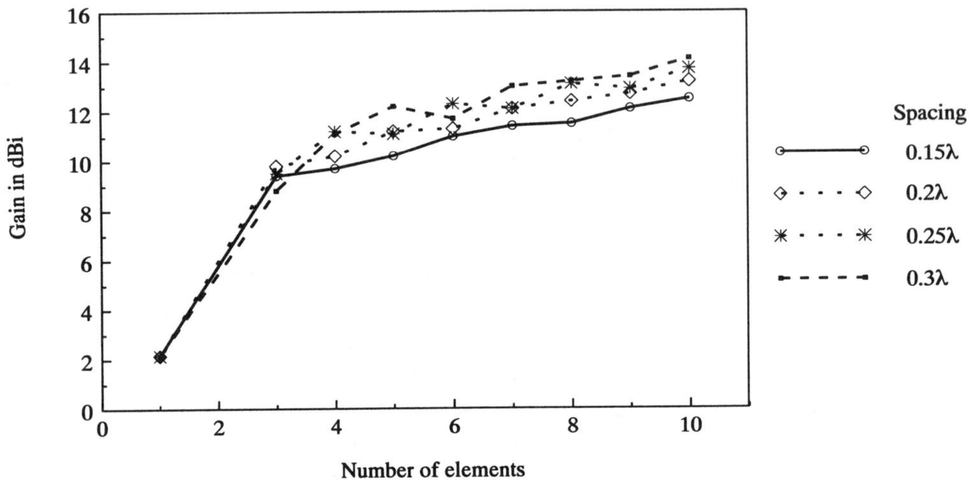

Figure 5.10 Variation of gain with number of elements for uniform Yagis.

around 6 dBi [5, p. 765]. If the antenna is optimized, its gain can be increased to 10 dBi but its bandwidth is reduced to about 9%.

5.9 FREQUENCY-INDEPENDENT (FI) ANTENNAS

In general, it can be said that the impedance and radiation pattern characteristics of an antenna are determined by its shape and dimensions in terms of wavelength. In 1954 Rumsey suggested that if an antenna could be defined entirely in terms of angles, for instance, instead of any characteristic length dimensions, then it would be possible for such an antenna to have properties that are independent of the frequency of operation. A change of frequency, however, results in rotation of the radiation pattern. This type of antenna that is based on the angle condition can be implemented by use of conical surfaces and equiangular spirals.

There are two types of frequency-independent antennas, namely, logarithmic conical spirals, called log spirals, and logarithmic periodic, called log periodics. It can be shown that they are both the same type of structure except that the log periodic has periodicity at discrete angles whereas the periodicity of the log conical spiral is continuous.

One of the characteristics of FI antennas is that the radiation is in the backfire direction.

5.10 LOG-PERIODIC ANTENNA

The log-periodic antenna consists of an array of dipole elements and displays fairly stable impedance and pattern characteristics over the design bandwidth. This antenna has self-similar figures with discrete similarity at discrete angles (ϕ) only. The *log-periodic dipole array* (LPDA) antenna is shown schematically in Figure 5.11(a).

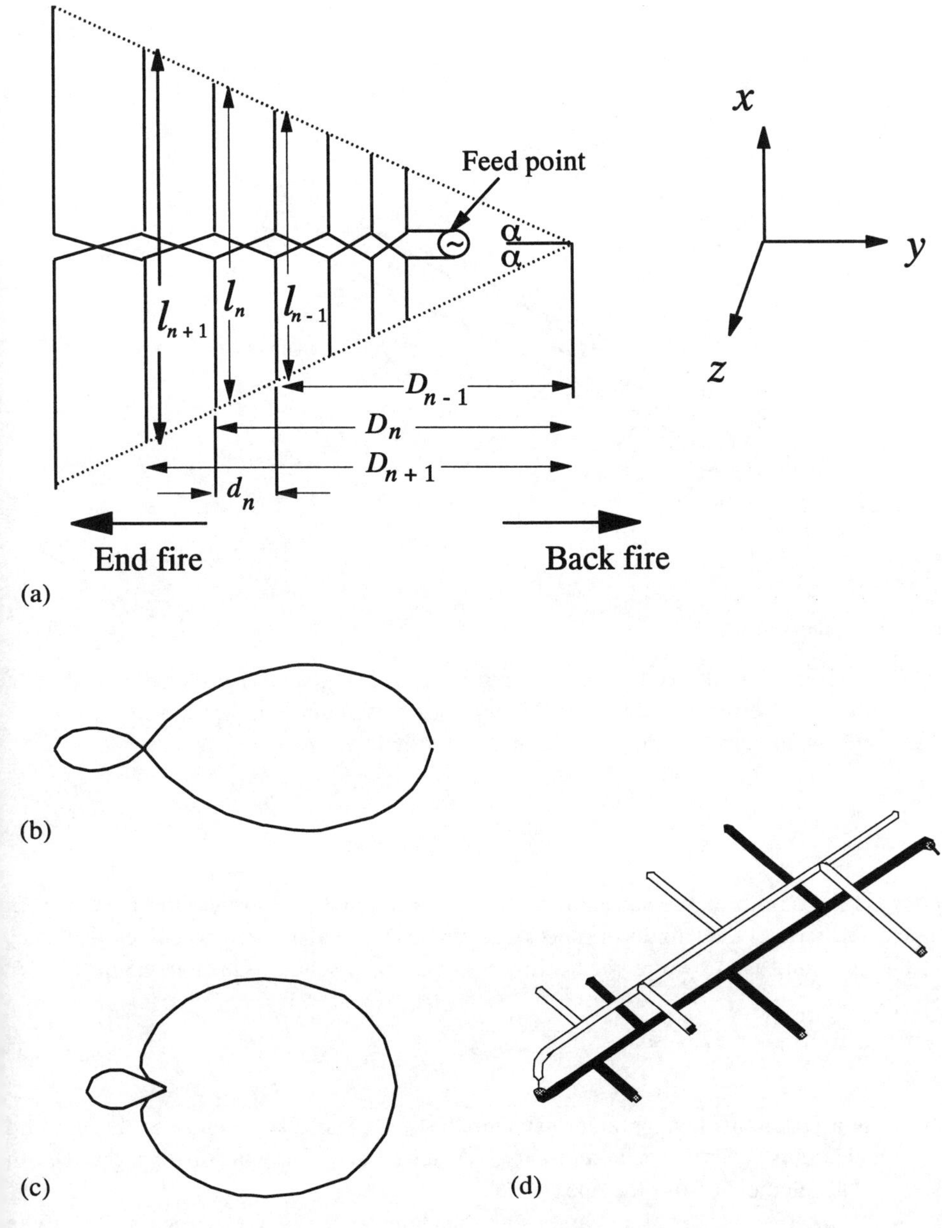

Figure 5.11 The log-periodic dipole array LPDA: (a) schematic of the log periodic dipole array; (b) radiation pattern in the x-y plane; (c) radiation pattern in the z-y plane; (d) coaxial line 50Ω or 75Ω feed; and (e) 300Ω twin wire feed for the LPDA.

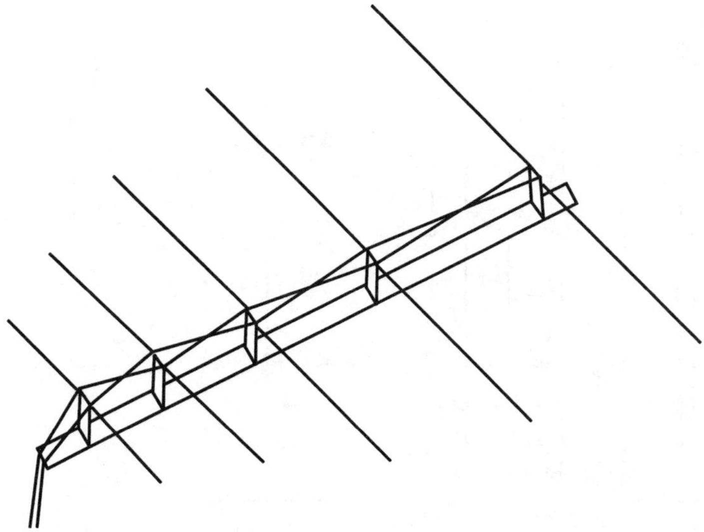

(e)

Figure 5.11 (continued).

It consists of dipoles of decreasing length and spacing such that the apex angle that is formed by the intersection of two imaginary lines drawn through the ends of the dipoles on each side of the array is 2α; and the tan of the half-angle α is given by

$$\tan\alpha = \frac{l_n}{2D_n} \tag{5.22}$$

where l_n is the length of the nth element and D_n is the distance of the nth element from the apex. The ratio of the lengths of successive dipoles is equal to the ratio of their distances from the apex and is known as the scaling factor s_f. The scaling factor is given by

$$s_f = \frac{l_n}{l_{n-1}} = \frac{D_n}{D_{n-1}} \tag{5.23}$$

where s_f is a constant of magnitude less than 1, l_n, l_{n-1} are the lengths of the nth and $(n-1)$th elements (where l_n is larger than l_{n-1}), and D_n, D_{n-1} are the distances of the nth and $(n-1)$th elements from the apex.

A spacing parameter S_p that varies with the element can also be defined and is given by

$$S_p = \frac{d_n}{2l_n} \tag{5.24}$$

where d_n is the distance between the nth and the $(n + 1)$th elements and l_n is the length of the nth element.

Since l_n is a half-wave dipole, it is equal to $\lambda/2$, and thus the spacing parameter S_p is also equal to d_n/λ; that is, it is equal to the spacing in wavelengths at the frequency of the resonant dipole.

Using the relationships of (5.22) and (5.23), the spacing parameter can be rewritten as

$$S_p = \frac{(1 - s_f)}{4} \cot \alpha$$

The properties of the antenna at frequencies f, rf, r^2f, . . . , r^nf are the same. When these frequencies are plotted on a logarithmic scale, these frequencies are equal spaced; this is the reason why this type of antenna is called a log-periodic antenna. If the frequencies are closely spaced, the antenna can provide uniform performance over a wide range of frequencies [5, Vol. 2, p. 767]. The center fed dipoles are connected to a twin transmission line so that adjacent elements are fed in antiphase as shown in Figure 5.11(a). This gives backfire radiation (in the direction of elements of decreasing length). If the phase reversal is not included, the radiation occurs in the end-fire direction so that the radiation from the shorter elements is obscured by the longer elements. Scalloped radiation patterns are also produced by end-fire arrays, and impedance behavior is erratic [6, p. 14–27]. The array is fed at its high-frequency (shortest element) end, and the transmission line is shorted at its low-frequency end. The upper frequency limit is determined by the length of the shortest dipole and occurs when this length is resonant at approximately half a wavelength. The lower frequency limit is determined by the length of the longest dipole. When the array is fed by a particular frequency, the wave will travel along the array from the shortest element until it encounters the resonant dipole element (i.e., the dipole whose length is approximately half a wavelength). This resonant element will have a lower impedance than any other element and, thus, will extract more energy than any other element. The energy extracted from the wave will progressively increase as the wave travels from the high-frequency end to the resonant element, and the energy extracted will decrease beyond this element. The elements that extract energy form the active region, which is restricted to a few elements on either side of the resonant element. As the frequency is increased the active region moves toward the shorter elements, and as the frequency is decreased it moves toward the longer elements. The width of the active region also reduces with increasing frequency, with the highest frequency having the smallest active region. At the low-frequency end, up to 50% of the length is used; but at the highest operating frequency, only about 15% may be used [11, p. 705]. The effective electrical aperture (the aperture in terms of wavelengths) remains essentially independent of frequency. The antenna can be considered as a cascade of cells, with each cell consisting of a section of transmission line shunt loaded at its center by the dipole element impedance. This dipole impedance also includes the mutual coupling between the elements. Shunt loading results in attenuation in the feeder line [6, p. 14–27]. This attenuation increases as the scaling factor s_f is increased, as long as the spacing parameter S_p does not become too small. In order to ensure that radiation does not occur in the end-fire direction, this attenuation through the active region should be about 20 dB.

The value of the constant s_f varies between 0.76 and 0.98 for values of S_p between 0.06 and 0.2. However, as the value of S_p is reduced, the VSWR tends to increase. Typically VSWR values of around 1.4 are achievable. As s_f approaches its upper limit of 0.98, the elements are more closely spaced and hence the active region contains more elements. The inclusion of more elements in the active region increases the directivity, and values between 8 dBi and 12 dBi are achievable. The apex angle 2α also affects the antenna directivity, and angles vary from between 2 deg and 40 deg. Small angles result in excessively long antennas, whereas large angles cause disruption of the phasing between elements and result in radiation pattern distortion [12, p. 18]. The antenna resistance has little effect on the directivity [5, Vol. 2, p. 769]. For thinner elements the resistance changes more rapidly with frequency, and there is a reduction in the directivity. Thicker elements reduce the Q and hence increase the bandwidth. As the spacing parameter S_p is reduced, the VSWR tends to rise.

At frequencies in the VHF range (30 MHz to 300 MHz) the antennas can be mounted high enough (in terms of wavelengths) to avoid ground effects. However, at frequencies in the HF range (3 MHz to 30 MHz) the proximity to the ground results in loss, which reduces gain, and the image of the antenna formed by reflection in the ground plane causes alteration of the radiation pattern. End loading is also associated with these low frequencies and makes the elements appear to be electrically longer; this results in the operating frequency being lowered. In the EMC field log periodics are mainly used at frequencies between 100 MHz and 1,100 MHz.

Half-power beamwidths (HPBWs) in the E plane, that is, the plane x-y of the dipoles, are of the order of 50 deg. The H-plane (y-z plane perpendicular to the plane of the dipoles) HPBWs are of the order of 90 deg. The radiation pattern of the log periodic in the x-y and y-z planes are shown in Figures 5.11(b, c), respectively. Typical feeding arrangements are shown in Figures 5.11(d, e). Phase reversal between adjacent elements is achieved either by connecting alternate monopoles to opposite sides of a transmission line or by physically crossing the wires of a twin transmission line feed for dipoles supported on insulators [11, p. 708].

Smith [12, p. 18] developed algorithms for selecting the optimum spacing parameter S_p and the scaling factor s_f for *log-periodic dipole antennas or array* (LPDAs) of specified gains. For a LPDA of linear gain g the spacing parameter is given by

$$S_p = -5.76909996 \times 10^{-4} \times g^2 + 0.0167208176g + 0.0602945516 \qquad (5.26)$$

For gains from 8 dBi to 12 dBi, spacing parameters vary between 0.1428 and 0.1804, as shown in Figure 5.12.

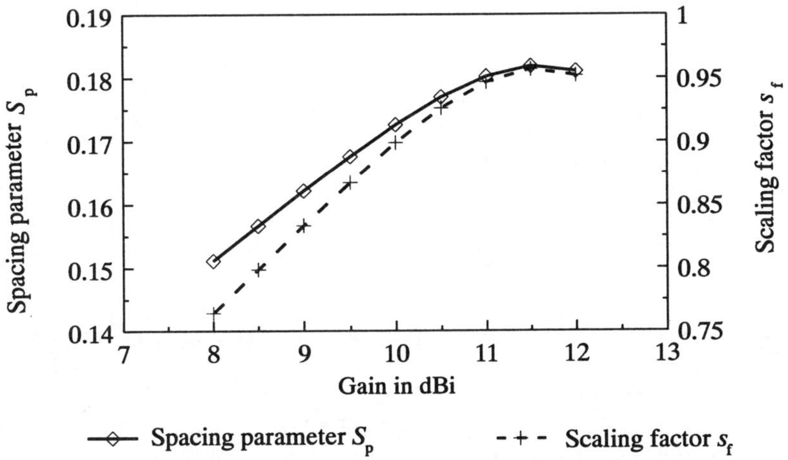

Figure 5.12 Variation of the spacing parameter and scaling factor with gain.

The optimum scaling factor s_f can calculated from the optimum spacing parameter S_p and is given by

$$s_f = 3.9866009 S_p + 0.236230336 \tag{5.27}$$

For the range of spacing parameters (S_p)from 0.1428 to 0.1804, the scaling factor varies between 0.806 and 0.955, which is shown in Figure 5.12.

The impedance of the antenna Z_a can be matched to the transmission line by selecting the correct length-to-diameter ratio (l/d_0) of the element. The impedance Z is given by

$$Z_a = \frac{Z_0^2}{8Z_1} \sqrt{s_f + \left(\frac{8Z_1}{Z_0}\right)} \tag{5.28}$$

where Z_0 is the characteristic impedance of the transmission line and Z_1 is given by

$$Z_1 = 276 \log_{10}\left(\frac{l}{d_0}\right) - 270 \tag{5.29}$$

Using the range of S_p and s_f in Figure 5.12, the antenna can be matched to a $75 - \Omega$ line by choosing values of the length-to-diameter ratio (l/d_0) between 10.501 and 10.554.

The number of elements depends on the bandwidth and the scaling factor s_f. A design bandwidth (B_s) is calculated that is greater than the desired bandwidth B_w, which is the ratio of the highest to the lowest frequency. This larger design bandwidth is required to accommodate the active region for the high-frequency end. The design bandwidth [6, p. 29–15] is given by

$$B_s = B_w[1.1 + 7.7(1 - s_f)^2] \cot \alpha \tag{5.30}$$

The number of elements is given by

$$N = \left[1 + \frac{\log_e B_s}{\log_e(1/S_f)}\right] \tag{5.31}$$

For an LPDA operating from 100 MHz to 1,100 MHz (i.e., $B_w = 11$) and an apex angle (2α) of 53 deg (which gives an LPDA of length equal to its width), the design bandwidth is 28 and this gives the number of elements as 21. For an LPDA operating from 200 MHz to 1,000 MHz (i.e., $B_w = 5$) and an apex angle of 53 deg, the design bandwidth is 13 and this gives the number of elements as 16.

The trapezoidal tooth antenna with curved teeth introduced by DuHammel in 1957 was the first successful log periodic. It consisted of a self-complementary structure, that is, if the two arrays shown in Figure 5.13(a) were to be folded along the center line of the metal sheet, they would form a single sector, with the metal section of double thickness. It had a bidirectional pattern, that is, it radiated in two directions, into and out of the paper. This was adapted by Isbell to a straight-toothed version as shown Figure 5.13(b). The teeth act as monopoles. The distance from the back of each tooth to the apex is D_n, and r_n is the distance from the apex to the front of the tooth. The ratio of r_n/D_n is a measure of the width of the teeth. The scaling factor s_f is defined as above, and the angle subtended by the extreme edges of the teeth (at the apex) is 2α. However, in this case there are two other angles to be considered. The angle 2β defines the angle of the supporting section, and the angle ψ is the angle between the two arrays. When ψ is equal to 180 deg, the antenna behaves like the one used by DuHammel and gives a bidirectional pattern. If the input impedance is plotted on a Smith's chart over several periods, the points lie on a circle whose center is on the zero-impedance line [6, p. 14–25]. The characteristic impedance of the antenna is the geometric mean of the maximum and minimum values of the resistance (real part of the impedance). The two arrays can be considered as a uniform biconical TEM transmission line. Since the monopoles are attached to the opposite sides of the center supporting strip, the required 180-deg phase delay is achieved.

This antenna radiates an unidirectional linearly polarized beam [5, Vol. 1, p. 376] with the electric field parallel to the edges of the teeth as shown in Figure 5.13(b). The variation of the -10-dB beamwidth in the *E* and *H* planes with the angle ψ (between the arrays) is shown in Figure 5.13(c) for a scaling factor of 0.707, $\alpha = 22.5$ deg and $\beta = 5$ deg. It can be seen that at around $\psi = 60$ deg the *E*- and *H*-plane beamwidths are equal. This feature is particularly useful when the log periodic is used as a feed for a reflector that requires circular symmetry.

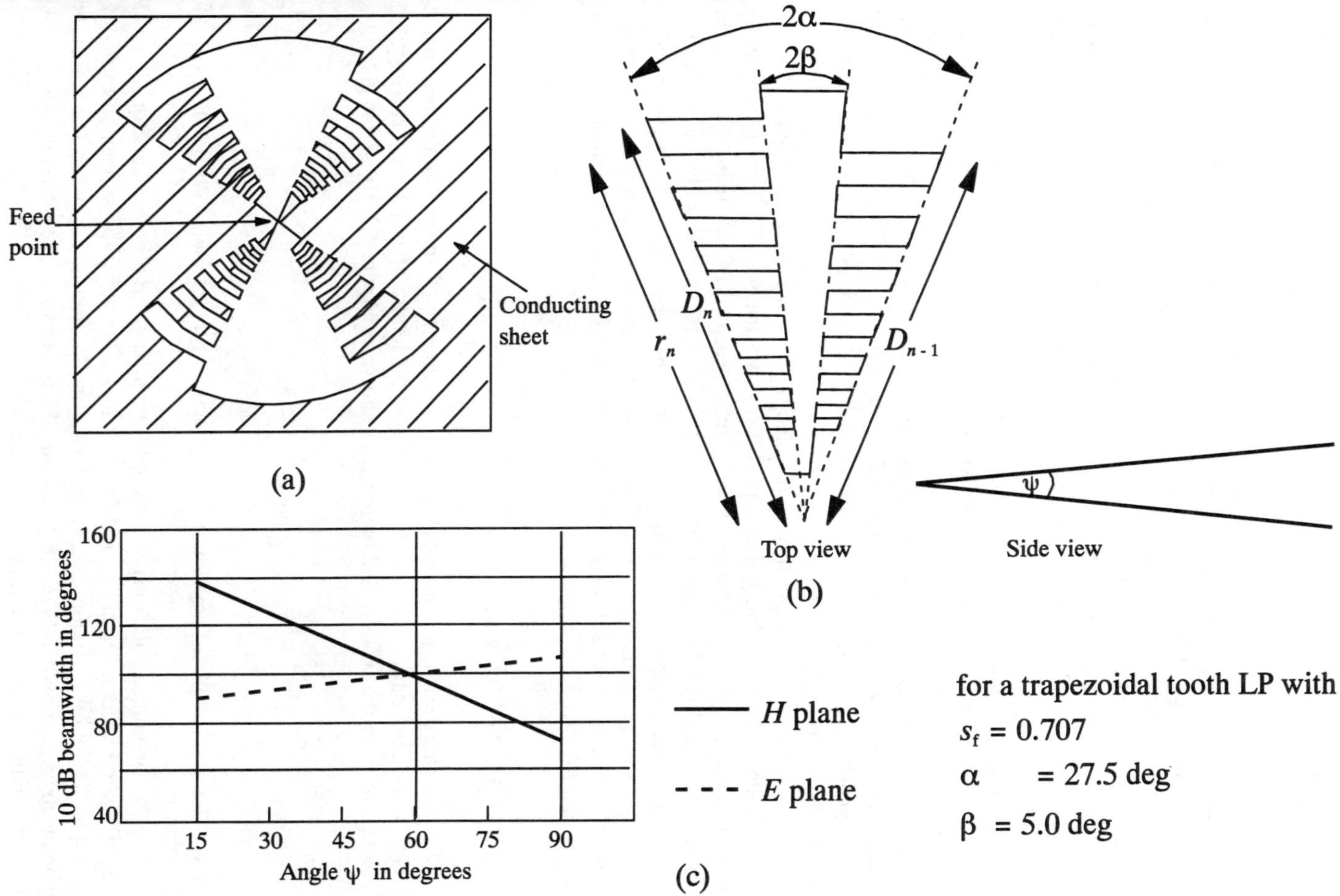

Figure 5.13 The trapezoidal tooth log-periodic antenna: (a) DuHammel's trapezoidal tooth LP; (b) straight-toothed trapezoidal tooth LP; and (c) beamwidths of the straight-toothed trapezoidal tooth LP.

5.11 BILOG

The bilog antenna [22] is a hybrid antenna comprised of a LPDA in front of a bowtie antenna. The advantage of using this antenna is that it can be used over a wider range of frequencies, thus obviating the need to use two antennas. Apart from the time saving, the test setup provides increased accuracy since the repositioning of antennas is not required. The bilog can be used over the frequency range 30 MHz to 1 GHz. At 30 MHz the wavelength is 10 m, and if a LPDA were to be used, the longest element would have to be about 5 m.

At the lower frequencies the antenna acts as a LPDA and does not require a balun when fed by a two-wire transmission line. At the lower frequencies the antenna acts as a broadband balanced dipole and a balun is required.

The radiation patterns for horizontal polarization of the bilog at 30 MHz and 150 MHz demonstrate the familiar figure eight of dipole patterns, as shown in Figure 5.14(a). As the frequency is increased, the pattern tends toward the unidirectional pattern of the LPDA, as shown in Figure 5.14(b).

For immunity and susceptibility testing the input power in watts required to give electric field strengths of 1 V/m, 3 V/m, and 10 V/m, at a distance of 3m from the antenna, is shown in Figure 5.15. It can be seen that at the lower end of the frequency range the input power required is higher but that above 250 MHz the power is almost constant. The antenna correction factor of the bilog is shown in Figure 5.16 and is compared to those for the LPDA and the biconical antenna. The magnitudes of the antenna factors are similar to the LPDA above 300 MHz, and between 100 MHz and 300 MHz it appears to be better than the biconical antenna.

5.12 HELICAL ANTENNAS

The helix antenna shown in Figure 5.17(a) has a diameter D; the circumference of the helix is $C = \pi D$. The spacing between turns is S, and the length of each turn is L. The pitch angle ψ is related to the spacing and diameter as shown in Figure 5.17(b) and is given by

$$\tan \psi = \frac{S}{\pi D} \tag{5.32}$$

where ψ is the pitch angle in degrees, S is the spacing in meters, and D is the diameter of the helix in meters.

A helical antenna could consist of a single conductor or multiple conductors and can radiate in several modes depending on the electrical diameter (the diameter in terms of wavelength). The most common mode is the axial mode of radiation, when the boresight is along the axis, as shown in Figure 5.17(c), and the minor lobes are relatively small. This mode occurs when the circumference C (of one turn) is of the order of one wavelength.

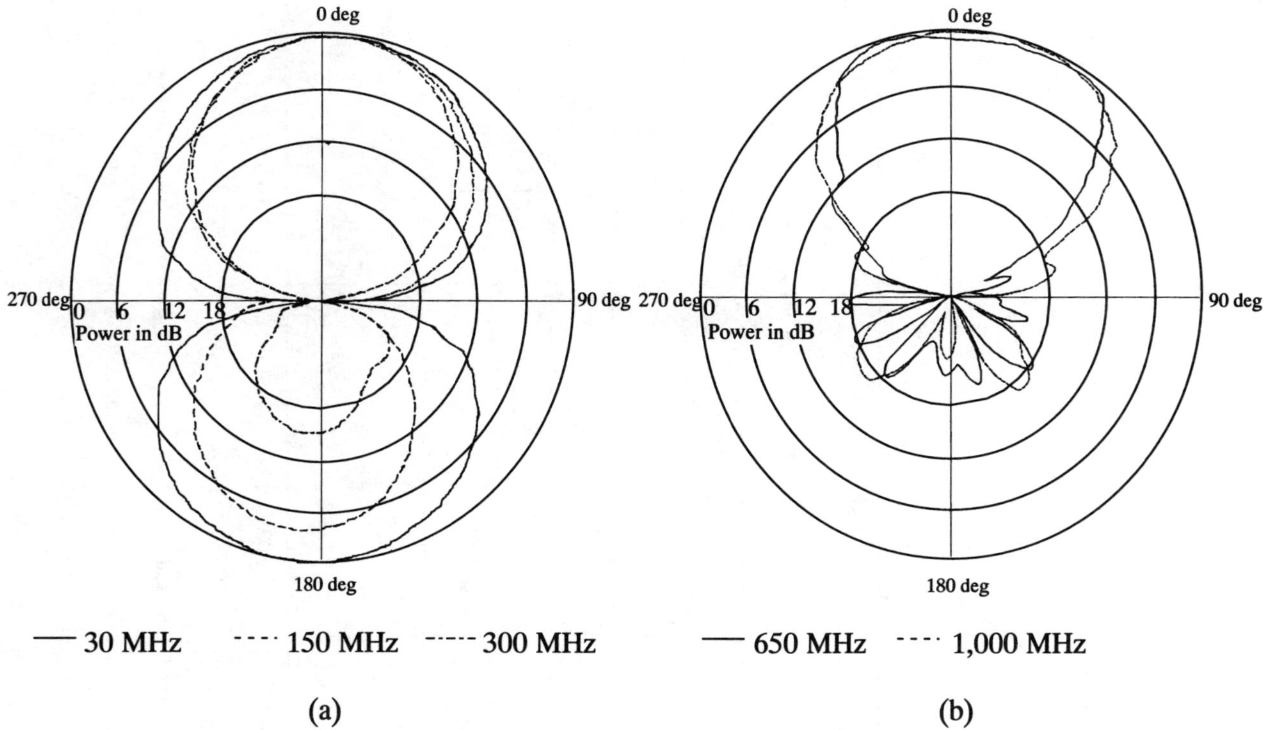

Figure 5.14 Radiation patterns of the bilog antenna for horizontal polarization. At (a) 30, 150, and 300 MHZ, and (b) 650 and 1,000 MHz. (After [22].)

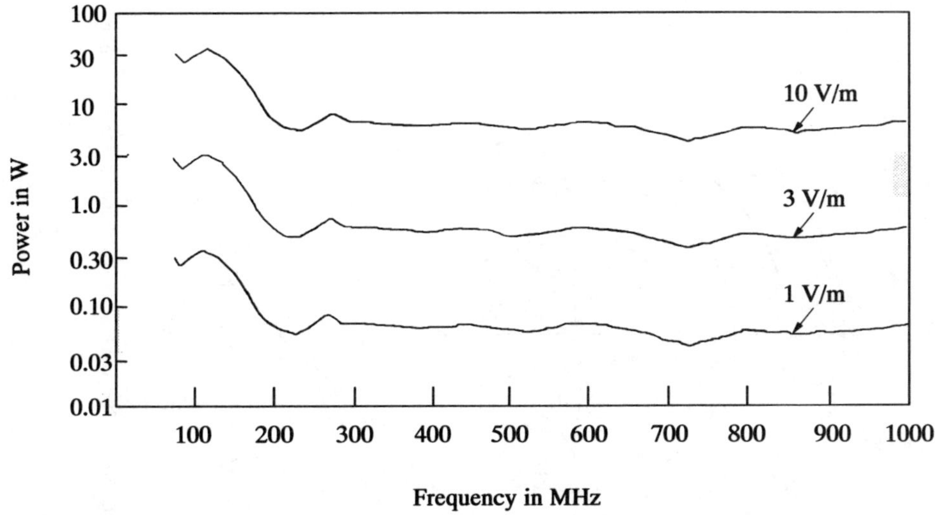

Figure 5.15 Power required for bilog antenna to produce a given electric field at 3m. (After [22].)

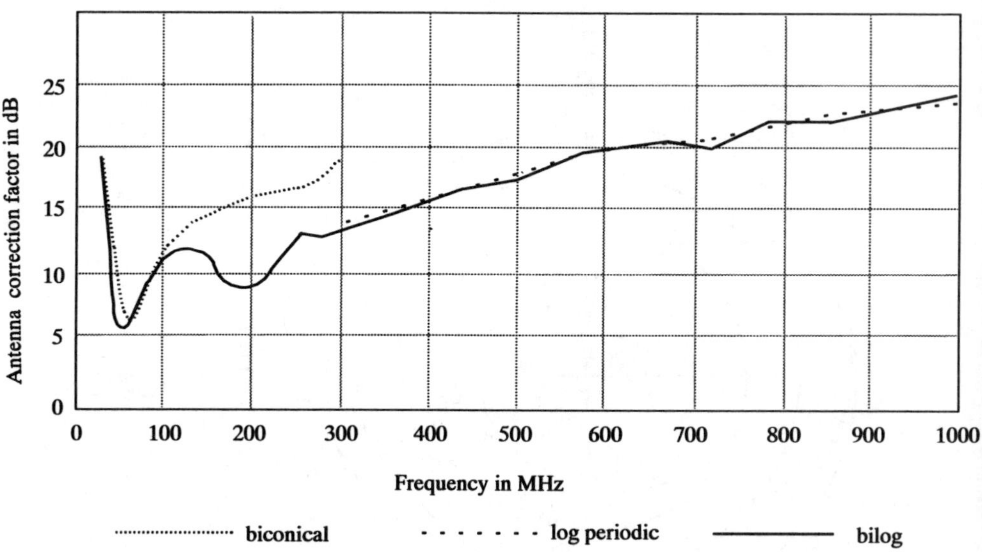

Figure 5.16 Antenna correction factors for the biconical, log-periodic, and bilog antennas. (After [22].)

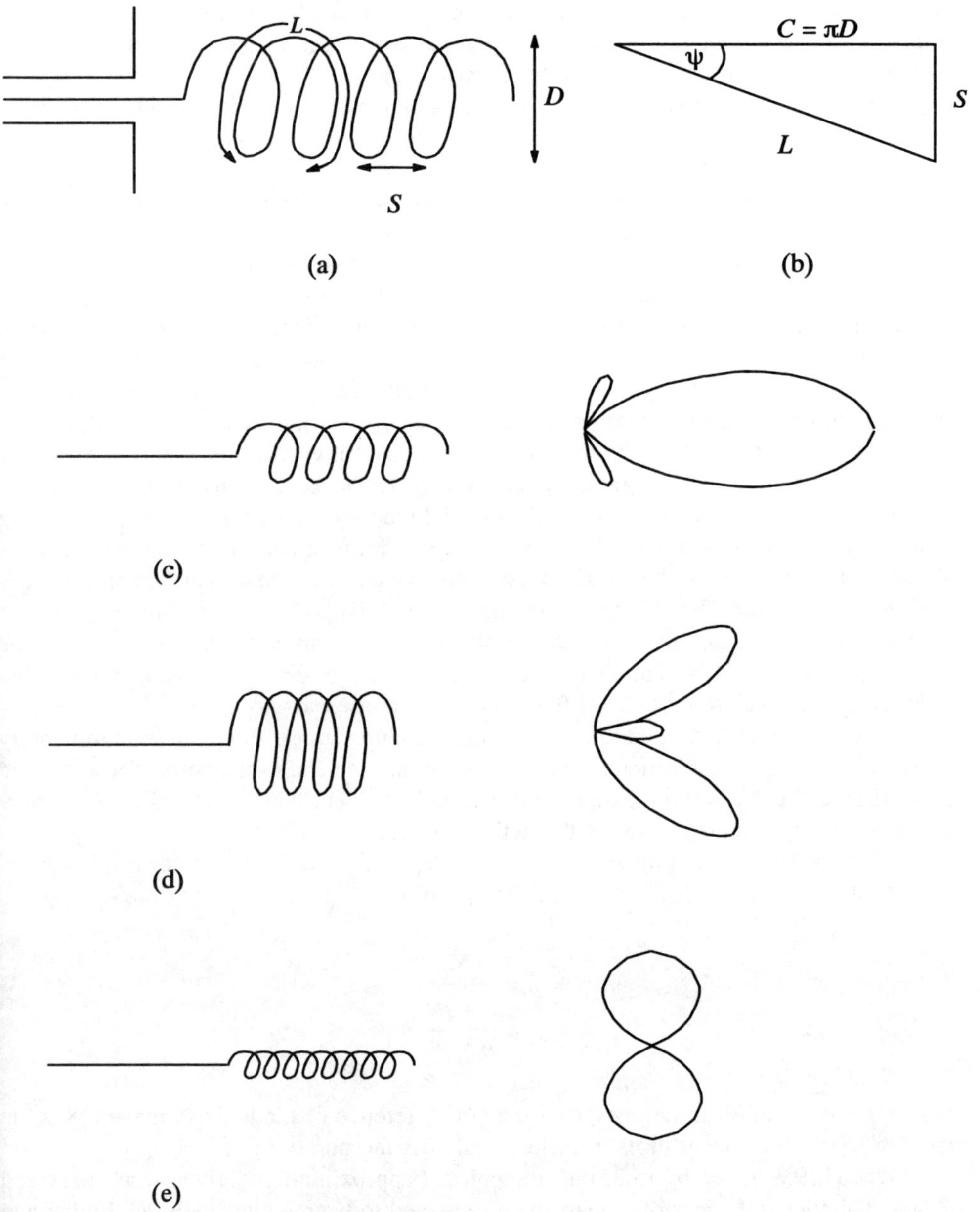

Figure 5.17 The helical antenna: (a) Characteristics, (b) relationship between diameter and length of one turn, (c) axial mode patten for circumferences of around one wavelength, (d) conical mode pattern for circumferences greater than one wavelength, and (e) normal mode pattern for circumferences much less than one wavelength.

When the diameter is much smaller than a wavelength, the helix antenna acts like a dipole and has an omnidirectional radiation pattern perpendicular to its axis, as shown in Figure 5.16(e). This is known as the normal mode.

Higher order modes occur when the helix diameter is greater than $\lambda/3$. The single lobe of the axial mode splits up into two lobes, known as the conical mode of radiation, as shown in Figure 5.16(d).

The axial-mode helix has wideband impedance characteristics, and the radiated electric field is circularly polarized [6, Chapter 13]. Helices wound like a left-hand screw receive *left-hand circularly polarized* (LHCP) waves, whereas those wound like a right-hand screw receive *right-hand circularly polarized* (RHCP) waves. Most axial-mode helices are of uniform diameter, but helices of nonuniform diameter are used to give wider bandwidth and improved radiation characteristics. The axial-mode helix can have a 1.7:1 bandwidth, and its directivity is close to the maximum attainable of an antenna of that size, over the whole of the bandwidth, because the phase velocity is automatically adjusted to the correct value to produce the maximum directivity for each frequency [13, p. 7–1].

The helix usually has a ground plane or is cavity-backed. Cavity-backed helices are preferred because they reduce back radiation and the forward gain is enhanced, resulting in an improved F/B (front-to-back) ratio. In a typical arrangement for an uniform helix, as shown in Figure 5.18(a), the coaxial feed is connected to the helix via a microstripline-matching transformer. For the frequency range 650 MHz to 1,100 MHz, the transformer is about 12-cm long and the cavity has a diameter of 26 cm and depth of 12.5 cm. The conductor from which the helix is wound could be tubing of circular cross section or a flat strip, and a lightweight cylindrical foam dielectric is used as the former.

The VSWR measured at the input of the matching transformer can be significantly reduced by adding two additional tapered turns at the free end of the helix. These tapered turns reduce the VSWR by suppressing the reflected currents. The cut-off frequency occurs when the circumference of the helix is 0.75λ.

The HPBW for pitch angles between 12 deg and 15 deg and helices of circumference C of values $0.67\lambda < C < 1.3\lambda$ can be given by

$$\theta = \frac{52}{\dfrac{C}{\lambda}\sqrt{\dfrac{NS}{\lambda}}} \tag{5.33}$$

where θ is the HPBW in degrees, C is the circumference of the helix in meters, S is the spacing between adjacent turns in meters, and N is the number of turns.

The HPBW given by (5.33) is an empirical approximation derived from measured data and applies to helices of at least three turns and to beamwidths between 30 deg and 60 deg. The optimum dimensions for a wideband helix in the frequency range 200 MHz to 500 MHz has a pitch angle of 14 deg, conductor diameter of 0.017λ, and axial length of 1.65 wavelengths at the center frequency [13, p. 7–3]). These dimensions give a value

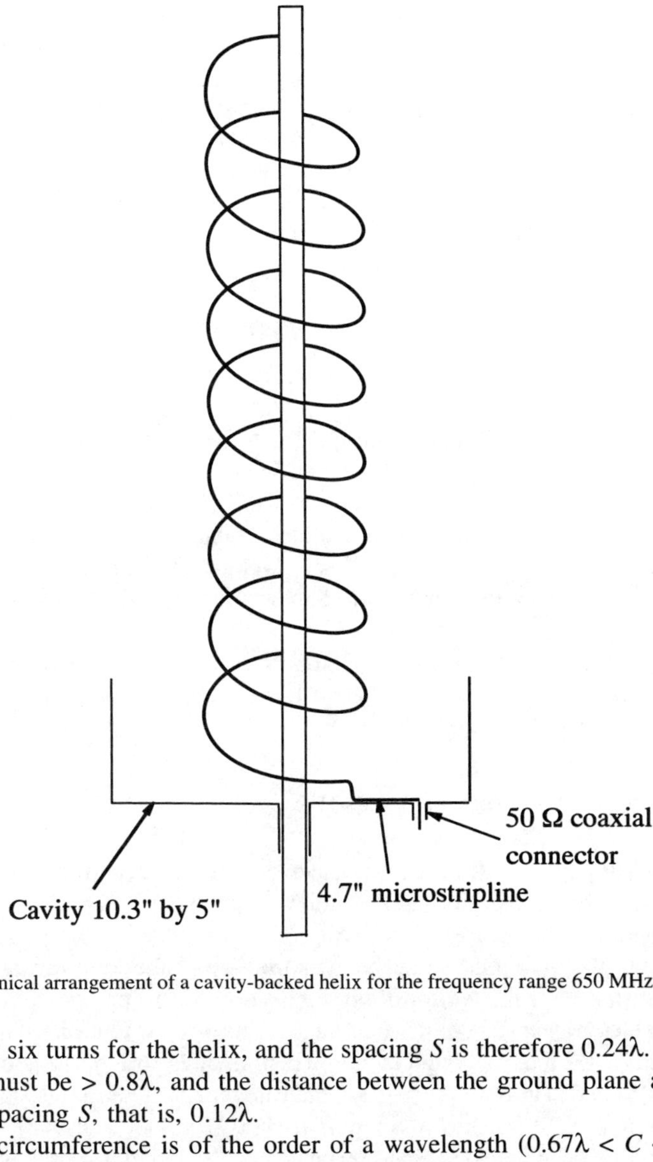

Figure 5.18 Mechanical arrangement of a cavity-backed helix for the frequency range 650 MHz to 1,100 MHz.

of approximately six turns for the helix, and the spacing S is therefore 0.24λ. The ground plane diameter must be $> 0.8\lambda$, and the distance between the ground plane and the first turn is half the spacing S, that is, 0.12λ.

When the circumference is of the order of a wavelength ($0.67\lambda < C < 1.3\lambda$), the terminal impedance of the helix is nearly resistive [13, p. 7–6] and the resistance R_h (within 20%) is given by the following empirical relation

$$R_h = \frac{140C}{\lambda} \tag{5.34}$$

where C is the circumference of the helix. This terminal impedance of the helix varies with frequency, but the variation will be less pronounced for helices with a larger number of turns than it is for helices with a few turns.

The radiation pattern can be obtained to a first approximation by considering the helix to be an array of N isotropic elements spaced a distance apart. The pattern of the array is the product of the pattern of each isotropic element and the array pattern. The isotropic element has a pattern of approximately cos θ, and the array has a radiation pattern that is the same as that of a uniformly illuminated aperture; thus, the resultant electric field radiation pattern is given by

$$\boldsymbol{E} = A\frac{\sin(N\psi/2)}{\sin(\psi/2)}\cos\theta \tag{5.35}$$

where $\boldsymbol{E}$ is the electric field in volts per meter, N is the number of turns, A is the normalization factor equal to $\sin(\pi/2N)$, and $\psi = kS\cos\theta - \delta$, where $k = 2\pi/l$ and δ is the progressive phase between turns and equals kLc/v; where L the length of one turn, v is the phase velocity along the helical conductor, and c is the velocity of light. When we increase the number of turns we get a narrower beam, which is expected since this is tantamount to increasing the number of elements in an array or increasing the aperture of an antenna.

Helical antennas can also be wound with more than one winding, such as bifilar, quadrifilar, and multifilar helices. The bifilar winding results in backfire radiation and does not generally require a ground plane.

5.13 LARGE AND RESONANT LOOPS

When the circumference of a loop is greater than 0.1λ the current distribution is significantly different from the uniform distribution assumed in the case of a small loop (Chapter 4). Assuming a uniform current distribution gives predicted radiation resistances that are much less than the measured values, and as the size of the loop increases the calculated radiation resistance (using uniform current distribution) departs even further from the measured values. For instance, if we assume a uniform current distribution for a loop of circumference 0.1λ, the calculated radiation resistance is 86% of the actual value, whereas if we have a loop of circumference 0.3λ, the calculated radiation resistance is 26% of the actual value [6, p. 5–9], as shown in Figure 5.19(a). The loop is usually fed by a twin wire transmission line. The resistance and reactance of the circular loop have been calculated as functions of the electrical circumference (the circumference in terms of wavelength) and are shown in Figure 5.19(b). When the circumference is approximately a whole number of wavelengths, the resistance has minimum values at points that are known as resonance points and at which the loop radiates the maximum power. Whereas when the circumference is an odd number of half wavelengths (0.5λ, 1.5λ, 2.5λ, . . . , $(2n - 1)\lambda/2$), the resistance

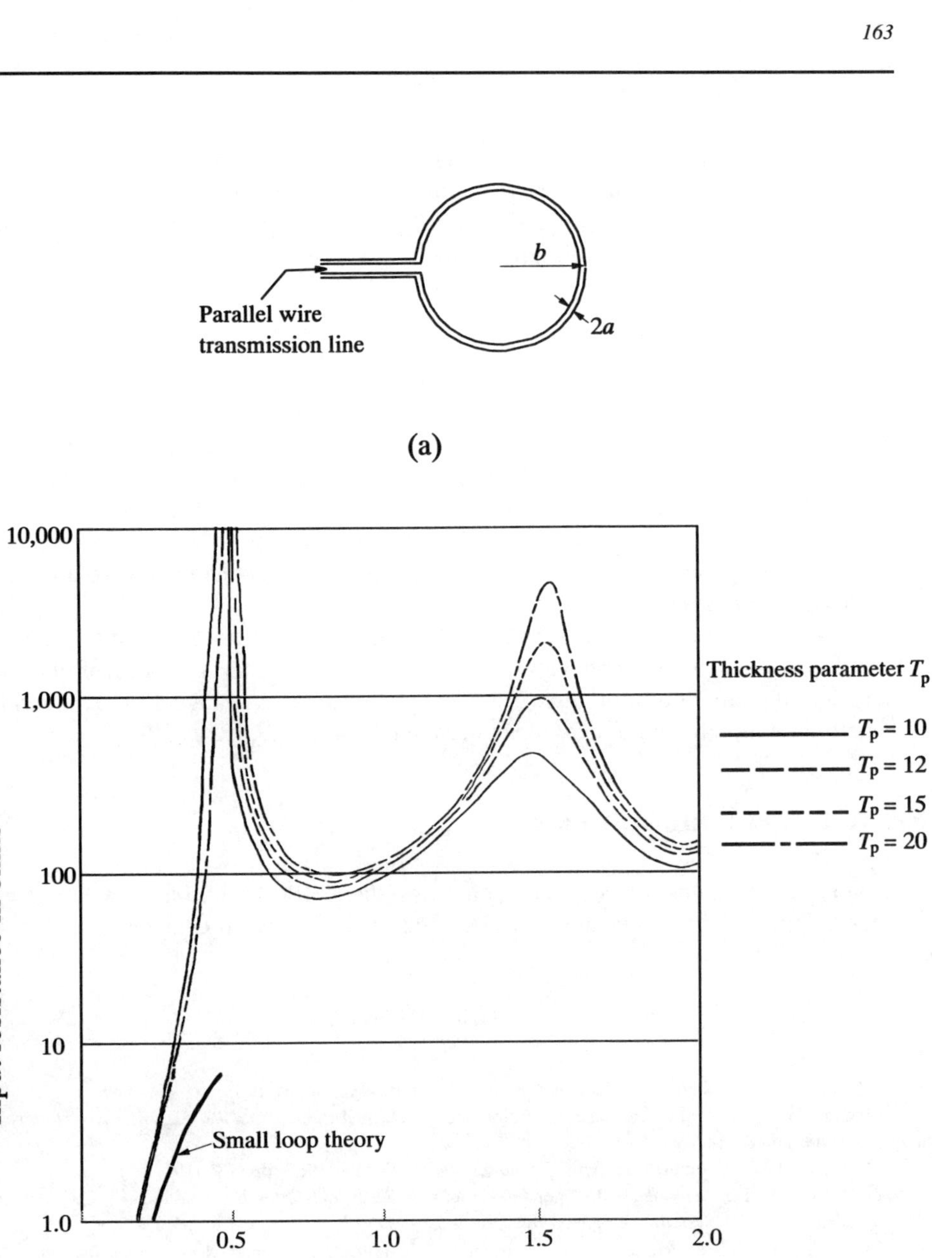

Figure 5.19 Characteristics of the single-turn loop antenna: (a) physical characteristics of a single turn loop, and (b) variation of input resistance with electrical circumference.

has maximum values points and these are known as antiresonance points. The actual values of the resistance at these resonance and antiresonance points are determined by the thickness parameter T_p of the wire forming the loop. The thickness parameter T_p is a function of the relative radii of the wire and the loop and is given by

$$T_p = 2\log_e\left(\frac{2\pi b}{a}\right) \quad (5.36)$$

where b is the radius of the loop and a is the radius of the wire forming the loop.

The resistance is relatively independent of the thickness parameter at the resonance points, but at antiresonance points it can be seen that the greater the thickness parameter, the higher is the resistance. This is to be expected since for a fixed-loop radius the larger the thickness parameter, the smaller is the radius of the conductor wire and hence the increase in resistance.

A single-turn loop can be optimized so that it can have a high gain over a frequency range where its length varies between 1λ and 1.7λ. For instance, a loop of diameter 20.3 cm has a directivity of more than 3.5 dB over the frequency range 470 MHz to 806 MHz where the loop circumference varies between 1λ and 1.7λ [6, p. 29–13].

5.14 DOUBLE-RIDGED HORN

Double-ridged horns are used at frequencies above 100 MHz, but they are more commonly used at frequencies above 1 GHz and are described in Chapter 6.

REFERENCES

[1] Nail, J. J., ''Designing a Discone Antenna,'' *Electronics*, August 1953, pp. 167–169.

[2] Rappaport, E., ''Antenna Designer's Notebook,'' *IEEE Antennas and Propagation Newsletter*, February 1988, pp. 12–14.

[3] Lucas, J. G., ''Antenna Systems,'' *Student Quart. J.*, June 1967, pp. 186–195.

[4] Mayes, P. E., ''Frequency Independent antennas: Birth and Growth of an Idea,'' *IEEE Antennas and Propagation Newsletter*, August 1982, pp. 5–8.

[5] Rudge, A. W., K. Milne, A. D. Olver, and P. Knight, *Handbook of Antenna Design, Vols.* 1 and 2, Stevenage, UK and New York: Peter Peregrinus Ltd., 1982.

[6] Johnson, R. C., and H. Jasik, *Antenna Engineering Handbook*, New York: McGraw-Hill Book Company, 1984.

[7] Green, H. E., ''Antenna Designer's Notebook,'' *IEEE Antennas and Propagation Newsletter*, June 1985, pp. 11–13.

[8] Aerospace Recommended Practice ARP 958, ''Broadband Electromagnetic Interference Antennas Standard Calibration Requirements And Methods,'' Society of Automotive Engineers Inc., 3–1-68.

[9] MIL-STD 461A, August 1968.

[10] Weiner, M. M., "Monopole Element at the Centre of a Circular Ground Plane Whose Radius is Small or Comparable to a Wavelength," *IEEE Trans. on Antennas and Propagation*, Vol. 35, No. 5, May 1987, pp. 488–495.

[11] Kraus, J. D., *Antennas*, New York: McGraw-Hill Book Company, 1988.

[12] Li, S. T., J. W. Rockway, J. C. Logan, and D. W. S. Tam, *Microcomputer Tools for Communications Engineering*, Norwood, MA: Artech House, Inc., 1968.

[13] Jasik, H., *Antenna Engineering Handbook*, New York: McGraw-Hill Book Company, 1961.

[14] Ramo, S., J. R. Whinnery, and T. Van Duzer, *Fields and Waves in Communication Electronics*, second edition, New York: John Wiley & Sons, 1984.

[15] Ramo, S., and J. R. Whinnery, *Fields and Waves in Modern Radio*, second edition, New York: John Wiley & Sons, 1953.

[16] Connor, F. R., *Introductory Topics in Electronics and Telecommunications—Antennas*, second edition, London: Edward Arnold, 1992.

[17] Jordan, E. C., *Electromagnetic Waves And Radiating Systems*, Constable and Company Ltd., 1953.

[18] Fewkes, J. H., and J. Yarwood, *Electricity, Magnetism And Atomic Physics, Vol.* 1, second edition, London: University Tutorial Press Ltd., 1962.

[19] Shepherd, D., "A Five-Decade System For Testing RFI Susceptibility," *RF Design*, September 1988, pp. 20–23.

[20] Balanis, C. A., *Antenna Theory—Analysis And Design*, New York: Harper and Row, 1982.

[21] *IEEE Standard Dictionary of Electrical and Electronics Terms*, fourth edition, (F. Jay, ed.), 1988.

[22] Marvin, A., S. J. Porter, and D. Riley, "30 MHz to 1 GHz in One Sweep—the Bilog."

Chapter 6
Antennas for Frequencies Above 1 GHz

Above 1 GHz wire antennas of resonant length are too small to be practicable. Frequency-independent antennas such as the log conical and planar spirals are popular, and aperture antennas such as the double-ridged horns are used. Smooth-walled horns are used where a wide frequency range is not required and for calibration purposes.

6.1 FREQUENCY-INDEPENDENT ANTENNAS

In 1954 Rumsey suggested that if an antenna could be defined in terms of angles instead of any characteristic length dimensions, then it would be possible for such an antenna to have properties that are independent of the frequency of operation. However, such a structure would be infinite in length and, when truncated, would not necessarily retain its frequency-independent characteristics because of the reflected fields from the truncation point.

The logarithmic spiral antenna is a true frequency-independent antenna [4, p. 698]. The first practical FI was constructed in 1958 by John D. Dyson at the University of Illinois (Kraus [4, p. 697]). It was a bidirectional planar spiral. The unidirectional planar spiral was achieved by placing a cavity behind the spiral. The cavity forms an image of the spiral that retransmits through the original spiral and, thus, modifies the radiation pattern. *Radar-absorbing material* (RAM) is placed in the cavity to prevent this reradiation, but this absorbs about 50% of the power input to the antenna.

6.2 BAND THEORY

The band theory is used to explain the radiation from log spiral as well as Archimedean spiral antennas. Although there is no mathematical basis for this theory, it is in good agreement with experimental results is easily understood and compatible with intuitive reasoning [7].

The spiral antenna behaves like a two-wire transmission line that is gradually transformed into a radiating structure. Allowable radiation bands exist for all circles that have a circumference equal to an integral number of wavelengths. For a more detailed explanation of the modes of radiation of a spiral see Section 6.3.1. In the case of the fundamental mode or mode 1, which has a main lobe along the axis, the radiation can be explained in the following way.

Consider a two-arm spiral, as shown in Figure 6.1(a), that is fed by currents that are in antiphase (i.e., 180 deg out of phase) but of equal amplitude. Consider two diametrically opposite points, A and A′, on the same arm of the spiral. If the current at A is in an upward direction, the direction of the current at A′ will depend on the phase delay caused by the length of the arc AA′. When the arc length is $\lambda/2$, the phase delay is 180 deg; so instead of the current at A′ being in the direction opposite to the current at A′ (as would be the case for dc), it is in the same direction because of this additional 180-deg phase delay. A similar situation occurs for the currents at B and B′ for the other arm of the antenna. Thus, all the currents are in the same direction at points A, A′, B, and B′. These points can all be considered to be at an average diameter of λ/π. The points C, C′, D, and D′ will similarly have currents in phase with each other but at right angles to the currents at A, A′, B, and B′. Since these currents are in time as well as spatial quadrature, the antenna radiates circularly polarized waves. This band of the spiral that has a mean diameter of λ/π is called the first radiation band. This gives the fundamental mode or mode 1 of radiation. At other frequencies where the circumference of the spiral is not equal to a wavelength (i.e., the diameter = λ/π), the phase relationship between the currents in adjacent conductors becomes random so that the net radiated energy is very small.

For mode 2, four arms are required, and the phase between the feed currents to opposite arms is zero or a multiple of 360 deg. As shown in Figure 6.1(b), the distance along the arc between points F and F′, which have in-phase currents, will be l, and thus the radiation band will have a circumference of 2λ or a diameter of $2\lambda/\pi$. The direction of the current vectors at F and F′ are in opposite directions, so the radiation on axis will be zero and split beams are obtained on either side of the boresight for this mode (see Figure 6.4).

Band theory can also be used to explain the deterioration in the axial ratio with decreasing spiral diameter for spirals whose outer peripheries are left open-circuited or short circuited. This occurs because the energy traveling outward toward the periphery is reflected back with an opposite sense. This reflected energy is at a lower level than the energy from the feed point, and thus the result is elliptical polarization. This can be seen in Figure 6.2, where the incident electric field E_i from the feed traveling toward the periphery is RHCP and the reflected electric field E_r is LHCP. Assuming that the two fields E_i and E_r are in phase at time $t = 0$, we can see that if $E_i > E_r$ the resultant electric field is right-hand elliptically polarized. The ellipticity (the ratio of the minor-to-major

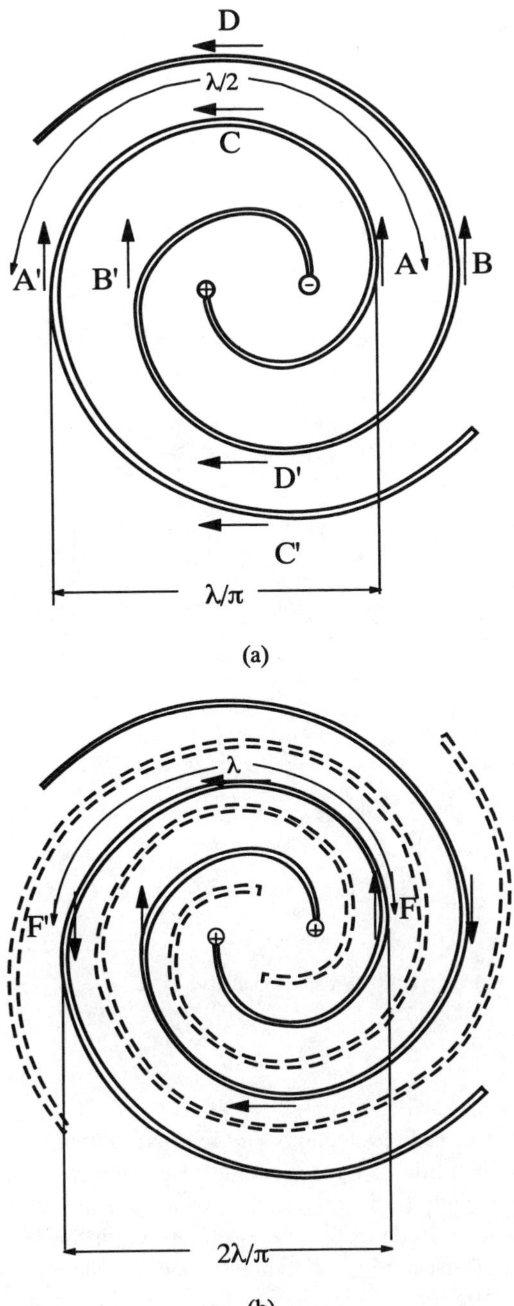

Figure 6.1 Band theory for spiral antennas: (a) mode 1 radiation with two arms, and (b) mode 2 radiation with four arms.

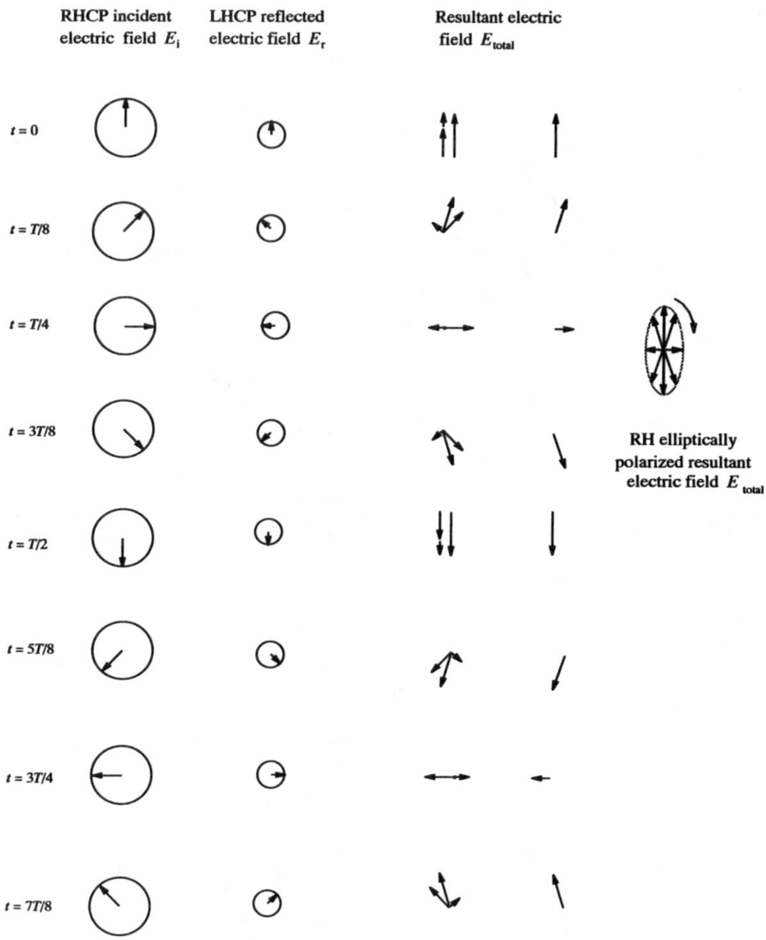

Figure 6.2 The resultant electric field from spiral antennas.

axis) of the resultant electric field depends on the relative magnitudes of E_i and E_r. As E_r becomes of similar magnitude to E_i the resultant field tends to linear polarization. This occurs when the spiral diameter is very small ($<\lambda/\pi$) and the reflected energy (from the edges) is of a similar magnitude to the initial energy from the feed [7]. Similarly, when the spiral diameter is very large, $E_i >> E_r$ and the resultant electric field is nearly circularly polarized and of the same sense as the incident electric field radiating from the center of the spiral toward the edges.

6.3 LOG SPIRAL

The logarithmic spiral is really an equiangular spiral but is commonly called a log spiral. The two main types of log spiral are those on flat surfaces and those conformed onto cones (conical log spirals). The planar log spirals may be circular or square; but the circular ones are more commonly used. In order to have radiation of an electric field we must have a change in the current with time. This can be achieved by moving a current element or by changing the rate of current flow with time. Since current is the rate of change of charge with time, the rate of change of current is the same as the acceleration (or deceleration) of a charge. In the spiral antenna the acceleration of a charge is achieved by the change of direction rather than the change of magnitude of the current. When the conductor (in which the charge is traveling) is bent into an arc normal to the direction in which the charge is traveling, the velocity vector $\boldsymbol{v}$ of the charge is changed to $\boldsymbol{v}'$, as shown in Figure 6.3(a). Using the triangle of forces, as shown in Figure 6.3(b), we can see that $\boldsymbol{v}'$ - $\boldsymbol{v}$ is $\delta\boldsymbol{v}$. The rate of change of $\delta\boldsymbol{v}$ with time gives us acceleration. The current is also attenuated in the case of a finite structure, as much as 20 dB in the first wavelength for planar log spirals [1, p. 185] so that the current is negligibly small at the point of truncation. The decay is an approximately constant function of the electrical arm length (i.e., the arm in terms of wavelengths). This has the affect of constantly shortening the active arm length as the frequency is increased. However, since the electrical arm length remains relatively constant, the HPBW is also constant. Even the truncated spiral with just one or two arms can provide frequency-independent (FI) operation over a wide bandwidth.

The decrease of currents is less rapid for narrow arms or narrow slot structures [8, p. 14–9]. The equations that determine the inner and outer edges of a planar log spiral are given in polar coordinate form by the radius vectors $\boldsymbol{r}_1$ and $\boldsymbol{r}_2$ from the center. Both edges of the spiral have the same value of k and a, but one of them is rotated through an angle of δ with respect to the other. The angle δ thus defines the angular width of the arms.

$$\boldsymbol{r}_1 = ke^{a(\alpha-\delta)} \tag{6.1}$$

$$\boldsymbol{r}_2 = ke^{a\alpha} \tag{6.2}$$

where r_1 and r_2 are the radial distances to the points P_1 and P_2 on the inner and outer edges of the spiral, as shown in Figure 6.3(a), $\alpha-\delta$ is the angle with respect to the x-axis for the inner spiral, α is the angle with respect to the x-axis for the outer spiral, k is a constant that determines the start of the spiral, and a is a constant that is known as the expansion coefficient. Both k and a are positive constants, and a determines the rate of spiral and usually takes values between 0.2 and 1.2 [1, p. 182]. Dividing (6.1) by (6.2), we get the ratio of the lengths r_1 to r_2, which is given by

$$K_2 = \frac{r_1}{r_2} = e^{-a\delta} \tag{6.3}$$

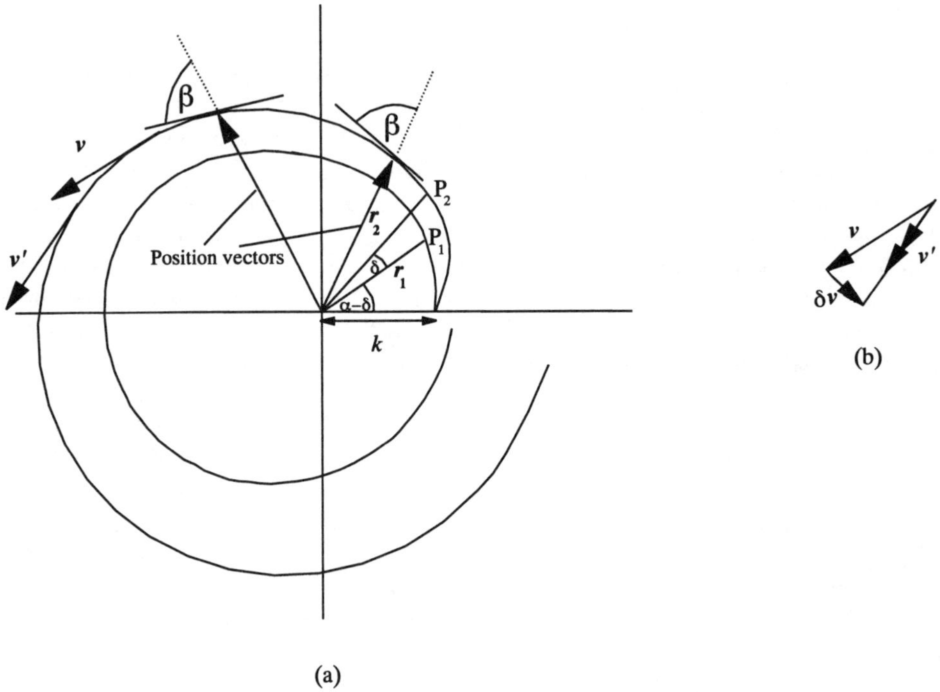

Figure 6.3 The equiangular or log spiral antenna: (a) velocity vectors for a charge moving in an arc; (b) addition of velocities using the triangle of vectors.

where K_2 is the ratio of r_1 to r_2.

When the angle δ is zero, the inner and outer edges are coincident; when $\delta = \pi$, the outer edge of the first turn touches the inner edge of the second turn of the spiral so that the spiral becomes a disk. Thus, in order to maintain a separation between the arms, the angle δ must lie between 0 and π; which means that K_2 must lie between 1 and $e^{-\pi a}$. When $\delta = \pi/2$ or 90 deg we have a self-complementary antenna. A self-complementary antenna has the metal area congruent to the nonmetallic area so that when the two are superimposed a solid metallic sheet is obtained. Alternatively, a rotation of the metallic area causes it to be superimposed on the nonmetallic area. The width of the conductors is equal to the space between the conductors.

Taking natural logs, that is, logs to the base e, for the general case of r, (6.2) can be rewritten as

$$\log_e r = \log_e k + a\alpha \log_e e \tag{6.4}$$

Since $\log_e e = 1$, (6.4), reduces to $\log_e r = \log_e k + a\alpha$. The rate of change of the radius with angle α is found by differentiating with respect to r and is given by

$$\frac{dr}{d\alpha}\frac{1}{r} = a \tag{6.5}$$

This gives the rate of change of the radius r with angle α as

$$\frac{dr}{d\alpha} = ra \tag{6.6}$$

Thus, the rate of change of r is proportional to the expansion coefficient a. The expansion coefficient a defines the tightness of the spiral. If a is large, then r increases rapidly with angle α and we have a loosely wound spiral, whereas when a is small r increases slowly and we have a tightly wound spiral.

The angle β that the spiral (or, strictly speaking, the tangent to the spiral) makes with the position vector at a point on the spiral is constant, and this gives the equiangular spiral its name. The angle β is sometimes called the wrap or winding angle of the arms and is given by

$$\tan\beta = \frac{1}{a} \tag{6.7}$$

$$\beta = \tan^{-1}\left(\frac{1}{a}\right) \tag{6.8}$$

For loosely wound spirals a is large and angle β is small, and for tightly wound spirals a is small and angle β is large.

An equiangular or log spiral can also be recognized by the fact that the width of the spiral as well as the distance between the turns increase as the radial distance from the center increases.

The spirals can have two or more arms. A rotation of 360/N deg leaves the structure unchanged. Thus, if the spiral has two arms a rotation of 180 deg will superimpose the spiral onto itself, whereas if the spiral has four arms a rotation of 90 deg achieves the same result [8, p. 14–4].

6.3.1 Modes of Radiation

There are a number of independent modes of radiation available that depend on the number of arms. For N arms there are (N - 1) modes available. The first or fundamental mode (which is also known as the sum pattern mode) has a single main lobe coincident with the boresight or main (z-) axis of the spiral. All the other modes have a null on the main axis and two lobes on either side of the axis, with the angle of the lobe increasing in proportion to the mode number as shown in Figure 6.4. These higher order modes are known as difference pattern modes.

For dual polarization more than two arms are required. Most of the radiation takes place in the first active region. For good spiral design this active region has a circumference of approximately M wavelengths, where M is the mode number. Thus mode 1 has an active region confined to a radius of $\lambda/2\pi$ or 0.16λ or a diameter d_1 of 0.32λ, as described in section 6.2. For loosely wound spirals, the attenuation in the first active region may not be large enough and, hence, the waves travel to the second active region.

In addition to these higher order positive modes, there are negative modes associated with cavity-backed spirals. These are due to the energy being reflected from the periphery of the spiral. In general, only the first-order negative mode ($M - 1$) is of significance.

In order to obtain radiation on boresight, the opposite arms of the spiral have to be in antiphase, that is, 180 deg out of phase. The difference pattern modes occur when the opposite arms are fed in phase.

The phase ϕ_p associated with a particular arm of the spiral [8, p. 14–4] is given by

$$\phi_p = \frac{2\pi M n}{N} \tag{6.9}$$

where ϕ_p is the phase in radians, M is the mode number, n is the arm number, and N is the total number of arms.

Thus, for mode 1 propagation, the total number of arms is 2, and it can be seen that arm 1 has a phase of π and arm 2 has a phase of 2π. For mode 2, the total number of arms

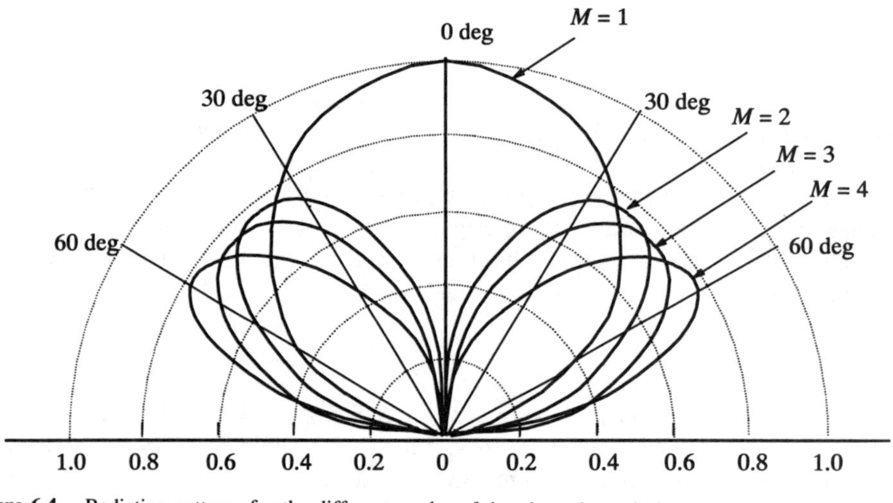

Figure 6.4 Radiation patterns for the different modes of the planar log spiral antenna.

is 4, and the phases obtained for arms 1, 2, 3, and 4 using (6.9) are π, 2π, 3π, and 4π, respectively. Thus it can be seen that the phases of opposite arms (e.g., 1 and 3) differ by 2π, that is, they are in phase.

6.3.2 Rotation of Radiation Pattern With Frequency

The radiation pattern of the spiral is rotated as the frequency is changed. When the frequency is changed from a frequency of say f to f/k_3 (where k_3 is a constant) the pattern is rotated through an angle of $\log_e k_3/a$. This rotation is not observable for tightly wound spirals (large angle β) since the radiation pattern has circular symmetry. However, for loosely wound spirals (small angle β) the beam is elliptical and the rotation of the beam can be observed [5]. Tightly wound spirals and those with wider arms tend to have smoother and more uniform radiation patterns that exhibit smaller variations in HPBWs.

Although the width of a log spiral should increase with its distance from the feed point, in practice, if a metallic strip of constant width (or a wire of constant diameter) is used for each arm, a moderately wide bandwidth of between 5:1 and 10:1 can be obtained for bidirectional spirals [8, p. 14–9].

6.3.3 Planar Log Spiral

The first planar log spiral was constructed by J. D. Dyson in 1958. It was a bidirectional antenna. The planar log spiral can be implemented by either cutting a metal spiral (usually on a dielectric substrate) or by cutting a spiral slot in a metallic ground plate, as shown in Figures 6.5 and 6.6.

The spiral is shown with less than two turns for clarity. The planar log spiral exhibits a decrease of current with distance that is faster than $1/r$ and, thus, the end effect becomes negligible at a distance that depends on the exact shape of the structure [8, p. 14–3].

6.3.4 Slot Planar Log Spiral

The planar spiral in slot form is formed, as shown in Figure 6.6(a), by removing the metallic spiral from an infinite ground plane. This is based on Babinet's principle, which states that a radiating flat strip element can be replaced by the equivalent shape slot in an infinite metallic sheet since they both have the same radiation characteristics. The slot spiral of Figure 6.6 is shown with less than two turns for sake of clarity.

6.3.4.1 Input Impedance

The log spiral antenna has a theoretical impedance at all frequencies that is equal to the intrinsic impedance ξ_0 of free space (120π or 377 Ω) divided by the number of arms [9,

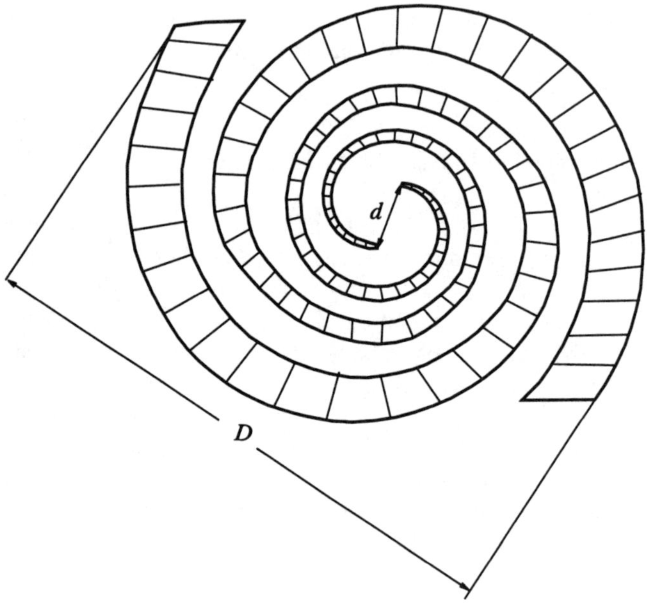

Figure 6.5 The metallic planar log spiral.

p. 520]. This applies to self-complementary spirals with opposite arms interconnected. Thus, for the two-armed spiral antenna the theoretical input impedance is 120π/2 (i.e., 188 Ω). However, the measured input impedance is typically between 50 Ω and 100 Ω [4, p. 700]. The experimental values are lower due to the presence of the cable feeding the structure. Narrow arm structures have higher impedances [8, p. 14–9].

6.3.4.2 The Arm Length of the Spiral

The length of the arm of a spiral is specified as the length along the center line of the arm. In order to calculate the length of an arm consider the elementary length δl as shown in Figure 6.7.

Using Pythagoras's theorem, we have

$$\delta l^2 = (r\delta\alpha)^2 + \delta r^2 = \delta r^2\left[\left(r\frac{\delta\alpha}{\delta r}\right)^2 + 1\right] \tag{6.10}$$

$$\delta l = \delta r\left[\left(r\frac{\delta\alpha}{\delta r}\right)^2 + 1\right]^{\frac{1}{2}} r$$

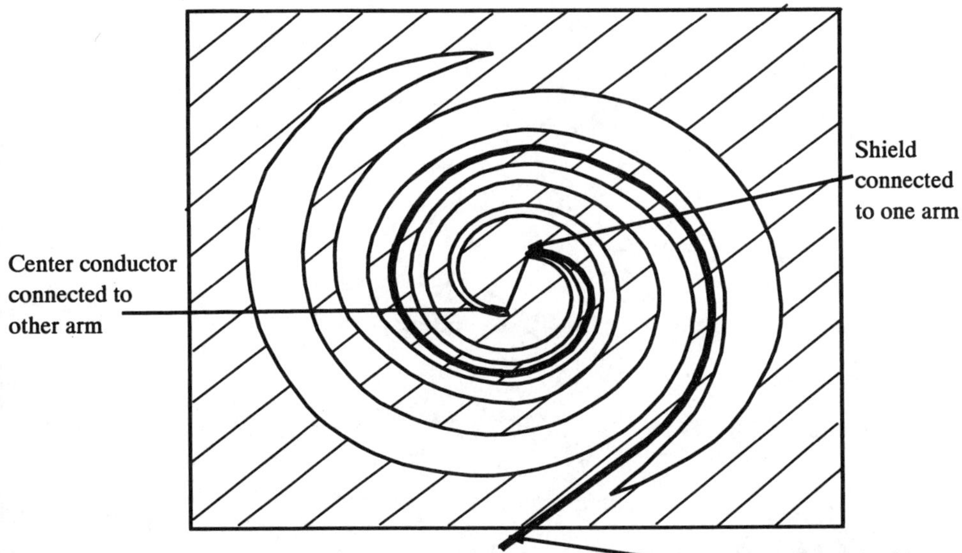

(a)

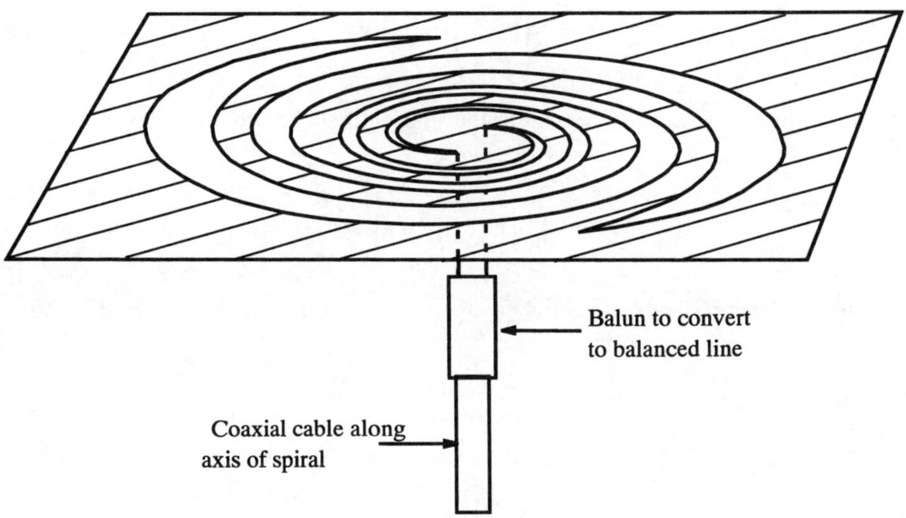

(b)

Figure 6.6 Feeding arrangements for the slot form of the planar log spiral: (a) infinite balun coaxial line feed; (b) axially fed coaxial line.

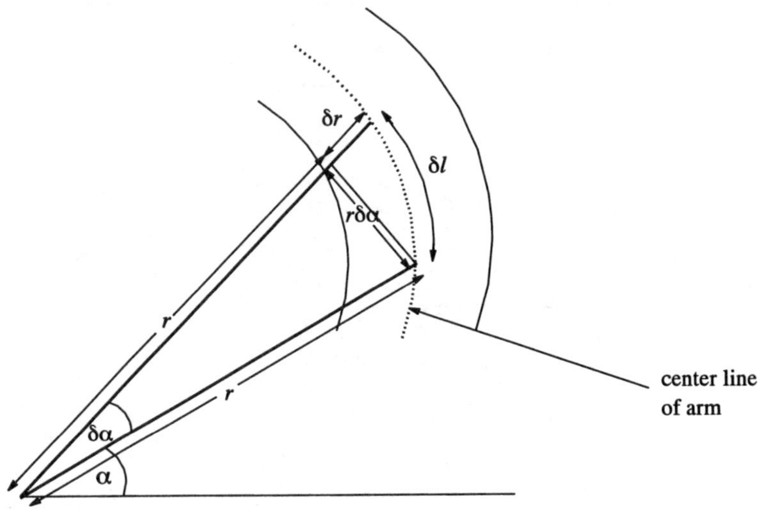

Figure 6.7 Calculation of the arm length of a log spiral.

and thus we can say the following approximation holds:

$$dl \approx \left[\left(r\frac{d\alpha}{dr}\right)^2 + 1\right]^{\frac{1}{2}} dr \tag{6.11}$$

The total length L is found by evaluating the integral between the limits of r_1 and r_2 as

$$L = \int_{r_1}^{r_2} \left[\left(r\frac{d\alpha}{dr}\right)^2 + 1\right]^{\frac{1}{2}} dr \tag{6.12}$$

From (6.6) we can see that $(rd\alpha/dr)$ is equal to $1/a$; thus, (6.12) can be rewritten as

$$L = \int_{r_1}^{r_2} \left[\left(\frac{1}{a}\right)^2 + 1\right]^{\frac{1}{2}} dr = \left[\left(\frac{1}{a}\right)^2 + 1\right]^{\frac{1}{2}} (r_2 - r_1) \tag{6.13}$$

6.4.3.3 *Figure of Merit*

The diameter d (see Figure 6.5) of the start of the spiral governs the high-frequency end of the antenna. This, in turn, is governed by the feed used. The low-frequency limit is determined by the overall diameter D of the spiral. If the diameter d of the feed region is, say, $\lambda_H/10$ (where λ_H is the wavelength at the high-frequency end), the overall diameter D is $\lambda_L/2$ (where λ_L is the wavelength at the low-frequency end), and the ratio D/d is 25, we have

$$\frac{D}{d} = \frac{25}{1} = \frac{\lambda_L}{2} \bigg/ \frac{\lambda_H}{10} \qquad (6.14)$$

Thus the ratio of the wavelengths is 1:5 and hence the frequency bandwidth is 5:1. The ratio of D/d is called the dimension ratio and λ_H/λ_L is called the wavelength ratio. The product of the dimension ratio and the wavelength ratio is taken as the figure of merit for a particular structure and is a measure of how compact an antenna can be made for a given bandwidth. This figure of merit is usually one or greater. In the above case, the dimension ratio is 25 and the wavelength ratio is 1/5, which gives a figure of merit of 5 [8, p. 14–10].

6.3.4.4 Feeding Arrangements

The feed for the spiral can be located axially and may consist of a coaxial cable, with the inner conductor connected to one arm and the outer conductor connected to the other arm (using a tapered balun for instance), as shown in Figure 6.6(b). Alternatively, a balanced feed can be used by bonding a cable to the underside of the metal between the arms [4, p. 700], as shown in Figure 6.6(a). Because of the rapid attenuation of the field with distance along the spiral, the current flow in the cable is negligible with this arrangement [8, p. 14–9]. The cable may be embedded in the ground plane, or soldered to it, and this arrangement is known as an infinite balun. This is the only type of balun that will allow the fullest use of the infinite impedance and pattern bandwidths of this type of antenna. [1, p. 183]. The main disadvantage of this type of feed is that it requires the use of sufficient ground plane to carry the feed cable; as a consequence, the feed region in the center is fairly large. This, in turn, restricts the upper frequency limit of the antenna. When the diameter of the cable approaches the width of the metal to which it is attached, it is necessary to attach a dummy cable to the opposite arm in order to maintain symmetry of construction and to prevent a tilt in the radiation pattern. Consistently good radiation patterns can be obtained with spirals between $1\frac{1}{4}$ and $1\frac{1}{2}$ turns.

The electric field components E_θ and E_ϕ in the θ and ϕ directions, as shown in Figure 6.8(a), can be measured separately by antennas oriented at right angles to each other. A practical bidirectional planar log spiral has an unsymmetrical radiation pattern that also varies with frequency as shown in Figure 6.8(b–i). These are radiation patterns measured by Dyson [1, Figures 8 and 9], at frequencies between 595 MHz and 12 GHz (i.e., a bandwidth of 20:1). These have been obtained for a slot spiral cut into a 1/32-inch-thick copper sheet of 14 in^2. This sheet is then bolted onto a larger ground plane, but

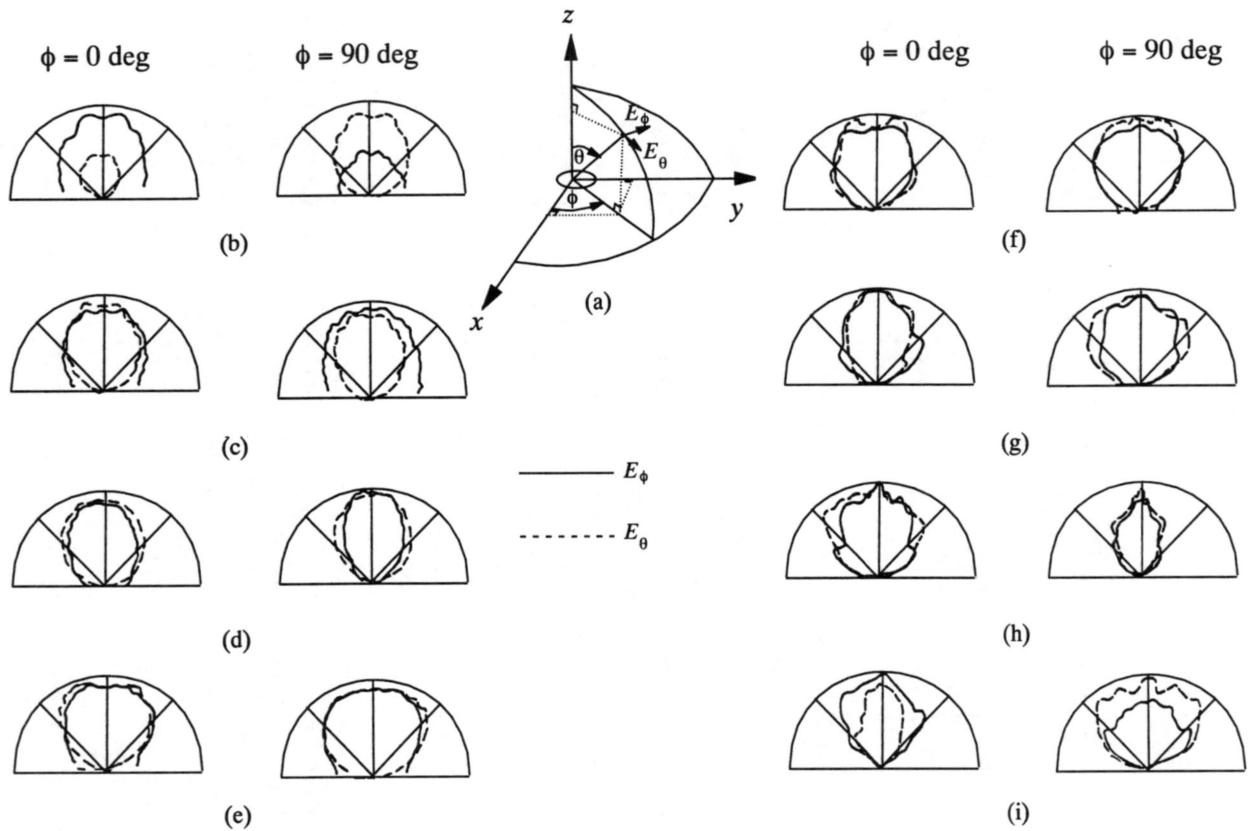

Figure 6.8 Radiation patterns for the bidirectional planar log spiral in the $\phi = 0$ deg and $\phi = 90$ deg planes: (a) E_θ and E_ϕ vectors; (b) frequency 595 MHz axial ratio 1.94; (c) frequency 1 GHz axial ratio 1.19; (d) frequency 1.4 GHz axial ratio 1.04; (e) frequency 2 GHz axial ratio 1.04; (f) frequency 3.95 GHz axial ratio 1.07; (g) frequency 5.6 GHz axial ratio 1.07; (h) frequency 9.027 GHz axial ratio 1.07; and (i) frequency 12 GHz axial ratio 1.24.

the size of this ground plane has not been stated. The spiral had $a = 0.30$, $k = 0.2$, $K_2 = 0.62$, the arm length was 38.7 cm, and the maximum diameter was 28.4 cm.

The variation of the HPBW with azimuth angle is quite marked. For instance, at 3.979 GHz the HPBW varies between 60 deg and 105 deg, as shown in Figure 6.9(a). If the frequency is varied and the HPBW is viewed at the azimuth angle $\phi = 90$ deg, the variation at spot frequencies in the range 2.472 GHz to 7.505 GHz is shown in Figure 6.9(b).

A spiral antenna can have any number of arms. However, the variation of the radiation pattern with the expansion coefficient a can only be calculated theoretically for an antenna that has an infinite number of arms [9, p. 520]. This is known as a sheath spiral. The variation of the radiation pattern for such a spiral, with the expansion coefficient a, is shown in Figure 6.10. It can be seen that when a is greater than one, the HPBW is proportional to a and the gain is inversely proportional to a. However, when the expansion coefficient a is less than one, there is no monotonic dependence on a [9, p. 520]. In practice, an antenna with only a few arms (even as few as two) is a reasonably good approximation for a sheath spiral.

6.3.5 Cavity-Backed Spiral

The planar log spiral, when placed in front of a cavity ($\lambda/4$ deep), can be used to give a unidirectional radiation pattern. An image of the original spiral is formed and, since it is in antiphase with the original spiral and appears to be $\lambda/2$ behind the original, constructively reinforces the field radiated by the original spiral. However, the image of the spiral radiates through the original spiral and distorts the pattern. In addition, since the cavity depth is around $\lambda/4$ for only a limited range of frequencies, the bandwidth is reduced to around 2:1 (i.e., an octave). If the cavity-backed spiral is loaded with RAM to absorb the radiation, the bandwidth can be increased to 10:1 [8, p. 14–9]. However, the absorber used in the cavity dissipates about 50% of the total input power to the antenna; thus, the efficiency of the antenna is reduced to less than 50% in the case of a transmitting antenna. In the case of a receive antenna, this dissipative loss raises the antenna noise temperature and thus reduces the system detection threshold [10].

The cavity diameter must be greater than or equal to the spiral diameter. If the cavity diameter is made too large, higher order coaxial or waveguide modes may be excited in the cavity. This would result in pattern deterioration at these cavity resonant frequencies. If the cavity diameter is made too small, the gain of the antenna is reduced, whilst the VSWR and ellipticity ratio are increased [7, p. 86].

The planar-backed spiral is also used as a feed for a parabolic reflector.

6.3.6 Conical Log Spiral

The conical log spiral antenna is an equiangular spiral projected onto a conical surface, as shown in Figure 6.11. It can also be considered as a tapered form of the helical antenna,

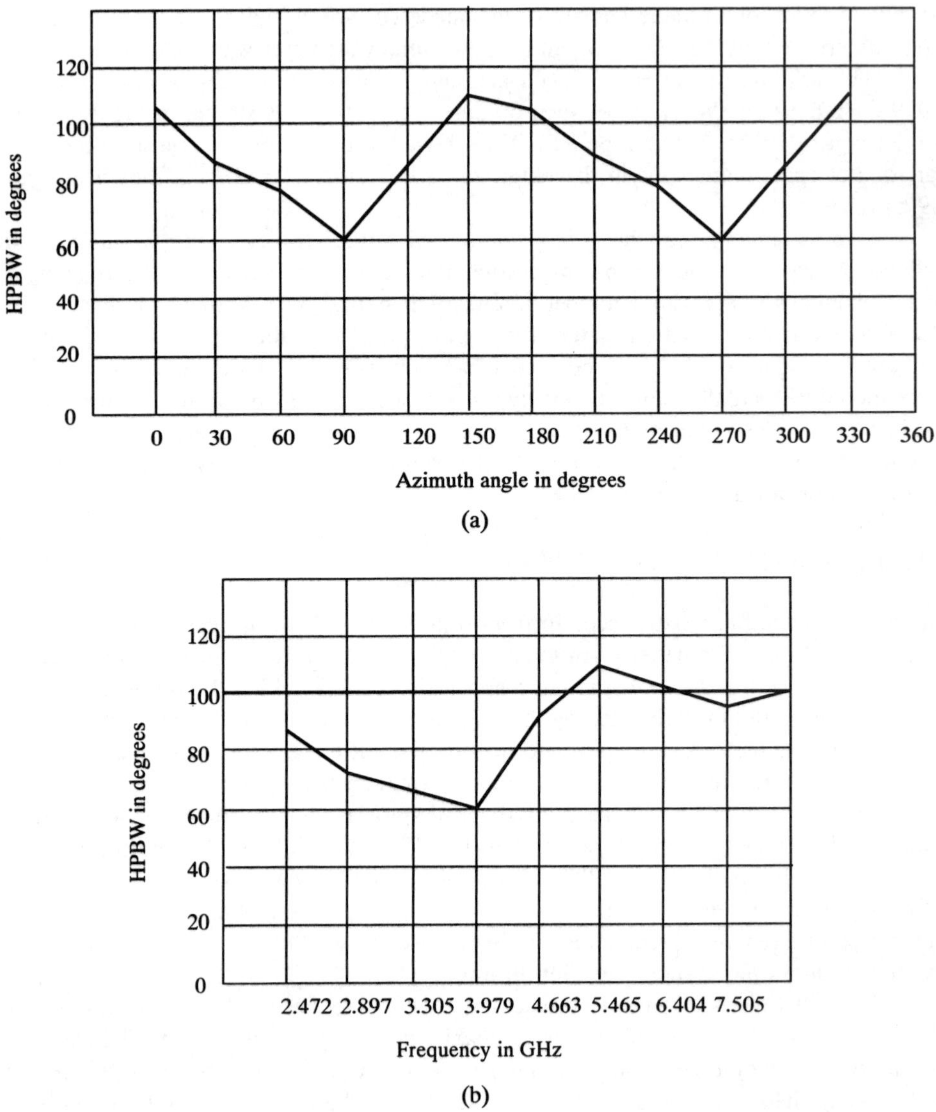

Figure 6.9 Variation of HPBWs of the radiated electric field with (a) azimuth angle in different planes at 3.97 GHz, and (b) frequency for a planar slot log spiral at ϕ = 90.

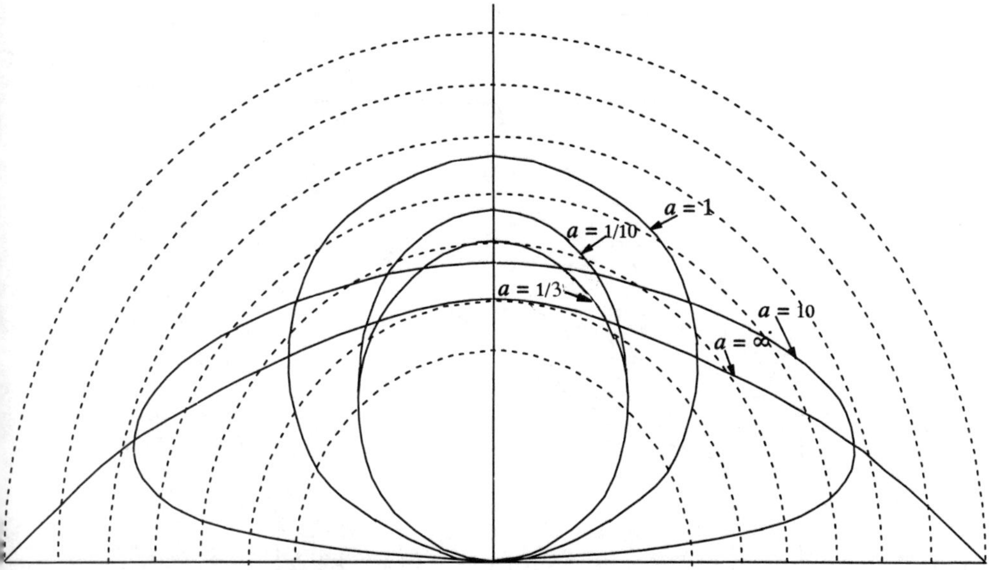

Figure 6.10 Variation of the radiation pattern with the expansion coefficient for a planar log spiral with an infinite number of arms.

and for small cone half-angles ($\theta_{hc} \leq 10$ deg), the conical log spiral behaves like a helical antenna. It does not require a cavity in order to produce a unidirectional radiation pattern. The conical log spiral can be used at frequencies below 1 GHz, and commercially available antennas, when made to conform to MIL-STD drawings, cover the frequency ranges from 100 MHz to 1 GHz, 200 MHz to 1 GHz, and 1 GHz to 10 GHz. However, they can be used from about 50 MHz to 18 GHz. Typical HPBWs are of the order of 60 deg in elevation and azimuth, and gains are of the order of 5 dBi to 8 dBi. They are used for radiated emissions as well as susceptibility measurements.

For conical log spirals that are excited at the apex, the field is transverse to the radial direction and the current decreases inversely as the radial distance r from the center. This decrease is not large enough to make the end effects negligible when the cone is truncated [8, p. 14–3]. Calculated magnitudes of the attenuation are in the range of 7 dB to 10 dB per wavelength along the spiral arm. End effects produce variations in the radiation pattern.

The constants that govern the shape of the conical log spiral are similar to those of the planar log spiral, but they are dependent on the cone half-angle θ_{hc}. The expansion coefficient a is equal to $a'/\sin\theta_{hc}$, where a' is the expansion coefficient for a planar log spiral. Thus the conical log spiral equations equivalent to (6.7) and (6.8) for the planar log spiral are given by

$$\tan\beta = \frac{1}{a} = \frac{\sin\theta_{hc}}{a'} \tag{6.15}$$

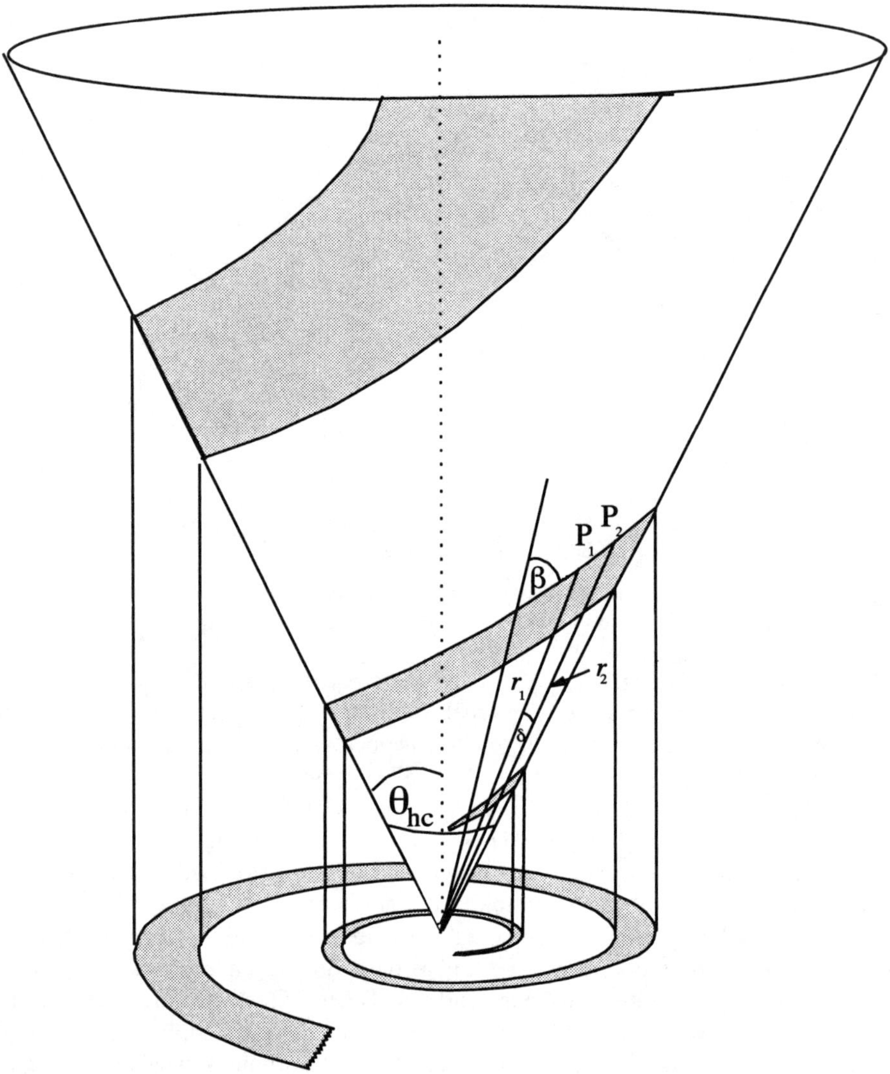

Figure 6.11 Conical log spiral.

$$\beta = \tan^{-1} \frac{\sin \theta_{hc}}{a'}$$

where θ_{hc} is the cone half-angle and a' is the expansion coefficient on the equivalent planar spiral. The equations that determine the inner and outer edges of a conical log spiral are given in polar coordinate form by the radius vectors $\boldsymbol{r}_1$ and $\boldsymbol{r}_2$ from the center. Both edges of the spiral have the same values of k and a, but one of them is rotated through an angle of δ with respect to the other. The angle δ thus defines the angular width of the arms.

$$\boldsymbol{r}_1 = ke^{a(\alpha - \delta)} \tag{6.16}$$

$$\boldsymbol{r}_2 = ke^{a\alpha} \tag{6.17}$$

where r_1 and r_2 are the radial distances to the point P_1 and P_2 on the inner and outer edges of the spiral, $\alpha - \delta$ is the angle with respect to the x-axis for the inner spiral, α is the angle with respect to the x-axis for the outer spiral, k is a constant that determines the start of the spiral, and a is a constant that is known as the expansion coefficient. Both k and a are positive constants, and a determines the rate of spiral. The ratio of the lengths r_1 to r_2 is given by

$$K_2 = \frac{r_1}{r_2} = e^{-a\delta} \tag{6.18}$$

As the cone half-angle decreases from $\pi/2$ (planar spiral), the spiral becomes more tightly wound. For small values of a' (of the order of 0.02) the main lobe is rotationally symmetrical [3, p. 331]. The pattern rotates with frequency as in the case of a planar log spiral, but this effect is not observable in the case of rotationally symmetrical beams. Like the planar log spiral, the conical log spiral can be constructed in either conductor or slot form.

The two-arm conical spiral is a symmetrical structure and can be fed by a balanced two-wire transmission line, brought along its axis, and a wire connected to each arm. This should preferably be a shielded line. A metallic shield or cylinder on the axis appears to have very little effect on the antenna characteristics as long as its diameter is less than one-third of the antenna diameter at any point on the axis [2, p. 495]. The balun to convert from an unbalanced to a balanced line may also be placed inside the antenna if it is nonradiating and not affected by the fields inside the cone. The tapered line balun, which is extremely wideband, can be used, although there appears to be some interaction between the fields around the balun and those due to the antenna [2, p. 495]. If the pattern degradation can be tolerated, then this balun is ideally suited to the conical log spiral. Alternatively, the feed may be a coaxial cable bonded to the spiral arm (or to the ground plane in the case of the slot form) as an infinite balun, in the similar manner to the planar case. However, the use of an infinite balun results in an effective increase in the width of the arms, and the impedance and other parameters will therefore be equivalent to those obtained for larger values of angular width δ.

If the arms are decreased to a narrow constant width they can be formed from cable alone. In this wire version the feed cable becomes one arm and at the apex the center conductor is connected to the outer braid of the dummy cable, which forms the opposite arm of the antenna. The apex diameter of approximately 4 cm is a practical limit of a feed cable of type RG8/U, with an upper frequency limit of around 1,700 MHz [3, p. 333]. If RG141/U cable is used, the apex can be reduced to 2 cm, and thus the upper frequency can be extended. However, this type of antenna constructed from cables results in an effective increase in the width of the arms, and the impedance and other parameters will therefore be equivalent to those obtained for arms of larger width.

The conical spiral is fed at its apex and, in common with other frequency independent antennas, provides backfire radiation. However, this backfire radiation occurs only for cone half-angles of 15 deg or less. As the cone angle is increased, there is a marked deterioration in the front-to-back ratio, until the value 90 deg is reached, when the antenna becomes a planar spiral and the pattern becomes bidirectional.

6.3.6.1 The Active Region

At the feed region of a two-arm spiral, out-of-phase traveling wave currents are excited. These currents do not radiate until they reach the active region, which is defined as that region that controls the primary characteristics of the radiated field [2]. As the smaller turns are eliminated, there is very little observable change in the near-field distribution until the antenna is truncated at a radius such that the near field is approximately 3 dB below the maximum recorded near field. This sets the upper edge of the active region. Similarly, as successive turns are removed from the base end, there is a negligible change until the antenna is truncated at a radius such that the near field is approximately 15 dB below the maximum recorded near field [2, p. 491]. This sets the lower edge of the active region. The active region is therefore defined as that region between the electrical radius r_3^- (at the highest frequency of operation), at which the electric field magnitude is 3 dB below the maximum on the apex side, and the electrical radius r_{15}^+ (at the lowest frequency of operation), at which the magnitude is 15 dB below the maximum on the base side. The axial ratio of the electric field polarization ellipse on the boresight axis, for this active region, is 3 dB or less at the lowest frequencies. The magnitude of the electric field is measured by using a near-field small shielded loop probe at a distance of 0.03λ from the surface of the antenna [2, p. 490]. The bounds of the active region vary for different cone half-angles θ_{hc} and wrap angles β. These bounds are based on the near-field measurements, and they predict the far-field pattern most accurately for cone half-angles θ_{hc} of approximately 20 deg. For smaller cone angles the bounds are larger, and for larger cone angles the bounds are smaller.

6.3.6.2 Bandwidth

The bandwidth of the structure B_s is defined as the ratio of the radius of the base $D/2$ to the radius of the apex $d/2$. The operating bandwidth B_{op} of the antenna is defined as the ratio of the structure bandwidth B_s to the bandwidth of the active region [2, p. 490] and denoted by $B_{op} = B_s/B_{ar}$. The bandwidth of the active region B_{ar} is given by

$$B_{ar} = \frac{r_{15}^{+}}{r_{3}^{-}} \tag{6.19}$$

where r_3^- is the electrical radius on the apex side at which the magnitude of the near field is 3 dB below the maximum at the highest frequency of operation and r_{15}^+ is the electrical radius on the base side at which the magnitude of the near field is 15 dB below the maximum at the lowest frequency of operation.

6.3.6.3 Radiation Characteristics

The narrow on-axis beam occurs for small cone angles and wrap angle β = 80 deg approximately, with all turns of the active region being phased for backfire radiation. As the active region increases, the beam broadens and tends toward a multiple beam. In general, as the aperture of an antenna increases, the beam gets narrower. However, in the case of log conical spirals, the phase distribution across the spiral results in radiation occurring in progressively different directions. Larger turns or more widely spaced elements are phased to radiate at an angle to the axis. At a cone half-angle of 15 deg the radiation is mainly on axis for angles of wrap β = 80 deg. However, as the angle of wrap is decreased to 45 deg, a major portion of the active region is phased for broadside radiation.

The directivity of the antenna can be calculated from the HPBWs in the orthogonal planes and is given by [2, p. 493]

$$D_{db} = 10 \log_{10} \frac{32000}{\theta_1 \phi_1} \tag{6.20}$$

where D_{dB} is the directivity in decibels with respect to a circularly polarized isotropic source and θ_1 and ϕ_1 are the HPBWs in degrees in the orthogonal planes.

The electric field components E_θ and E_ϕ in the θ and ϕ directions can be measured separately by antennas oriented at right angles to each other. This variation is due to radiation caused by higher order modes that do not have rotationally symmetrical beams. The difference in HPBWs varies between 3 deg and 20 deg depending on the spiral wrap angle β and the cone half-angle θ_{hc}. The difference in the HPBWs for the E_θ and E_ϕ

components [2, Figure 12] is shown in Figure 6.12. In general, it can be said that the difference increases with decreasing wrap angle and increasing cone half-angle. It can be seen that for very small cone angles of 2 deg (θ_{hc} = 1 deg) the difference in the HPBWs is around 3 deg for a wrap angle of 80 deg and increases to 9 deg for a wrap angle of 45 deg. For large cone half-angles of 15 deg and a wrap angle of 55 deg, the difference is as high as 20 deg.

For cone half-angles of 10 deg, and β = 73 deg, the measured HPBWs for E_ϕ and E_θ polarization are 70 deg and 90 deg, respectively [3, p. 331]. HPBWs of the order of 40 deg to 50 deg can be obtained with small cone antennas. However, the coverage of very wide bandwidths would entail the use of very long antennas for these small cone angles. The length of the antenna in terms of the longest wavelength λ_L is given by

$$\frac{L}{\lambda_L} = \frac{1}{2 \tan \theta_{hc}} \left[\frac{D}{\lambda_L} - \frac{d}{B\lambda_L} \right] \tag{6.21}$$

where D is the diameter of the base, d is the diameter of the apex, λ_L is the wavelength of the lowest operating frequency, θ_{hc} is the cone half-angle, and B is the operating bandwidth. As the active region moves toward the truncated tip of the cone, the HPBW may at first become narrower and then the pattern may become rough. Any lack of precision in the structure of the antenna will cause the pattern to tilt or become distorted. As the frequency is increased further, the beam broadens and tends to break up into a number of lobes [2, p. 493]. Cone half-angles of 15 deg or less have higher front-to-back ratios. For K_2 between 0.85 and 0.9, and β = 73 deg, an apex diameter of $\lambda_H/4$ and a base diameter of $3\lambda_L/8$ provide a F/B ratio of 15 dB or more.

The axial ratio of the polarization ellipse for angles off boresight (from 0 deg to 90 deg) varies between 2 dB and 4 dB for a wrap angle β = 45 deg and θ_{hc} = 10 deg [8, p. 14–12]. The axial ratio is 2:1 (3 dB) up to about 60 deg off axis.

In the case of wrap angles $\beta \leq 45$ deg, and $\theta_{hc} \geq 7.5$ deg, as the antenna deviates from the self-complementary structure of δ = 90 deg, the front-to-back (F/B) ratio also may deteriorate and the pattern breaks up into many lobes. A large portion of the energy is also radiated in the direction of the base, that is, endfire instead of backfire radiation. For β = 80 deg, the F/B ratio increases from 6 dB to 25 dB as θ_{hc} decreases from 22.5 deg to 7.5 deg [8, p. 14–12].

For self-complementary antennas, δ = 90 deg, the variation of the average HPBW (for E_ϕ and E_θ polarization) with wrap angle β is shown in Figure 6.13 with the cone angles as a parameter. The HPBWs are much larger for self-complementary antennas.

The approximate directivity can be calculated from these HPBWs by using the formula of (6.20). These directivities have been calculated and are plotted in Figure 6.14 for the HPBWs of Figure 6.13.

The typical variation of gain and antenna correction factor with frequency for a commercially available conical log spiral is shown in Figure 6.15 for the frequency range

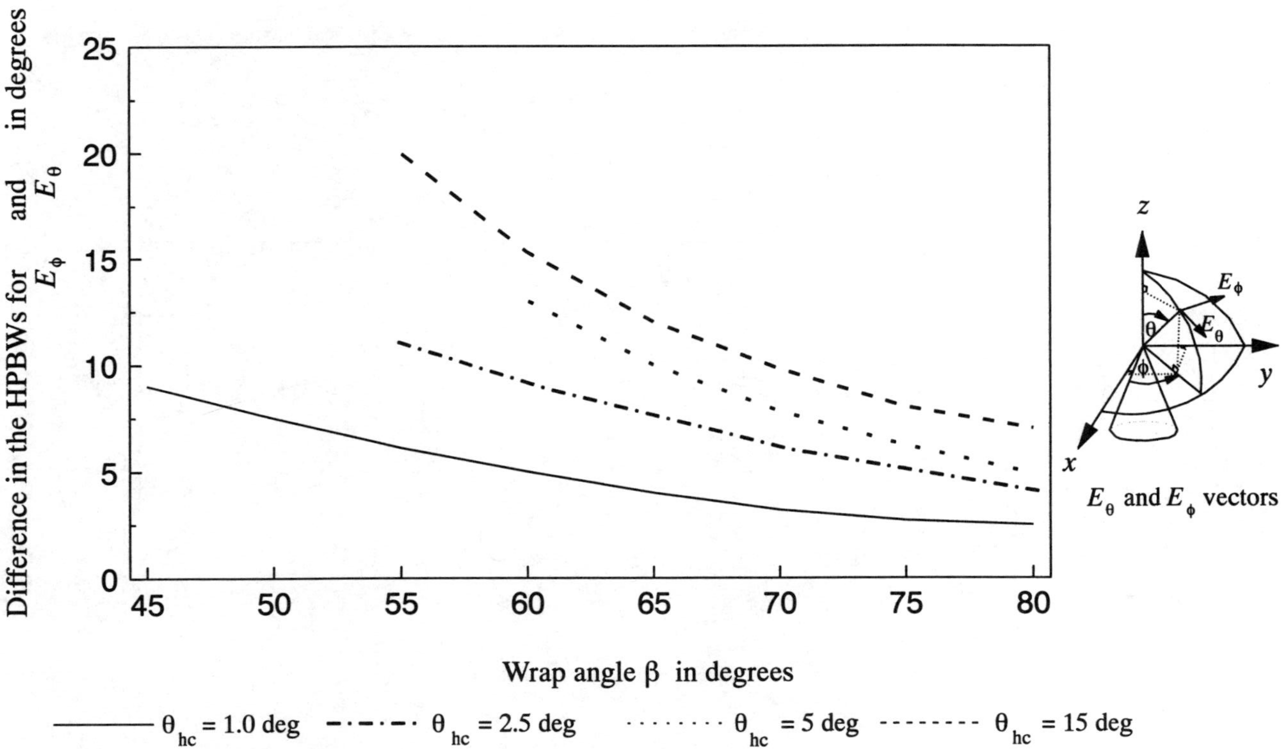

Figure 6.12 Variation of the difference in the HPBWs for the E_θ and E_ϕ components, with spiral wrap angle β and cone half-angles as parameters.

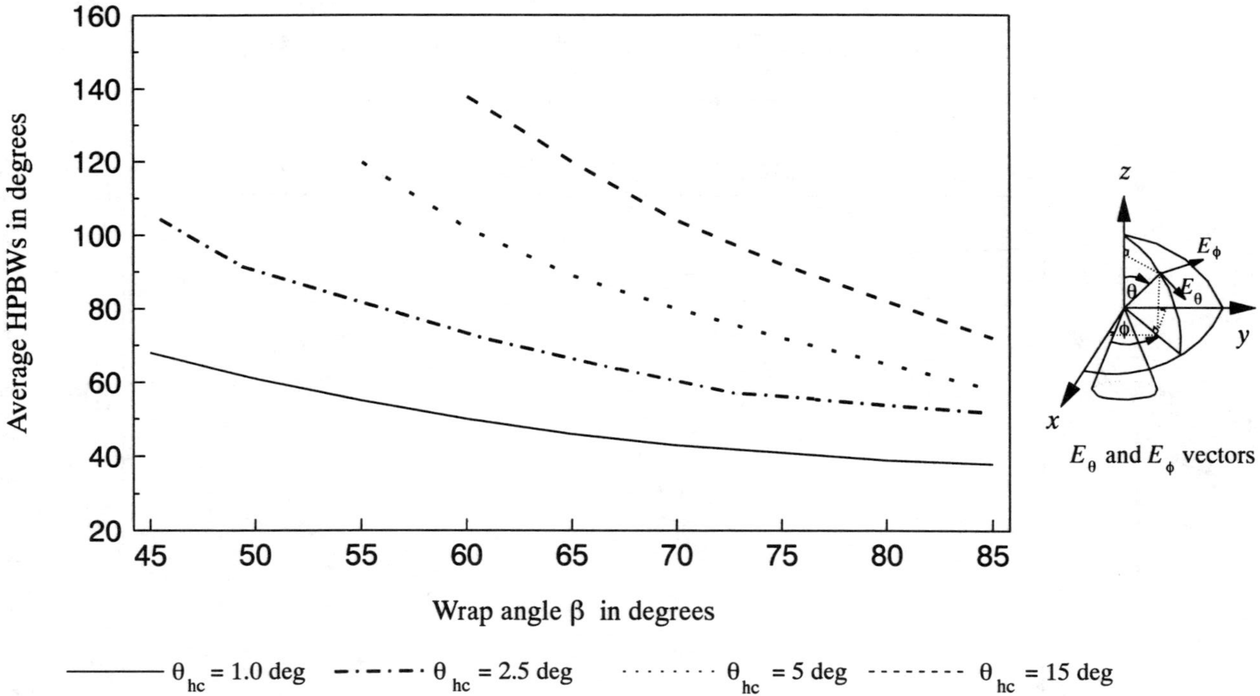

Figure 6.13 Variation of the average HPBWs (for E_ϕ and E_θ) with spiral wrap angle β, with the cone half-angle as a parameter.

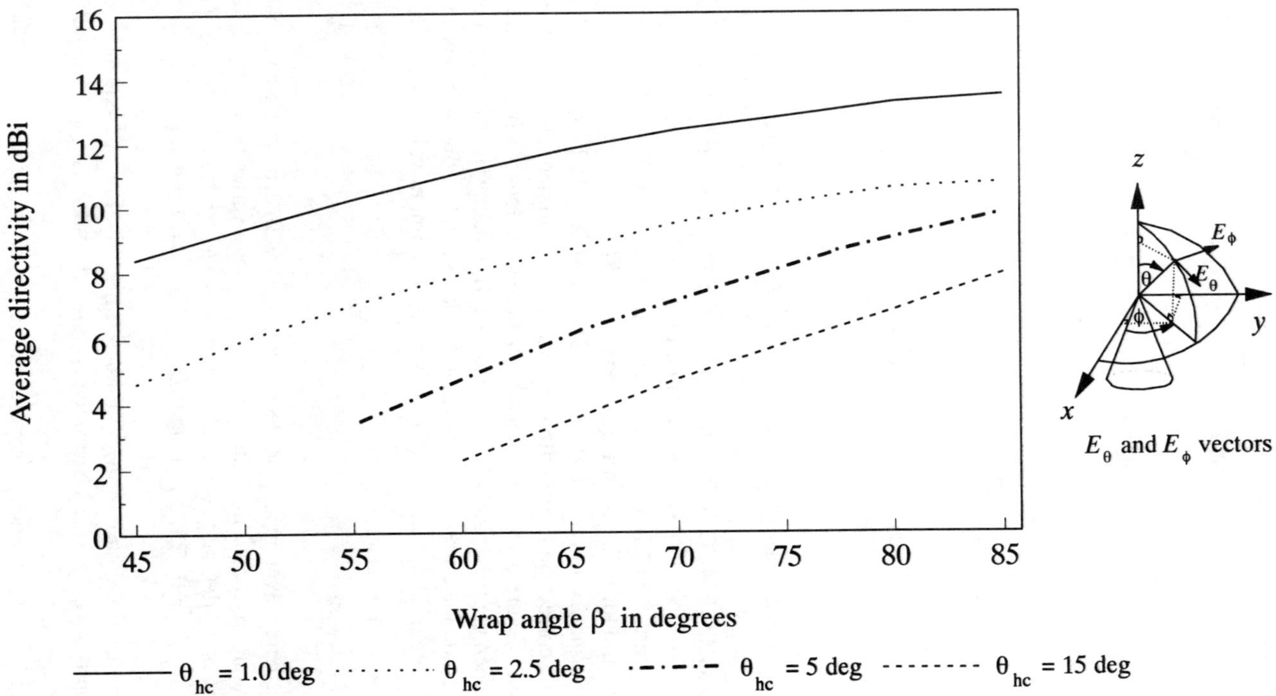

Figure 6.14 Variation of the average directivity for E_ϕ and E_θ with spiral wrap angle β, with the cone half-angle as a parameter.

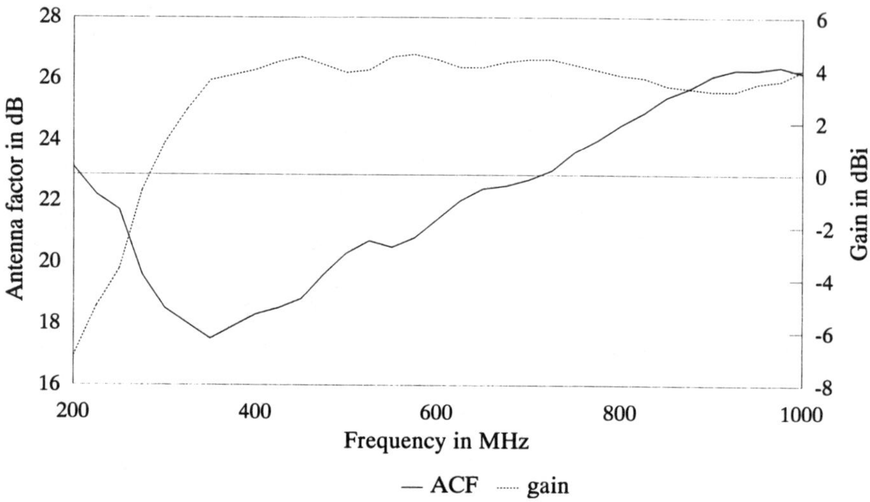

Figure 6.15 Variation of the gain and ACF for a conical log spiral in the frequency range 200 MHz to 1 GHz.

200 MHz to 1 GHz (EMCO 3101). It can be seen that the gain increases rapidly from approximately -6 dBi at 200 MHz to a value of 3.6 dBi at 350 MHz and then oscillates between 3.5 dBi and 4.5 dBi. The antenna correction factor varies between 17 dB and 27 dB, with the minimum value occurring around 350 MHz.

For the frequency range 1 GHz to 10 GHz (EMCO 3102) the typical variation of gain and antenna correction factor with frequency for a commercially available conical log spiral is shown in Figure 6.16. It can be seen that the gain does not increase monotonically but varies between approximately 0.3 dBi and 4 dBi, reaching its lowest value of 0.3 dBi at 10 GHz. The antenna correction factor, on the other hand, shows a steady increase from its minimum value of about 27 dB at 1 GHz to its maximum value of 50 dB at 10 GHz.

6.3.6.4 Input Impedance

The mean impedance level appears to decrease slowly with decreasing cone half-angle θ_{hc} [3, p. 333]. The approximate measured input impedance varies between 129Ω and 153Ω for pitch angles of 17 deg and cone half-angles between 10 deg and 30 deg. These values were obtained for $K_2 = 0.925$, arm length $L = 1.5$m, $a = 0.303 \sin \theta_{hc}$, and wrap angle $\beta = 73$ deg.

The diameter of the base of the spiral determines the lowest frequency, at which the diameter is approximately λ/2. The truncated apex diameter is approximately λ/4 at the highest frequency [4, p. 703].

The impedance is inversely proportional to the width of the arms. The impedance can be varied between 80 Ω and 320 Ω by reducing the angular width δ of the arms from

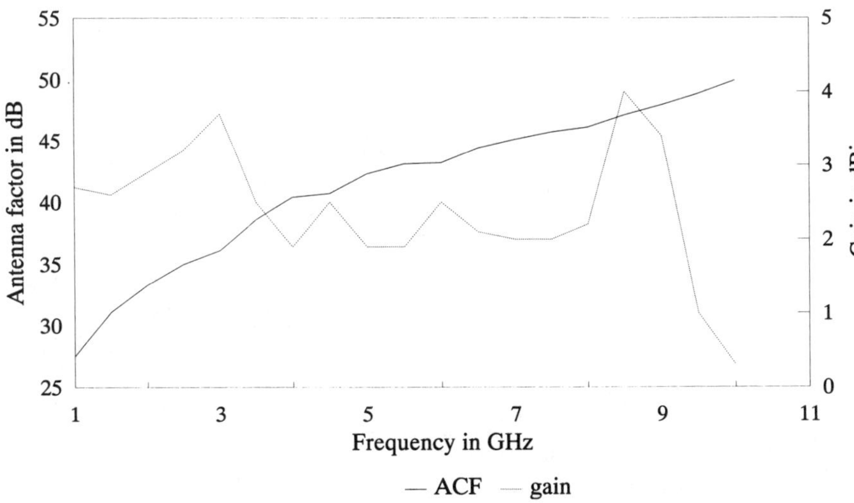

Figure 6.16 Variation of the gain and ACF for a conical log spiral in the frequency range 1 GHz to 10 GHz.

160 deg to 20 deg (narrow width). This variation with the angular width is shown in Figure 6.17, with the cone half-angles θ_{hc} of 7.5 deg, 10 deg, and 15 deg as parameters. The impedances obtained for a cone half-angle θ_{hc} of 10 deg and an infinite balun feed are appreciably lower and vary between 150 Ω and 60 Ω.

If the arm width is kept constant at 90 deg and the cone angle is increased, the impedance increases until the cone becomes a plane at θ_{hc} = 90 and the impedance approaches the theoretical value of 60π or 188 Ω [2, p. 496]. As the angle of wrap β increases, the impedance also increases, although the change is very small.

When the antenna operates at a frequency such that there is no distortion due to the truncated tip or base, the variation of the E_θ and E_ϕ radiation patterns with cone half-angles θ_{hc}, spiral wrap angle β, and angular arm width δ is shown in Figure 6.18. It can be seen that as the wrap angle is increased, the patterns become more even and tend toward the single lobe pattern. There is also a smaller difference between the E_θ and E_ϕ electric field patterns. At the small cone half-angle θ_{hc} = 7.5 deg, the patterns are smoother for the smaller wrap angles (β = 70 deg) than they are when θ_{hc} is larger (15 deg). For θ_{hc} < 10 deg and wrap angles > 75 deg it is possible to use thin arms or even constant-width arms [2, p. 498]. The constant-width arms can be constructed of coaxial cable, and this is used in many commercially available conical log spirals.

In the plane perpendicular to the axis of the cone, the pattern is almost omnidirectional. As in the case of other FI antennas, the radiation is in the backfire direction.

Dual polarization can be obtained by having another set of spirals wound in the opposite sense on the inside of the cone.

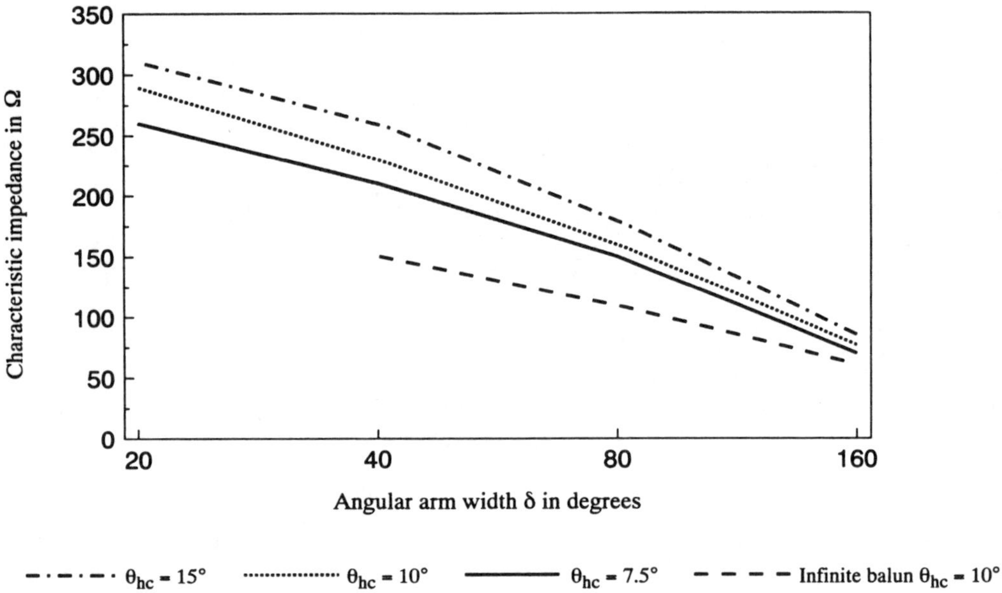

Figure 6.17 Variation of the characteristic impedance of a conical log spiral with the angular arm width δ, with the cone half-angle θ_{hc} as a parameter.

6.3.6.5 Phase Center

The phase center of an antenna is the point from which the EM waves appear to be emanating. The conical log spiral does not have a unique phase center, although an apparent phase center can be defined over most of the main beam [2, p. 493]. The apparent phase center in terms of the cone radius R_c/λ is given by

$$\frac{R_c}{\lambda} = \frac{1.2 \sin \beta}{2\pi(1.4 + \cos \beta)} \tag{6.22}$$

Because the fields have higher amplitude nearer the apex, the phase center tends to remain in this region, although the radiation region moves farther away from the apex at lower frequencies [2, p. 494].

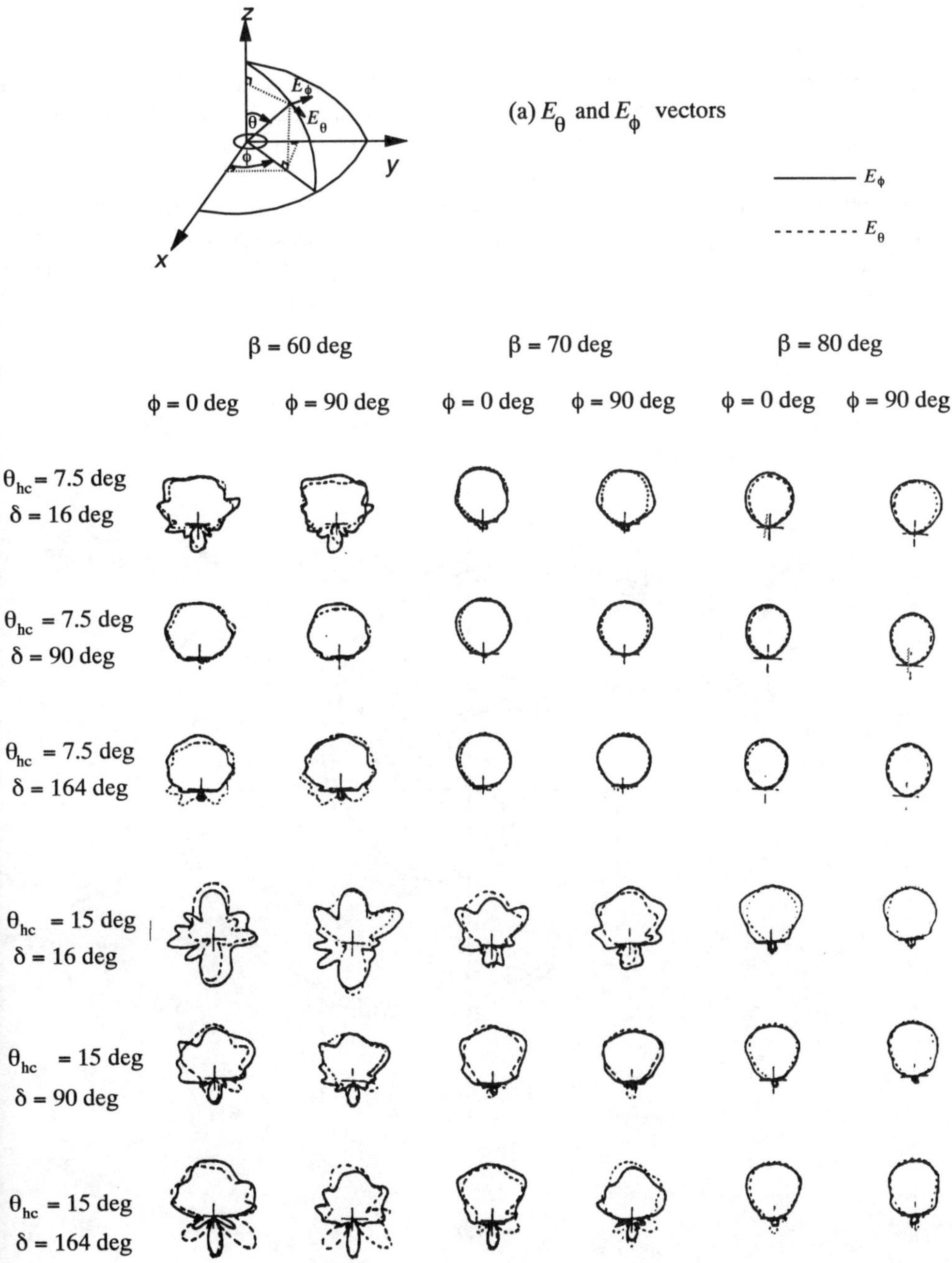

Figure 6.18 The variation of the E_θ and E_ϕ radiation patterns with cone half-angles θ_{hc}, spiral wrap angle β, and angular arm width δ for a conical log spiral. (After [2].)

6.4 ARCHIMEDEAN SPIRAL

The Archimedean spiral, as shown in Figure 6.19(a), is sometimes called an arithmetic spiral. Although the Archimedean spiral is a broadband antenna, it is not a true FI antenna. The equations that determine the inner and outer edges of the Archimedean spiral are given by $r_1 = a\alpha - \delta$ and $r_2 = a\alpha$. The width of the arm is given by $r_2 - r_1 = a\alpha\delta$. Thus, the width of the arm is constant for a fixed value of the expansion coefficient a and angular arm width δ. The adjacent turns of a single spiral are separated by a distance of $2a\pi$.

The angle β, which the radius makes with the tangent to the spiral, is not constant with the angle α, as in the case of log spirals. The angle β is given by

$$\tan \alpha = \beta \tag{6.23}$$

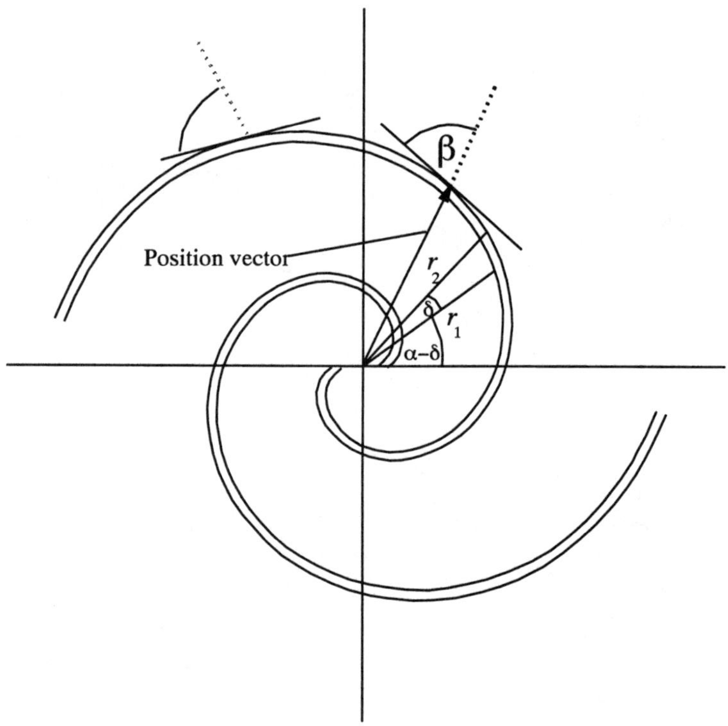

(a)

Figure 6.19 (a) The Archimedean spiral and (b) printed circuit balun.

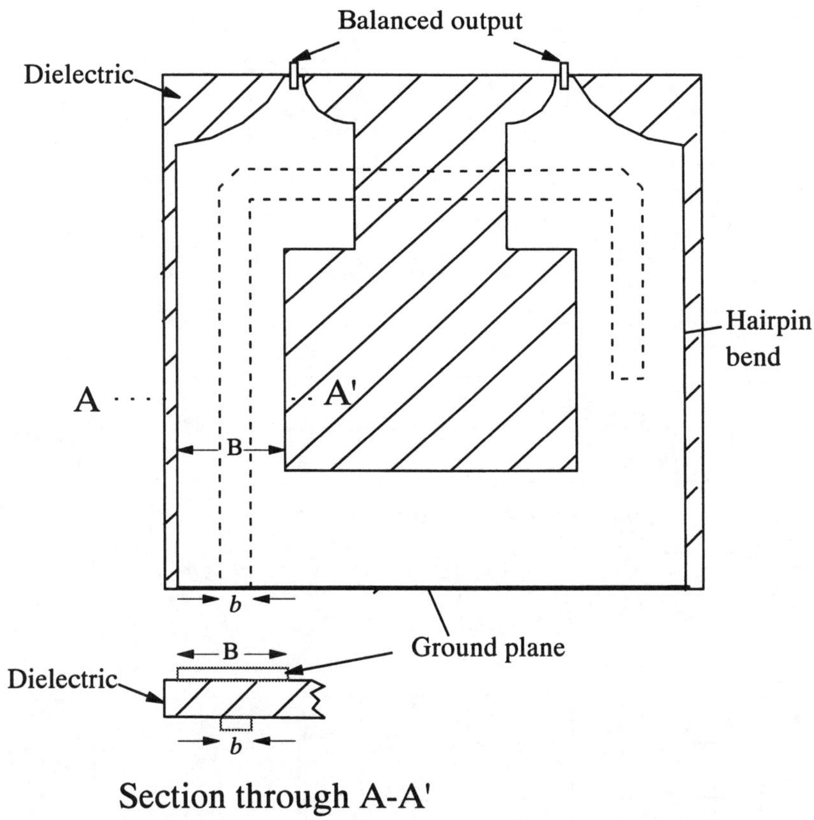

(b)

Figure 6.19 (continued).

Archimedean spirals must have many turns and be tightly wound to operate over wide bandwidths, whereas equiangular spirals can be constructed with just one or two turns that are loosely wound.

For a tightly wound spiral, at distances remote from the origin, the angle β varies slowly with the angle α and the Archimedean spiral closely approximates a tightly wound log spiral [4, p. 698].

6.4.1 Cavity-Backed Archimedean Spiral

An empirical estimate of the gain and axial ratio variation with the cavity diameter in terms of wavelengths is shown in Figure 6.20 [8, p. 14–16]. The results were obtained for

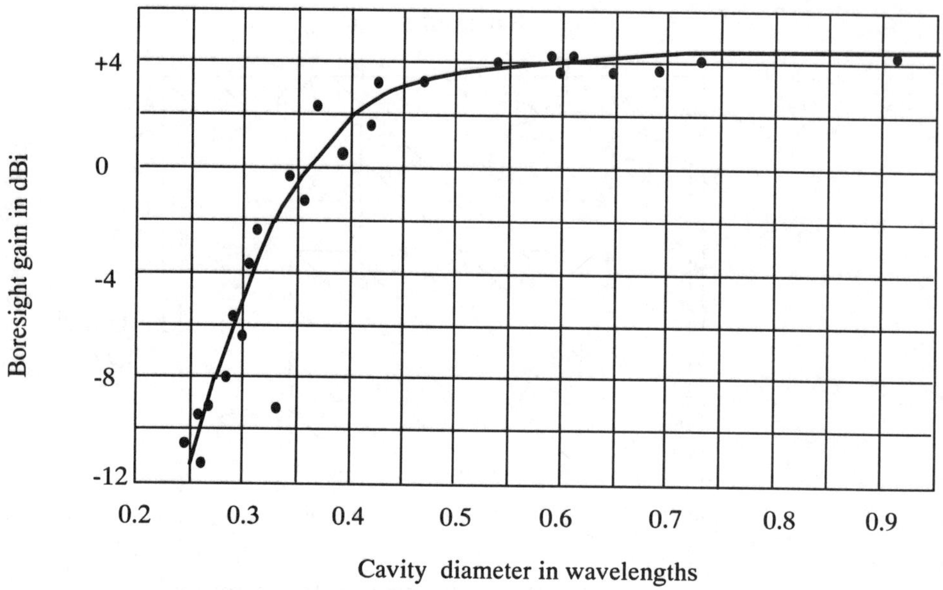

(a)

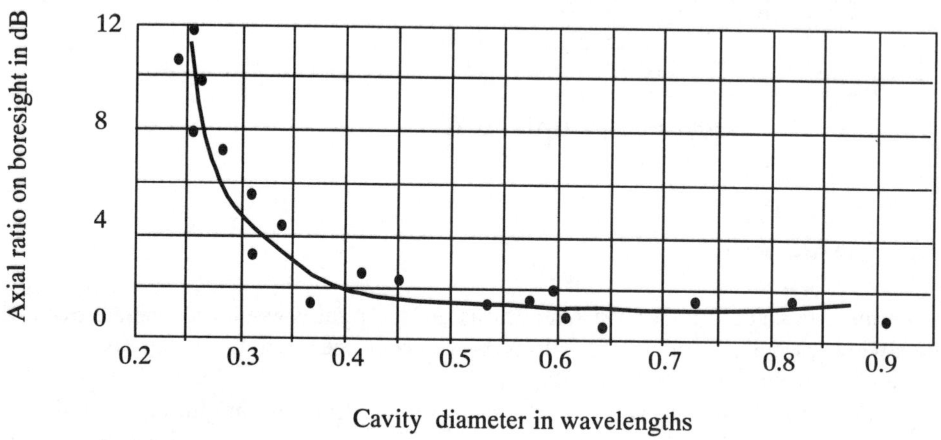

(b)

Figure 6.20 Variation of (a) the boresight gain and (b) axial ratio of a cavity-backed Archimedean planar spiral with cavity diameter.

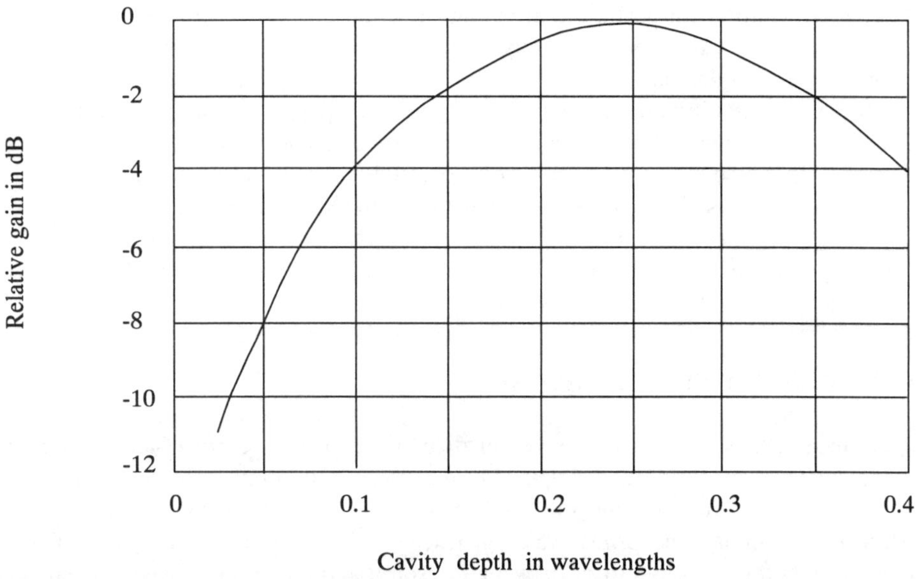

Figure 6.21 Variation of gain with cavity depth for an Archimedean planar spiral.

a large number of spirals in the 0.2-GHz to 4-GHz frequency band. The gain curve was normalized to the average axial ratio. The gain relative to an isotropic antenna, that is, the gain in dBi, increases almost linearly until a knee is reached at around $\lambda/3$, and then the gain curve starts to asymptotically approach the maximum gain at 0.5λ.

The boresight axial ratio variation with the cavity diameter is shown in Figure 6.20(b). The deterioration in the axial ratio obtained with the smaller cavity diameters is to be expected since the energy reflected from the edges of the spiral has polarization of the sense opposite to the energy radiated at the center of the antenna.

The variation of gain (normalized to the maximum gain attainable) with the cavity depth is shown in Figure 6.21. It can be seen that the maximum gain is obtained when the cavity depth c_d is $\lambda/4$. Increasing the number of turns increases the energy radiated since the effective aperture is now increased.

6.4.1.1 Feeding Arrangements

The feed and/or balun should be symmetrical and nonradiating. The most common feeding method used for the Archimedean spiral is a rigid coaxial line that is placed axially within the cavity. The center conductor is connected to one arm and the outer conductor is connected to the other arm. The coaxial line is tapered to obtain a good match over the frequency range of an octave or more [7, p. 89]. However, the feed itself radiates, giving rise to an unsymmetrical beam and affecting the ellipticity of the electric field polarization. The beam may also be squinted by a feed that radiates or is not well balanced and, thus, gives rise to boresight error.

A printed circuit balun, as shown in Figure 6.19 [6], is another popular balun that may be used. This consists of a hairpin bend used on one side of the board and a balanced line on the other side. The spiral arms are soldered directly to the balanced line. This type of balun has been used with antennas of octave bandwidth, in the 200-MHz to 4-GHz frequency range, and ellipticity ratios of less than 3 dB have been obtained, up to 60 deg off boresight. Boresight errors have also been virtually eliminated, and VSWRs of less than 2:1 have been obtained [7, p. 89].

6.5 MICROSTRIP PLANAR SPIRAL

Microstrip antennas are usually fairly narrowband. However, a *spiral-mode microstrip* (SMM) designed by Wang and Tripp [11] has bandwidth similar to the cavity backed spiral and, at the same time, has the smaller volume and conformability associated with a microstrip antenna. The SMM was constructed with a 1/2-inch-wide ring of RAM, which was just thick enough to fit between the spiral and the ground plane, and such that it is half underneath and half outside the truncated spiral. The RAM prevents the generation of higher order phase modes as well as the generation of the negative first-order mode (M = -1), which is the first-order mode being reflected from the outer periphery of the spiral. The gap widths (between the printed spiral and the ground plane) of 1.5 mm and 7.6 mm were used for Archimedean and equiangular spirals, and bandwidths of 6:1 were achieved in the frequency range 2 GHz to 12 GHz. The spiral diameters were 3 inches, and the diameter of the ground plane was 18 inch (46 cm). The smaller gap widths give lower gains and efficiencies at the lower operating frequencies. The axial ratio of the SMM antennas are larger than those obtained for the cavity spiral. They also suffer from pattern ripples and beam asymmetries, which are thought to be due to the imbalance of the balun feed and the edges of the finite ground plane. When compared with a 2.5-inch-diameter cavity-loaded spiral, the gains of 3 inch diameter SMMs (with gap widths of 7.6 mm) were between 1 dB and 3 dB better over the frequency range 2 GHz to 12 GHz. SMM offers several advantages over the cavity-backed spiral. The cavity-backed spiral has a cavity whose depth has a comparable dimension to the diameter of the spiral. For an antenna with a bandwidth of 2 GHz to 18 GHz, the SMM antenna has a thickness of 0.38 cm (0.15 inch) compared with 5.08 cm (2 inch) of a cavity-backed spiral.

6.6 DISCONE ANTENNA

The principle of the discone antenna was described in Chapter 5, and its characteristic parameters are shown in Figure 6.22. The discone antenna is often supplied with a ''snap-on'' N-type connector that provides the RF feed as well as the mechanical support

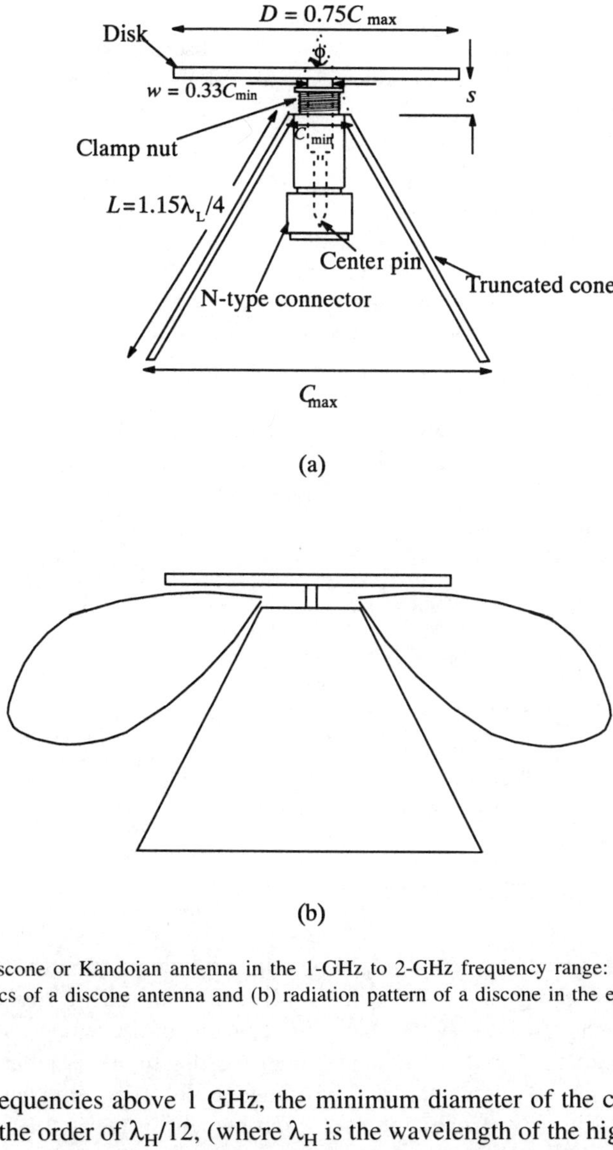

Figure 6.22 The discone or Kandoian antenna in the 1-GHz to 2-GHz frequency range: (a) physical characteristics of a discone antenna and (b) radiation pattern of a discone in the elevation plane.

for the disk. At frequencies above 1 GHz, the minimum diameter of the cone C_{min} must of necessity be of the order of $\lambda_H/12$, (where λ_H is the wavelength of the highest operating frequency) in order to accommodate the connector. At frequencies between 1 GHz and 2 GHz, Rappaport [12, p. 37] has found that for discone antennas mounted on the end of N-type connectors (and thus having values of $C_{min} = \lambda_H/12$ approximately) the diameter w of the feed pin must be taken into account. The dimension s (disk-to-cone spacing) also has a significant impact on antenna loading for large electrical dimensions of C_{min} and must be taken into account. The optimum separation s for a cone flare angle of 45 deg was found to be $0.41C_{min}$, but for larger flare angles $s = 0.5C_{min}$ was the optimum spacing. At lower frequencies (below 1 GHz) $s = 0.3C_{min}$ was used, as stated in Chapter 5.

The best flare angles (to provide a low VSWR) range between 45 deg and 75 deg, and the diameter of the feed conductor w should be one-third of the diameter of the top of the cone ($w = 0.33C_{min}$) at frequencies for which C_{min} is of the order of $\lambda_H/12$ or greater. At lower frequencies ($C_{min} \leq \lambda_H/75$), the optimum value of w is $0.3C_{min}$, as stated in Chapter 5. The impedance of the connector is a function of w/C_{min}; and for a ratio equal to 0.3, the characteristic impedance of 66.5Ω is obtained, which presents a good match to a 50-Ω transmission line. These values achieve a VSWR of 1.3:1 over an octave bandwidth. The disk diameter should also be $0.75C_{max}$ and the slant height L should be $1.15\lambda_L/4$ (where λ_L is the wavelength at the lowest operating frequency) instead of values of $0.7C_{max}$ and $\lambda_L/4$, respectively, for lower frequencies.

Four cones were constructed with C_{min} = 19 mm (0.75 in) and flare angles ϕ of 45 deg, 60 deg, 75 deg, and 90 deg. The body of each cone was soldered to a UG-21D/U N-type connector, and the rear end of the connector was made flush with the top of the cone. Disks of varying diameters were soldered to the rods that were, in turn, soldered to the center pins of the connectors.

The dimensions for the four antennas are shown in Table 6.1.

Table 6.1
The Parameters of Four Antennas With Different Flare Angles

Flare Angle	*Antenna 1*	*Antenna 2*	*Antenna 3*	*Antenna 4*
ϕ	45 deg	60 deg	75 deg	90 deg
C_{max}	74.2 mm	95.2 mm	108 mm	120.5 mm
C_{min}	19.1 mm	19.1 mm	19.1 mm	19.1 mm
L	73 mm	73 mm	73 mm	73 mm
D	55.65 mm	71.4 mm	81 mm	90.4 mm

The antenna can also be optimized by adjusting the position of the clamp nut. This clamp nut tuning changes the dimension s (disk-to-cone spacing) and also reduces the impedance mismatch created within the connector. An improvement of around 5 dB was obtained in the antennas with larger cone angles (65 deg to 90 deg), but for the cone flare angle of 45 deg the disk diameter had to be increased to $1.1C_{max}$ and the feed pin diameter reduced to $0.25C_{min}$ in order to obtain an improvement in the VSWR.

6.7 DOUBLE-RIDGED HORNS

Double-ridged horns can be used from around 50 MHz to above 40 GHz, but commercially available horns are usually found in the frequency ranges 200 MHz to 2 GHz (Eaton

96000) and 1 GHz to 18 GHz (Eaton 96001). The gains of double-ridged horns vary between 2.5 dBi and 8 dBi for the frequency range 200 MHz to 1 GHz and 4 dBi to 42 dBi for the frequency range 1 GHz to 18 GHz. HPBWs vary between 40 deg and 70 deg in the *H* and *E* planes, respectively. Double-ridged horns are particularly useful for immunity and susceptibility testing since they can handle powers of up to 800W in the lower frequency range and up to 500W in the higher frequency range.

In order to understand the behavior of double-ridged horns we must consider doubled-ridged waveguides, which are smooth-walled rectangular waveguides with ridges. The basics of waveguide theory are explained in Section 6.7.1.

6.7.1 Waveguide Theory

Waveguides may be of rectangular, circular, or elliptical cross section. The rectangular waveguides might have smooth inner walls or may have one, two, or four ridges, in which case they are known as ridged waveguides. Waveguides are used at frequencies above about 1 GHz since coaxial lines are subject to attenuation (which increases with frequency), and they also cannot withstand high power. At lower frequencies the size of the waveguide becomes too large to be practical, although waveguides are sometimes used for specialized applications at frequencies down to 300 MHz (in the case of smooth-walled waveguides) and down to 100 MHz in the case of ridged waveguides. Waveguides cannot support a TEM wave since one of the requirements for its propagation is two separate conductors.

The EM wave propagates in the waveguide in a similar manner to the way light waves travel in an optic fiber. We can imagine the waves reflected off the metal walls as they travel down the guide. When we draw rays of light in optics, the wavefront moves perpendicular to these rays. Consider a rectangular waveguide of cross-section dimensions a and b, as in Figure 6.23(a). The guide is excited at a point between the two side walls, and waves reflect off the side walls. In Figure 6.23(b) the rays are shown reflecting off the walls, whereas in Figure 6.23(c) the wavefronts are shown. The dotted lines indicate troughs, and the solid lines represent the crests. The waves reflected off the side walls cross, and where two troughs cross at point Q for instance, the resultant is a trough of twice the magnitude of the trough of each individual wave. Similarly, where crests cross (at P for instance) the resultant crest is twice the magnitude of the individual crests. Thus the resultant wave traveling down the guide is of twice the amplitude of the individual waves. This is the guided wave, and the distance between crests or between troughs is the guide wavelength λ_g. The distance between P and Q is half this value, and PS is a quarter of this value. The relationship between the guide wavelength and the wavelength of the individual incident waves can be derived using geometry and trigonometry. The wavelength of the individual waves is the same as the wavelength of the waves in free space, and thus this is known as the free-space wavelength λ_0. In Figure 6.23(c) the distance denoted by TQ is that between a crest and a trough and, therefore, is half this free-space wavelength. The width of the guide is a, and the distance RS is $a/2$.

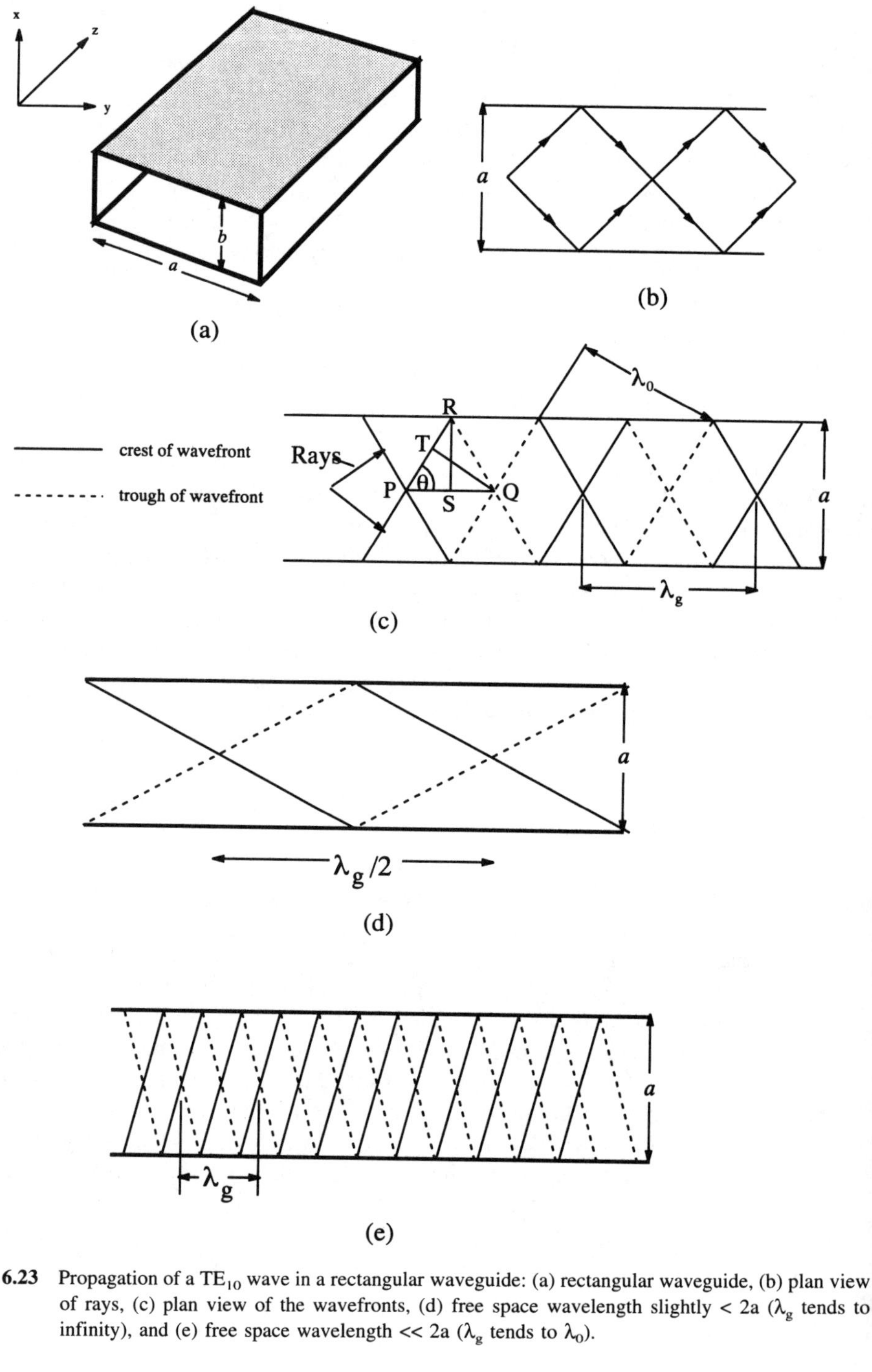

Figure 6.23 Propagation of a TE_{10} wave in a rectangular waveguide: (a) rectangular waveguide, (b) plan view of rays, (c) plan view of the wavefronts, (d) free space wavelength slightly < 2a (λ_g tends to infinity), and (e) free space wavelength << 2a (λ_g tends to λ_0).

In triangle *PTQ* the sine of angle *TPQ* is given by $\sin\theta = QT/PQ$, that is,

$$\sin\theta = \frac{\lambda_0/2}{\lambda_g/2} \tag{6.24}$$

In triangle *PRS*, using Pythagoras's theorem,

$$PR^2 = RS^2 + PS^2 \qquad PR = \sqrt{\left(\frac{a}{2}\right)^2 + \left(\frac{\lambda_g}{4}\right)^2}$$

$$\sin\theta = \frac{RS}{PR} = \frac{a/2}{\sqrt{(a/2)^2 + (\lambda_g/4)^2}} \tag{6.25}$$

Equating (6.24) and (6.25) we get

$$\frac{\lambda_0}{\lambda_g} = \frac{a/2}{\sqrt{(a/2)^2 + (\lambda_g/4)^2}}$$

Squaring and cross multiplying gives

$$\lambda_0^2\left[\left(\frac{a}{2}\right)^2 + \left(\frac{\lambda_g}{4}\right)^2\right] = \frac{\lambda_g^2 a^2}{4}$$

Dividing throughout by $\lambda_0^2\lambda_g^2 a^2/4$ gives

$$\left(\frac{1}{\lambda_g}\right)^2 = \left(\frac{1}{\lambda_0}\right)^2 - \left(\frac{1}{2a}\right)^2 \tag{6.26}$$

Let us now consider what happens as we vary λ_0. We will consider four cases: when it is equal to $2a$, slightly less than $2a$, much less than $2a$, and greater than $2a$.

1. If $\lambda_0 = 2a$, then the right-hand side of (6.26) becomes zero, and thus the guide wavelength $\lambda_g \to \infty$. In this case, the frequency tends to zero, that is, the wave does not propagate. The free-space wavelength at which this occurs is known as the cut-off wavelength λ_c (= $2a$), and the corresponding frequency is known as the cut-off frequency f_c. The waveguide acts as a high pass filter, and propagation only takes place at frequencies above this cut-off frequency.

2. When λ_0 is slightly < $2a$, it can be seen that the difference between the two terms on the right-hand side of (6.26) is very small, so the reciprocal of the square of the guide

wavelength is also very small. This makes the guide wavelength very long, as depicted in Figure 6.23(d).

3. When λ_0 is << (much smaller than) $2a$, it can be seen from (6.26) that the first term on the right-hand side of the equation is the only significant term, which means that the guide wavelength tends to the free space wavelength, as depicted in Figure 6.23(e).

4. When λ_0 is > $2a$, that is, the free-space wavelength is greater than the cut-off wavelength, an interesting situation occurs. The right-hand side of (6.26) is negative, and this makes its square root an imaginary value. This is analogous to the imaginary part of an impedance and a similar condition applies in this case. The guided wave has a purely reactive impedance, and the guide supports what are known as evanescent modes. These modes store energy in a manner similar to the way capacitors store energy; they do not propagate or dissipate energy. Use is made of this property (of waveguides below cut-off) in the design of ventilation panels for shielded rooms.

6.7.2 Modes in Square and Rectangular Waveguides

Now let us consider what happens at the metal side walls of the waveguide. We apply what are known as boundary conditions. One of these boundary conditions states that at a perfect conductor the tangential electric field is zero. This is in effect a generalized form of Ohm's law, which states that the voltage is equal to the product of the current and resistance. If we have a perfect conductor, the resistance is zero and thus the voltage is also zero. Since the electric field is the voltage per unit length, the electric field along or tangential to a perfect conductor is also zero. It can be shown that the normal (perpendicular) electric field varies across the broadside dimension a of the waveguide in a sinusoidal manner, from zero at the ends to a maximum at the center.

Other variations of the electric field are also possible across this dimension, but one half-sinusoidal variation is the lowest configuration that can be supported by the waveguide. This is known as the dominant or fundamental mode of propagation. For rectangular waveguides this is the TE_{10} mode where TE stands for transverse electric and the subscripts 1 and 0 stand for the numbers of half-wave variations along the y- and x-axes, respectively. In this mode the electric field only has a component in the x-y plane, transverse to the direction of propagation of the wave. As shown in Figure 6.24(a), the electric field only varies in the y-direction, that is, across the broadside or a dimension of the waveguide; there is no electric field perpendicular to the b dimension of the waveguide and hence no variation in the x-direction. Since there is one half-wave variation in the y-direction, the first subscript is one, and since there is no variation in the x-direction, the second subscript is zero. If the electric field had been perpendicular to the b dimension of the waveguide, the mode would be called the TE_{01} mode. The magnetic field has a component in the transverse plane as well as in the direction of propagation of the wave, and for this reason, in some older textbooks and papers the TE modes are called H modes. The magnetic field lines are shown in Figure 6.24(a) with the top surface of the guide

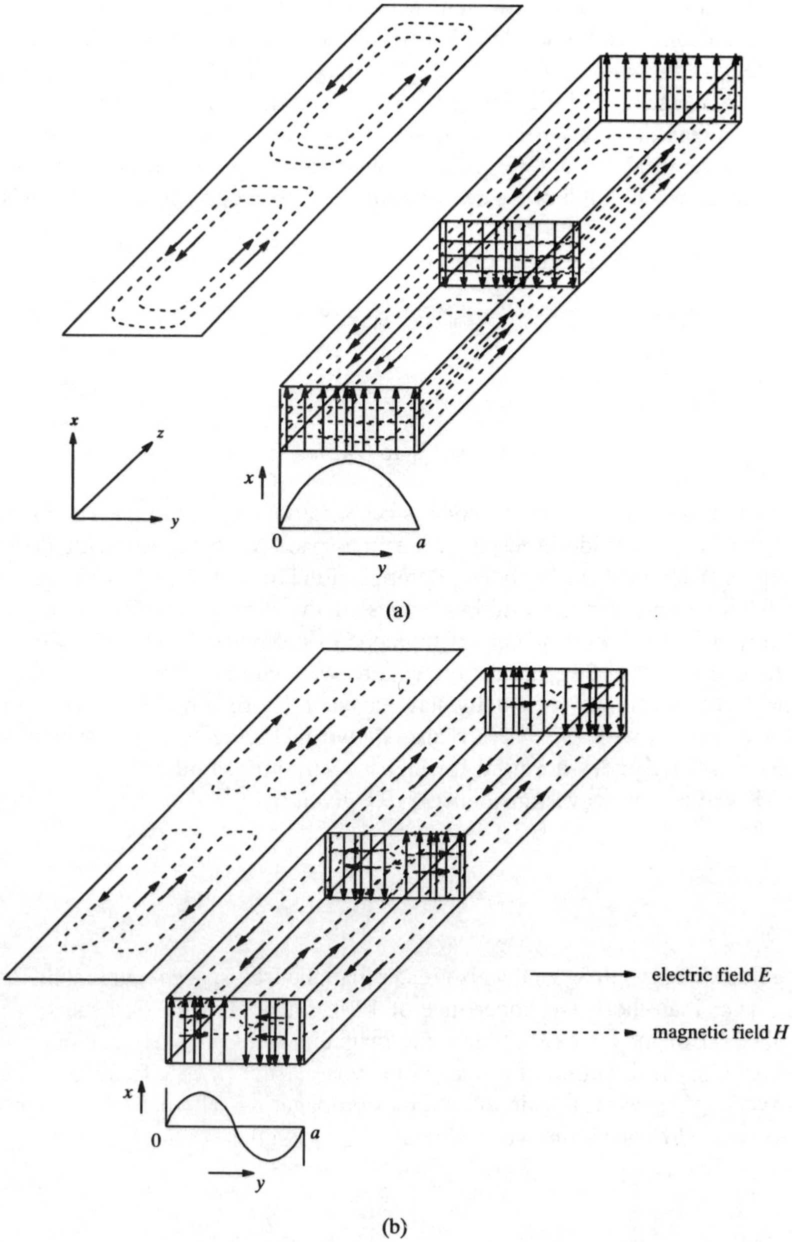

Figure 6.24 Propagation of (a) TE_{10} and (b) TE_{20} modes, showing the fields in rectangular waveguides.

detached, for clarity. If there are two half-wave variations of the electric field across the broadside dimension, then the mode is known as the TE_{20} and the electric field variation for this mode is as shown in Figure 6.24(b). The magnetic field lines are not shown on the top and bottom surfaces, and the top surface of the guide is shown detached, for clarity. Rectangular waveguides are normally chosen for use at certain frequencies so that they only propagate the fundamental TE_{10} mode (which also presents the lowest attenuation) and the free-space wavelength of the center frequency for propagating this mode is around $4a/3$.

These waveguides are usually chosen so that their dimensions lie between the limits

$$a = 0.6\lambda_0 \text{ to } 0.9\lambda_0 \tag{6.27}$$

and

$$b = 0.3\lambda_0 \text{ to } 0.45\lambda_0 \tag{6.28}$$

This ensures that only the dominant mode is propagated. Figure 6.25(a) shows the a and b dimensions of a waveguide in terms of the free-space wavelength for the dominant as well as some of the higher order modes of propagation [16, Fig. 7-8]. The curves and lines in Figure 6.25(a) represent the limiting values of the dimensions for the higher order modes. Figure 6.25(b) shows the cut-off frequencies normalized to (as a multiple of) the cut-off frequency of the TE_{10} mode for square waveguides ($b/a = 1$) as well as for rectangular waveguides, which usually have a b/a ratio of around 1/2. The modes in oversized waveguides with $b/a = 2$ are also shown in Figure 6.25(a) for completeness. These oversized waveguides are used for high-power applications.

For TE waves, the wave impedance Z_w is given by

$$Z_w = 120\pi\left(\frac{\lambda_g}{\lambda_0}\right) \tag{6.29}$$

Since the guide wavelength λ_g is always greater than the free-space wavelength, Z_w for TE modes is greater than the wave impedance of 120π (377) ohms in free space.

Another set of modes exists where the magnetic field only has components in the plane transverse to the direction of propagation, whereas the electric field has components in the transverse plane as well as in direction of propagation. These are known as the TM or E modes. For TM waves the wave impedance Z_w is given by

$$Z_w = 120\pi\left(\frac{\lambda_0}{\lambda_g}\right) \tag{6.30}$$

Since the guide wavelength λ_g is always greater than the free-space wavelength, Z_w forTM

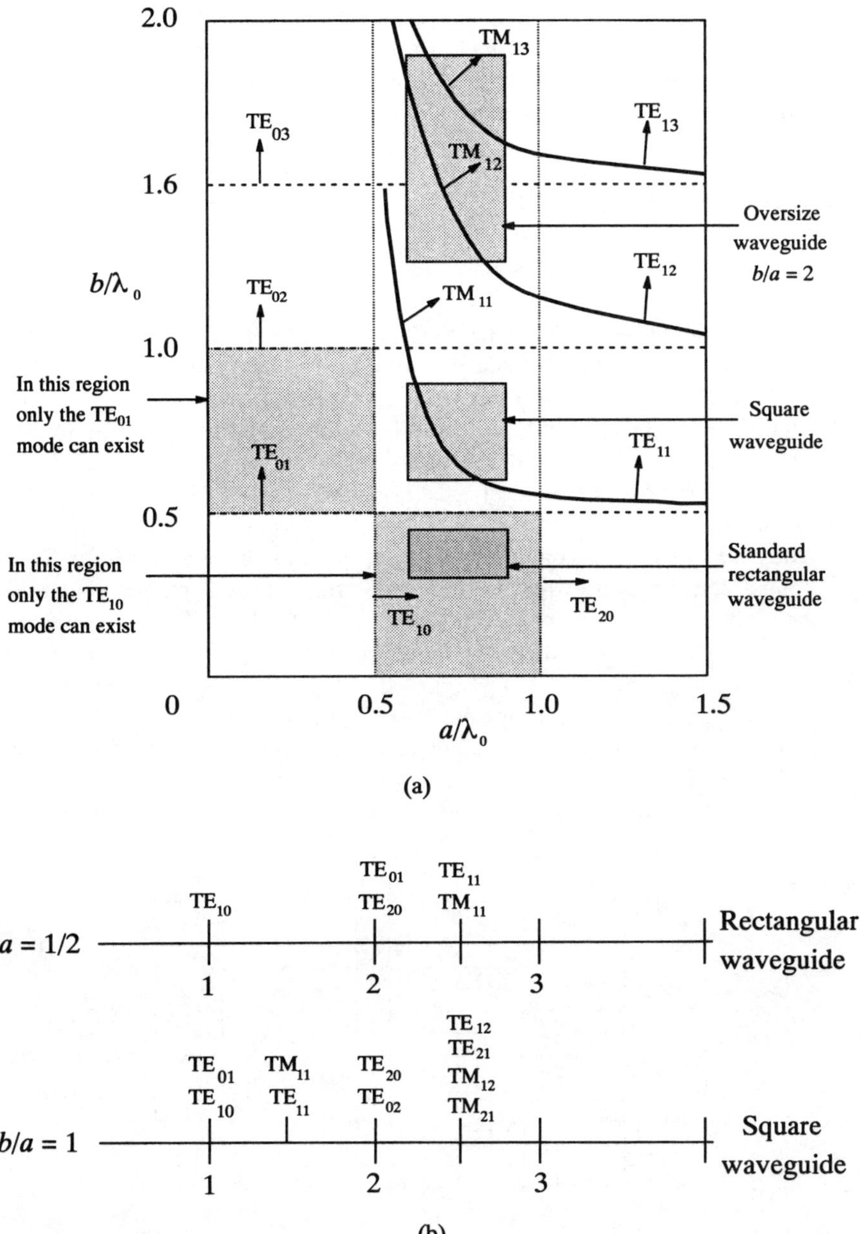

Figure 6.25 Dimensions and cut-off frequencies for some modes in rectangular waveguides: (a) dimensions of square and rectangular waveguides for various modes and (b) cut-off frequencies normalized to the cut-off frequency of TE_{10} mode. (After [16], Fig. 7.8.)

modes is less than the wave impedance of 120π (377) ohms in free space. Attenuation as a function of frequency for some of the lower order modes in a rectangular waveguide are shown in Figure 6.26, together with the attenuation in a coaxial cable for comparison [16, Fig. 8-2].

6.7.2.1 Rectangular Waveguide Designations

Standard rigid waveguides are chosen so that they only propagate the dominant or TE_{10} mode. There are two main types of designations used for rectangular waveguides. The DEF 5351 designation uses the letters WG, and the EIA RS 216A designation uses the letters WR. In the first case the letters WG are followed by a number that decreases with frequency, that is, increases with decreasing dimensions of the waveguide. For example, WG 10, which is used in the NATO E and F bands (2.3 GHz to 4.2 GHz), has inside dimensions of 7.214 cm (2.84 in) by 3.404 cm (1.34 in), whereas WG 16, which is used in the NATO bands I and J (formerly X-band, 8 GHz to 12.4 GHz), is 2.29 cm (0.9 in) by 1 cm (0.4 in).

In the case of the other commonly used designation, the letters WR are followed by numbers that decrease with increasing frequency; that is, the numbers increase with the size of the waveguide. For example, WR 284 is equivalent to WG 10, whereas WR 90 is equivalent to WG 16. We should note that with the WR designation the number following the WR number is approximately 100 times the larger dimension of the waveguide in inches. Table 6.2 shows both designations over the 400-MHz to 135-GHz frequency range. By measuring the inside dimensions of a waveguide we can look up its operating frequency range.

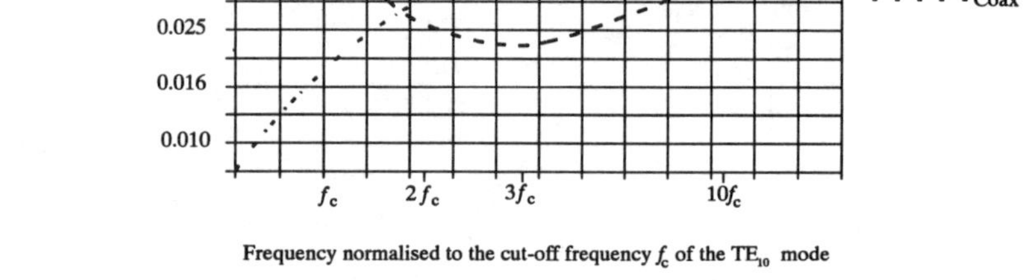

Figure 6.26 Attenuation for some of the modes in a rectangular waveguide as a function of the cut-off frequency of the dominant TE_{10} mode. (After [16], Fig. 8.4.)

Table 6.2
Rectangular Waveguide Designations and Dimensions

DEF 5351 No		*EIA RS216A No*	*Dimensions (in)*		*Dimensions (mm)*		*Operating Freq. Range (GHz)*
			Width	*Height*	*Width*	*Height*	
WG	1	WR 1800	18	9	457.2	228.6	0.41–0.61
WG	2	WR 1500	15	7.5	381	190.5	0.50–0.75
WG	3	WR 1150	11.5	5.75	292.1	146.05	0.61–0.96
WG	4	WR 975	9.75	4.875	247.65	123.82	0.75–1.12
WG	5	WR 770	7.7	3.85	195.58	97.79	0.96–1.45
WG	6	WR 650	6.5	3.25	165.1	82.55	1.12–1.70
WG	7	WR 510	5.1	2.55	129.54	64.77	1.45–2.20
WG	8	WR 430	4.3	2.15	109.22	54.61	1.70–2.60
WG	9	WR 340	3.4	1.7	86.36	43.18	2.20–3.30
WG	10	WR 284	2.84	1.34	72.13	34.036	2.60–3.95
WG	11A	WR 229	2.29	1.145	58.166	29.083	3.30–4.90
WG	12	WR 187	1.87	0.872	47.498	22.148	3.95–5.85
WG	13	WR 159	1.59	0.795	40.386	20.193	4.90–7.05
WG	14	WR 137	1.37	0.622	34.798	15.798	5.85–8.20
WG	15	WR 112	1.12	0.497	28.448	12.624	7.05–10.05
WG	16	WR 90	0.9	0.4	22.86	10.168	8.20–12.4
WG	17	WR 75	0.75	0.375	19.05	9.525	10.0–15.0
WG	18	WR 62	0.62	0.31	15.748	7.874	12.4–18.0
WG	19	WR 51	0.51	0.255	12.954	6.476	15.0–22.0
WG	20	WR 42	0.42	0.17	10.668	4.318	18.0–26.5
WG	21	WR 34	0.34	0.17	8.6360	4.318	22.0–33.0
WG	22	WR 28	0.28	0.14	7.112	3.556	26.5–40.0
WG	23	WR 22	0.22	0.11	5.588	2.794	33.0–50.0
WG	24	WR 19	0.188	0.094	4.775	2.387	40.0–60.0
WG	25	WR 15	0.148	0.074	3.759	1.8796	50.0–75.0
WG	26	WR 12	0.122	0.061	3.098	1.5494	60.0–90.0
WG	27	WR 10	0.10	0.05	2.54	1.27	75.0–110.0
WG	28	WR 8	0.08	0.04	2.032	1.016	90.0–135

6.7.3 Double-Ridged Waveguides

We have seen that the cut-off wavelength λ_c for the TE_{10} of a particular smooth-walled rectangular waveguide is twice its broad dimension a (i.e., $\lambda_c = 2a$). However, when a single or double ridge or septum is used, the cut-off wavelength of the dominant or TE_{10} can be increased to more than 2.5 times this value (i.e., $\lambda_c > 5a$). This means that the lower cut-off frequency of the ridged waveguide is less than two-fifths of that of the smooth-walled guide.

The bandwidth for the single-mode transmission can be 4:1 [8, p. 42–34]. The *maximum usable bandwidth* (MUB) is given by the ratio of the cut-off wavelength of the

TE_{10} mode to that of the TE_{20} mode. It should be noted that this is not the useful bandwidth since the guide is normally operated about 15% to 25% above the cut-off frequency. The ridges can also raise the cut-off frequency of the TE_{30} mode slightly [8, p. 40–5].

The bandwidth of the double-ridged waveguide can be extended further by suppressing the TE_{20} so that the MUB is given by the ratio of the cut-off wavelength of the TE_{10} mode to that of the TE_{30} mode. The ratio of the cut-off wavelength of the TE_{30} mode to that of the TE_{10} mode can be > 12:1 for double-ridged waveguides. Dielectric loading is often used for horns of bandwidths > 5:1 to reduce the phase error, as in the case of smooth-walled horns. It should be noted that the cut-off frequency can be reduced by the use of these dielectric materials but that this does not increase the bandwidth and usually results in greater loss.

The change in the cut-off frequency depends both on the width of the ridge and its height relative to the dimensions of the waveguide. The addition of the double septum or ridge also alters the amplitude distribution across the mouth of the horn, with a corresponding reduction in the sidelobe level down to -30 dB [4, p. 655]. The electric field lines are shown in Figure 6.27(a).

6.7.4 Double-Ridged Waveguide Horns

In the smooth-walled pyramid waveguide horn the electric field in the H plane (parallel to the broadside) tends to zero at the edges. This gives reduced sidelobes of around 23 dB. In the E plane, however, the amplitude is nearly uniform, resulting in a sidelobe level of 13 dB. The introduction of septum plates with heights equal to a quarter of the broadside dimension (so that the gap $b' = 0.5$) gives a stepped amplitude distribution, which is almost a binomial series distribution of 1:2:1. This can reduce the sidelobe level to 30 dB below boresight by accurate positioning of the ridges at the throat of the horn [4, p. 655].

The taper of the ridge height and width must be such that a smooth impedance transition is provided from the characteristic ridge impedance of $Z_0 = 50\Omega$ to the intrinsic impedance of free space, $\xi_0 = 377\Omega$. It has been found experimentally that an exponentially impedance taper for the following form is quite satisfactory:

$$Z = Z_0 e^{kx} \text{ for } 0 \le x \le l/2$$

$$Z = 377 + Z_0[l - e^{k(l-x)}] \text{ for } l/2 \le x \le l$$

where l is the length of the flare section and k is a constant, such that the impedance of the midpoint of the flared section is the average of the end point impedances.

The simplest type of ridge to fabricate is one of constant width. The length of the flared section should be fairly long so that the impedance transformation is accomplished over a length of the order of half a wavelength at the lowest operating frequency. However, if the ridge width is too large near the open mouth of the horn aperture, the H-plane

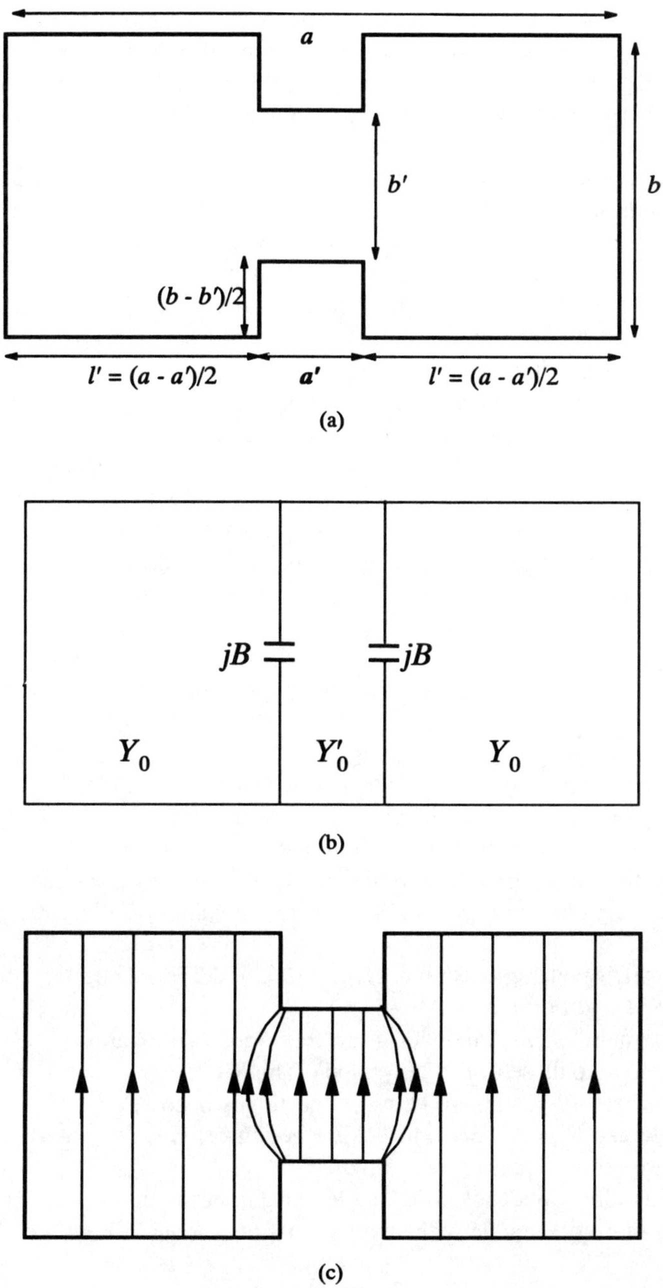

Figure 6.27 Double-ridged rectangular waveguide: (a) ridge dimensions in a double ridged waveguide, (b) equivalent circuit for cut-off wavelength computation, and (c) electric field for the TE_{10} mode in a double ridged waveguide.

distribution may have a large peak in the center and a large phase error occurs in the E plane [15]. Both of these factors result in a reduced gain for the horn. This can be alleviated by allowing the width of the ridges to increase in the flared section.

Consider the double-ridged waveguide shown in Figure 6.27(a). The width of each ridge is a' and the gap is b', so the height of each ridge is $(b - b')/2$. The transverse section of the waveguide can be considered to be a junction capacitance (at the central reduced height waveguide section) in parallel with two TE-mode transmission lines on either side with open and short-circuit terminations, respectively [12, p. 399]. This equivalent circuit model holds as long as the gap b' is smaller than the cut-off wavelength λ_c and much smaller than half the width of the normal height guide $(a - a')/2$. At cut-off, the susceptance looking into the shorted parallel plate guide to the right of the ridge is equal in magnitude but opposite in phase to the susceptance of the guide looking to the left [12, p. 399]. The susceptance at the center is zero at cut-off. The equivalent circuit is valid for $b/\lambda_c < 1$ in the case of the double-ridged waveguide.

It is also assumed that no higher mode interaction exists between the ridge and the narrow side of the guide. This would be the case when the width of the unreduced guide $(a - a')$ is much larger than twice the guide height, that is, $(a - a') >> 2b$.

The guide wavelength λ_g is given by the following resonance condition at the reference plane T:

$$\frac{Y'_0}{Y_0}\tan\frac{\pi}{\lambda_c}a' + \frac{B}{Y_0} - \cot\frac{\pi}{\lambda_c}(a-a') = 0 \tag{6.29}$$

where Y'_0 is the susceptance of the waveguide with reduced height, and Y_0 is the susceptance of the waveguide of normal height propagating the TE_{10} mode. $Y'_0/Y_0 = b/b'$ and λ_c is the cut-off wavelength of the guide. The numerical solution to (6.29) yields the cut-off wavelength for the TE_{10} normalized to the broadside a of the waveguide, that is, λ_c/a [12, p. 399]. These are plotted in Figure 6.28 for different values of a'/a and for values of b'/b between 0.1 and 0.5.

The cut-off wavelengths for the TE_{10}, TE_{20}, and TE_{30} modes are shown in Figure 6.29 for b'/b=0.1 and a/b = 2.

The maximum usable bandwidths are the ratios of the cut-off wavelengths of the TE_{20} or TE_{30} mode to those of the TE_{10} mode. The MUBs are shown in Figure 6.30. The MUB increases rapidly as the gap between the ridges becomes small.

When the gap b' is increased to $0.25b$, the cut-off wavelengths (normalized to the broadside a) are reduced as shown in Figure 6.31.

The bandwidths obtainable with the larger gap between the ridges (shorter ridges) are lower than those obtainable with smaller gaps (as shown in Figure 6.32), which is to be expected.

The ridged waveguide is much smaller than the smooth-walled guide with the same bandwidth; however, the losses are higher.

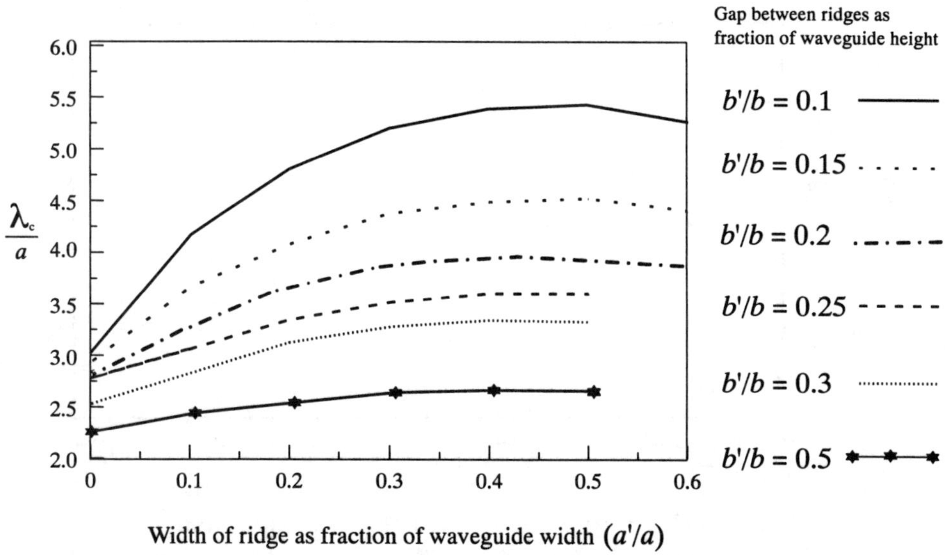

Figure 6.28 Variation of the TE_{10}-mode cut-off wavelength for double ridged waveguides with ridge width a'/a and for values of b'/b between 0.1 and 0.5 and $a/b = 2$.

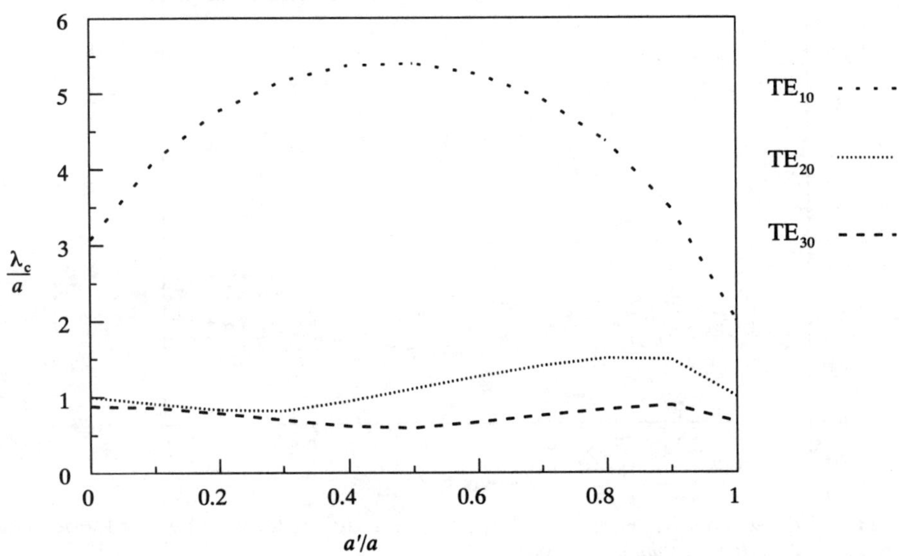

Figure 6.29 The cut-off wavelengths for the TE_{10}, TE_{20}, and TE_{30} modes for double ridged waveguides for $b'/b = 0.1$ and $a/b = 2$.

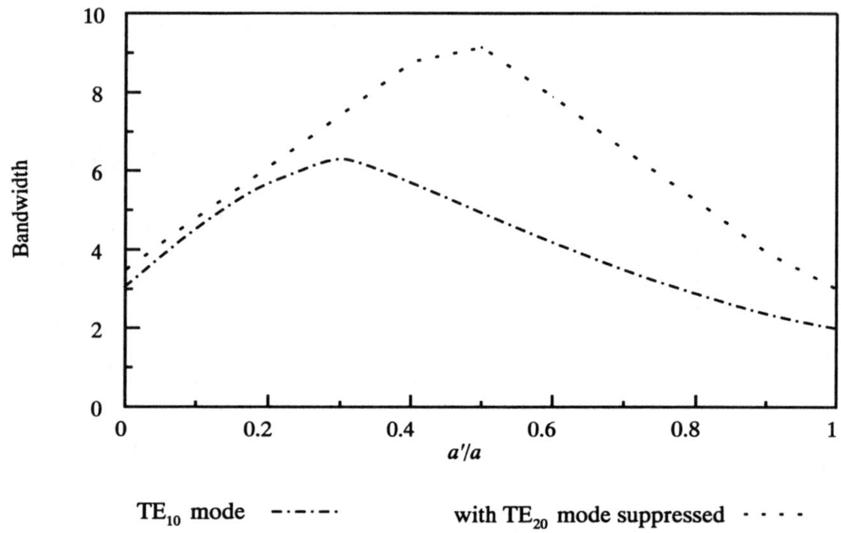

Figure 6.30 The maximum usable bandwidths of double-ridged waveguides with $b'/b = 0.1$ and $a/b = 2$.

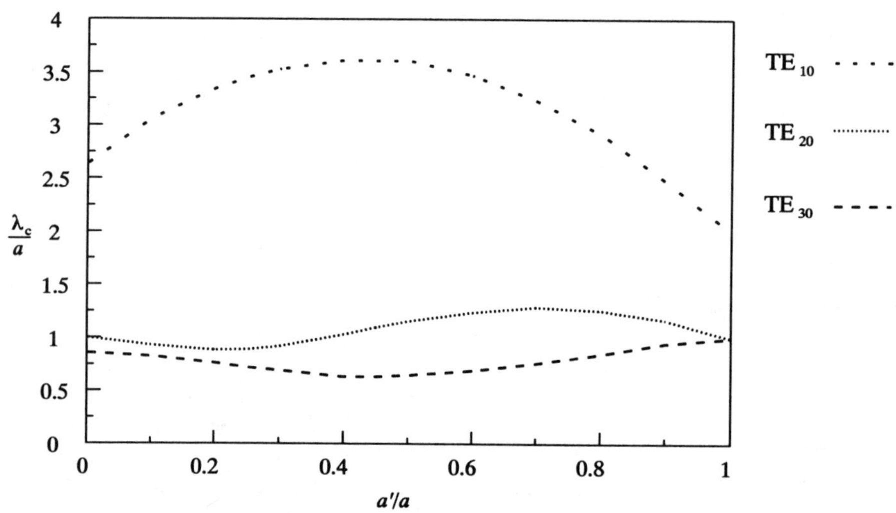

Figure 6.31 The cut-off wavelengths for the TE_{10}, TE_{20}, and TE_{30} modes of double ridged waveguides for $b'/b = 0.25$ and $a/b = 2$.

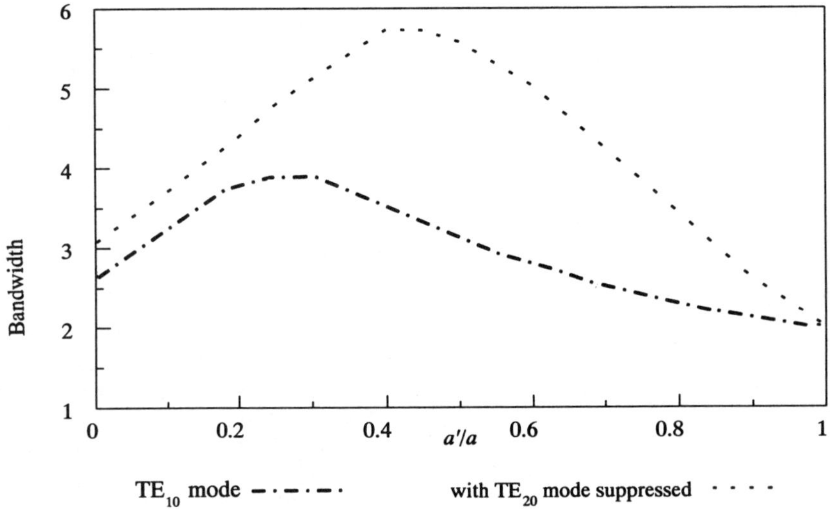

Figure 6.32 The maximum usable bandwidths of double-ridged waveguides with b'/b = 0.25 and a/b = 2.

The ACF, gain, and HPBWs for a commercially available (EATON 96000) double-ridged waveguide horn in the frequency range 200 MHz to 2 GHz are shown in Figure 6.33. The HPBWs shown are the average of those in the *E* and *H* planes.

In the frequency range 1 GHz to 10 GHz, the ACF, gain, and HPBWs for a commercially available (EATON 96001) double-ridged waveguide horn are shown in Figure 6.34.

Double-ridged horns can handle high powers and are therefore used as transmit antennas in, for example, susceptibility/immunity and SE measurements. The far-field distance is normally taken as $2D^2/\lambda$ where D is the diagonal in this case. However, the broadbanding of the horn results in a loss of gain, and this reduces the effective aperture D_e. This effective aperture is the diameter of a reflector antenna with the equivalent gain. The HPBWs of a 1-GHz to 10-GHz double-ridged horn, for example, vary between about 72 deg and 30 deg. The diameter D_e of an equivalent hypothetical circular reflector with an HPBW of θ_H is given by

$$D_e = \frac{58\lambda}{\theta_H} \tag{6.30}$$

where θ_H is the HPBW of the equivalent reflector in degrees, λ is the wavelength in meters, and D_e is the diameter in meters. The effective aperture D_e can be calculated at each frequency if we know the HPBWs. The far-field distance is then calculated at each frequency, and these values are listed in column 4 of Table 6.3. The far-field distance

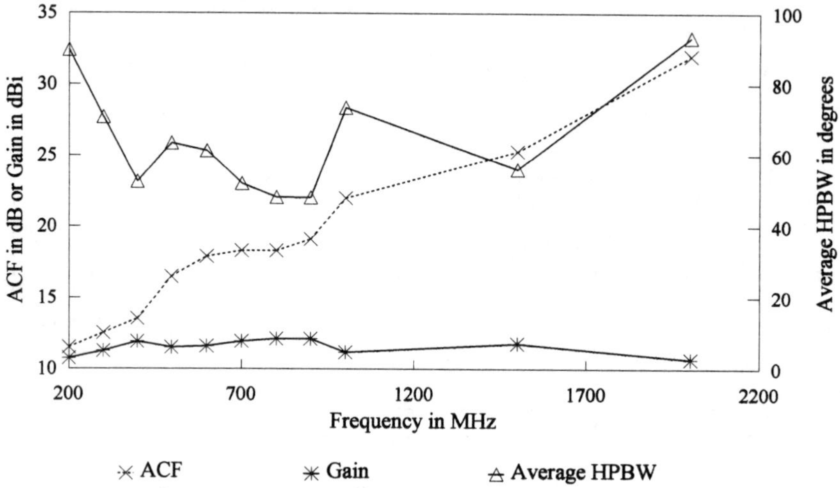

Figure 6.33 The ACF, gain, and HPBWs for a commercially available double-ridged waveguide horn in the frequency range 200 MHz to 2 GHz.

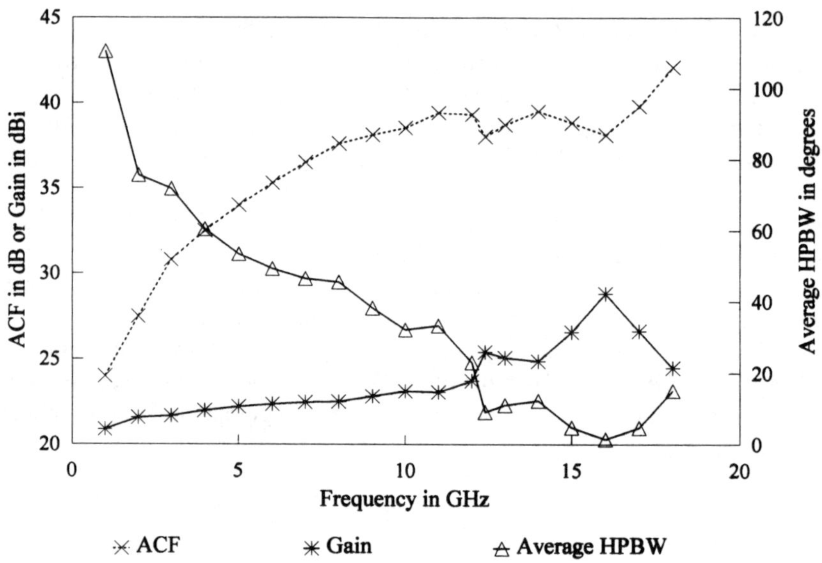

Figure 6.34 The ACF, gain, and HPBWs for a commercially available double-ridged waveguide horn in the frequency range 1 GHz to 10 GHz.

Table 6.3

Freq. (GHz)	*HPBW in H* Plane	*Effective Aperture* D_e	*Far-Field Dist. (m) Using* $2D_e^2/\lambda$	*Far-Field Dist. (m) Using* $2D^2/\lambda$
1	72	0.242	0.389	0.523
2	62	0.141	0.262	1.045
3	65	0.089	0.159	1.568
L	43	0.081	0.218	2.613
10	38	0.046	0.139	5.226

using the physical diagonal of the horn of 28 cm is shown in column 5 of Table 6.3. It can be seen that, using these distances, the horn would have to be placed much farther away (especially at the higher frequencies), with a subsequent reduction in the dynamic range of the whole measurement system. For instance, at 10 GHz the distance would be 5.2m instead of 13.9 cm!

REFERENCES

[1] Dyson, J. D., "The Equiangular Spiral Antenna," *IRE Trans. on Antennas and Propagation*, April 1959, pp. 181–187.

[2] Dyson, J. D., "The Characteristics and Design of the Conical Log-Spiral Antenna," *IEEE Trans. on Antennas and Propagation*, July 1965, pp. 488–499.

[3] Dyson, J. D., "The Unidirectional Equiangular Spiral Antenna," *IRE Trans. on Antennas and Propagation*, October 1959, pp. 329–334.

[4] Kraus, J. D., *Antennas*, International Edition. New York: McGraw-Hill Book Company, 1988.

[5] Mayes, P. E., and J. D. Dyson. "A Note on the Difference Between Equiangular and Archimedes Spiral Antennas," *IRE Trans. on Microwaves Theory and Techniques*, Vol. 9, March 1961, pp. 203–204.

[6] Bawer, R., and J. J. Wolfe, "Rebuttal," *IRE Trans. on Microwaves Theory and Techniques*, Vol. 9, March 1961, pp. 204–205.

[7] Bawer, R., and J. J. Wolfe, "The Spiral Antenna," *IRE Int. Conv. Rec.*, No. 8, 1960, pp. 84–95.

[8] Johnson, R. C., and H. Jasik, *Antenna Engineering Handbook*, New York: McGraw-Hill Book Company, 1984.

[9] Rudge, A. W., K. Milne, A. D. Olver, and P. Knight, *The Handbook of Antenna Design, Vol.* 1, Stevenage, UK and New York: Peter Peregrinus Ltd., 1982.

[10] Wang, J. J. H., and V. K. Tripp, "Spiral Microstrip Antenna Suits EW/ECM Systems," *Microwaves and RF*, December 1993, pp. 147–150.

[11] Wang, J. J. H., and V. K. Tripp, "Design of Multioctave Spiral-Mode Microstrip Antennas," *IEEE Trans. on Antennas and Propagation*, Vol. 39, No. 3, March 1991, pp. 332–335.

[12] Marcuvitz, N., "Waveguide Handbook," *MIT Rad Lab Series, Volume* 10, New York: McGraw-Hill Book Company, 1957, pp. 66–72.

[13] Rappaport, T. S., "Wide-Band Test Antenna," *RF Design*, April 1988, pp. 37–41.

[14] Nail, J. J., "Designing a Discone Antenna," *Electronics*, August 1953, pp. 167–169.

[15] Walton, K. L., and V. C. Sundberg, "Broadband Ridged Horn Design," *Microwave J.*, March 1964, pp. 96–101.

[16] Moreno, T., "Microwave Transmission Design Data," New York: Dover Publications, Inc., 1948.

Chapter 7
Calibration of Antennas

This chapter details the calibration of antennas by describing the measurement and calculation of gain and the calculation of antenna correction factors.

In the measurement of radiated emissions from a DUT or in-site surveys, antennas provide the interface between the free-space radiated wave and the guided waves (in the transmission line) whose characteristics are measured by the receiver. In susceptibility testing the antenna is the source of the radiated field and the antenna converts the guided wave from the transmission cable to the free-space wave used to test the DUT.

The field detected by the antenna depends on the spatial orientation (e.g., the direction in which it is pointing) as well as the polarization orientation of the antenna. An antenna (apart from an ideal isotropic one) can receive different proportions of the electric field depending on the direction in which it is pointing. Along its boresight (direction of maximum radiation when it is transmitting) it will receive the maximum signal. If a power density of 1 W/m^2 is present at a particular location and we put an antenna with a linear power gain of 12 at this position, then we will actually measure 12 W/m^2. Thus we must know that the linear power gain is 12 in order to deduce that the true power density present is only 1 W/m^2.

We can either measure or calculate the gain of antennas. Manufacturers usually quote the gains at spot frequencies only and these may not coincide with the measurement frequencies. The measurement may entail the comparison of the antenna's gain with that of a standard reference antenna. These reference antennas are usually standard gain horns, at frequencies above about 1 GHz. At lower frequencies the reference antennas are usually resonant dipoles.

The antenna detects the electric or magnetic field, but the receiver or analyzer measures the voltage across its input terminals. The signal from the antenna suffers losses between the antenna and the receiver, and thus the conversion between electric field, for example, and the voltage is not a simple matter of relating them by the well-known formula of $\mathbf{E} = V/h_e$ (where h_e is the effective height of the antenna). A correction factor, which is known as the antenna correction factor (or commonly just called the antenna factor), has to be applied in order to relate the voltage measured by the receiver to the electric field at the antenna location.

7.1 GAIN

First let us consider the difference between directivity and gain. The directivity or directive gain is the comparison between a lossless antenna and a fictitious isotropic antenna; whereas the gain (sometimes called the effective gain) is the comparison between an actual antenna and a particular reference antenna [1, p. 34–21]. The directivity in the case of a transmitting antenna is defined as

$$\text{directivity} = \frac{\text{maximum radiation intensity from a lossless antenna}}{\text{average radiation intensity from an isotropic antenna}} \tag{7.1}$$

The maximum radiation intensity is the maximum power per unit solid angle, and the average radiation intensity from an isotropic antenna is the same as would be obtained if the radiation from an antenna is averaged over a sphere. A similar definition applies to a receiving antenna since an antenna is a reciprocal device and has the same properties for transmitting as it has for receiving. The directivity does not take into account the heating losses in the antenna, and reflection losses due to the impedance mismatch between the antenna and the generator (or receiver). The gain is defined as the product of the directivity and the efficiency and is given by

$$G = \eta D \tag{7.2}$$

where G is the gain in linear terms or numeric gain, η is the efficiency as a percentage or fraction, and D is the directivity in linear terms. The gain and directivity are usually quoted in decibels, in which case the following relation applies

$$g_{\text{dB}} = D_{\text{dB}} + \eta_{\text{dB}} \tag{7.3}$$

where the subscript dB is used to denote the quantities in decibels. Note that since the efficiency is always less than one, η_{dB} is negative.

The efficiency depends on the ohmic (heating) losses in the antenna and the losses due to impedance mismatch between the antenna and the transmission line feeding it. Efficiencies of antennas vary between about 50% and 90%.

Although strictly speaking the gain and directivity could apply to any angular position, in practice these terms are only used for boresight or maximum values, unless otherwise specified.

7.1.1 Measurement of Gain

The gain of an antenna can be measured using one, two, or three antennas. The single-antenna method requires a smooth perfectly reflecting metal sheet, and the two-antenna method requires two identical antennas. The gain of antennas used as standards are measured, and these standard gain antennas are then compared with the antennas used for testing. These standard gain antennas are not used in test measurements, thus ensuring that they are not damaged. Standard gain antennas are usually smooth-walled pyramidal rectangular horns fed by standard rectangular waveguides. The gain of a pyramidal horn depends mainly on the flare angle, the length of the flare, and the frequency. At lower frequencies, where the horns are too large, resonant dipoles are used and the gains of unknown antennas are measured by substitution methods.

7.1.1.1 Friis's Transmission Formula

Let us consider the two antennas depicted in Figure 7.1 at a distance R apart. The antenna A is transmitting, and the antenna B is the receive antenna.

It can be shown that the effective area A_r presented by a receiving antenna to a plane wave [2, Chapter 6] is given by

$$A_r = \frac{G_r \lambda^2}{4\pi} \tag{7.4}$$

where A_r is the area of the antenna in square meters, G_r is the linear (or numeric) gain of the antenna, and λ is the wavelength of the radiation in meters. If the transmitting antenna has a gain of G_t and is radiating a power of P_t watts, then the power density P_d (the power per unit area) at a distance R from the receiving antenna is given by

$$P_d = \frac{P_t G_t}{4\pi R^2} \tag{7.5}$$

The power P_r received by the antenna B is the power density times the area A_r that it presents to the incident wave and is given by

$$P_r = P_d A_r = \frac{P_t G_t}{4\pi R^2} \frac{G_r \lambda^2}{4\pi} \tag{7.6}$$

This equation can be rewritten as

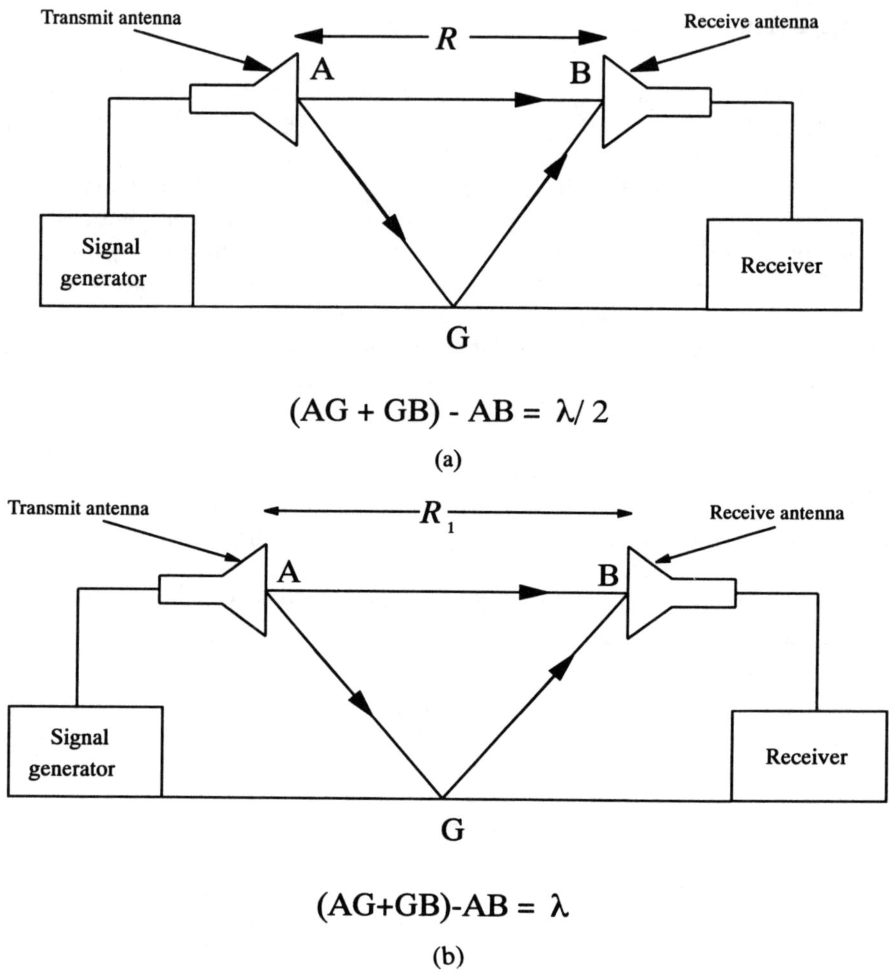

Figure 7.1 Measurement of gain using two antennas: (a) condition for destructive interference, and (b) condition for constructive interference.

$$P_r = P_t G_r G_t \left(\frac{\lambda}{4\pi R} \right)^2 \tag{7.7}$$

It can also be written in decibels as

$$10 \log_{10} P_r = 10 \log_{10} P_t + 10 \log_{10} G_r + 10 \log_{10} G_t + 20 \log_{10} \left(\frac{\lambda}{4\pi R} \right) \tag{7.8}$$

Equation (7.7) is known as Friis's transmission formula. The last term in (7.8) is dependent on the wavelength of the propagating frequency and on the distance between the antennas. It is known as the space attenuation or loss factor L_e. It is always negative in the far field ($R > \lambda/\pi$ for wire antennas or $> 2D^2/\lambda$ for aperture antennas, where D is the largest dimension of the antenna) but is usually written as a positive quantity since it is taken to represent a positive loss rather than a negative gain. Figure 7.2 shows the variation of this space loss or attenuation in decibel L_s versus distance, with frequency as a parameter. The space loss between antennas placed 1λ apart is approximately 22 dB.

The power P_t transmitted by the transmit antenna is not the same as the power provided by the generator P_g feeding the antenna. There is a loss of power between the antenna and the generator, which depends on the impedance mismatch between the two. This loss of power can be quantified by measuring the complex reflection coefficient of each individually. The complex reflection coefficient of each should be measured to the

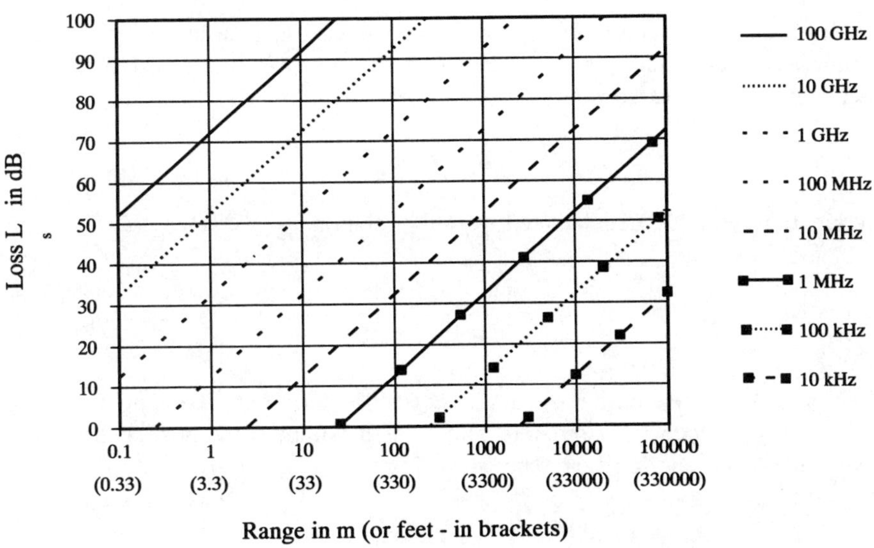

Figure 7.2 Space attenuation versus distance with frequency as a parameter.

same reference plane and a correction term M_t applied to the power output of the generator. The complex reflection coefficient gives the phase as well as the magnitude of the wave reflected by an impedance that is not matched to the characteristic impedance of the system. It can be measured over the whole frequency band of interest by a vector network analyzer such as a Hewlett-Packard 8510 or a Wiltron 360. The magnitude of the correction term [3, p. 649] is given by

$$M_t = \frac{(1 - |\tau_g|^2)\,(1 - |\tau_t|^2)}{|1 - \tau_g \tau_t|^2} \tag{7.9}$$

where τ_g and τ_t are the complex reflection coefficients of the generator and transmit antenna respectively to the same reference plane. The magnitude (modulus value) of the reflection coefficient is related to the magnitude of the *voltage standing wave ratio* (VSWR) S by the formula

$$|\tau| = \frac{|S - 1|}{|S + 1|} \tag{7.10}$$

The power transmitted by the antenna P_t and that supplied by the generator P_g is given by the relation

$$P_g = P_t M_t \tag{7.11}$$

where M_t is defined by (7.9). If a power meter is used to measure the power from the generator, a correction term has to be applied to the power P_m measured by it. The power P_m is related to the true transmitted power P_g by the relation

$$P_g = P_m M_m \tag{7.12}$$

The correction term M_m is calculated from the relation

$$M_m = \frac{(1 - |\tau_m|^2)\,(1 - |\tau_t|^2)}{|1 - \tau_m\, \tau_t|^2} \tag{7.13}$$

The received power also requires a similar correction term M_s to be applied to the power measured by the receiver in order to allow for the impedance mismatch between the receive antenna and the receiver. The correction term is given by

$$M_s = \frac{(1 - |\tau_r|^2)\,(1 - |\tau_s|^2)}{|1 - \tau_r \tau_s|^2} \tag{7.14}$$

where τ_r and τ_s are the complex reflection coefficients of the receive antenna and the receiver, respectively, at the same reference plane.

The correction terms M_t and M_s can be replaced by a combined correction term C, which is in decibels and is given by

$$C = 10 \log_{10} M_t + 10 \log_{10} M_s \tag{7.15}$$

If the transmit and receive antennas do not have the same polarization, then a correction for the polarization mismatch must be used in addition to the correction term of (7.15). If the polarization mismatch is P_{tr} in linear terms, then the correction term C_p in decibels is given by

$$C_p = 10 \log_{10} P_{tr} \tag{7.16}$$

Other factors to be taken into consideration are the errors due to near-field effects and multipath effects. The near-field effects are due to: (1) the fact that the near-field components may not have decreased to negligible values at the far-field distance of $2D^2/\lambda$ or λ/π. The magnitude of this correction term is of the order of 0.05 dB. And (2) here is a variation of the electric field across the aperture of the receive antenna. This is known as an amplitude taper and a variation of 0.25 dB can give rise to an error of about 0.1 dB in the magnitude of the gain measurement.

Thus the total errors due to these near-field or proximity effects are of the order of 0.15 dB [3, p. 650].

The multipath effects are caused by interference between the direct wave (from the transmit to receive antennas) and the waves that arrive at the receive antenna after reflection. Depending on the relative phases of the waves, constructive or destructive interference can result. When the antennas are separated by less than $2D^2/\lambda$ or λ/π the multipath interference is primarily due to multiple reflection between the two antennas, whereas at greater distances the multipath interference is mainly the result of reflections from the ground and other objects. The presence of multipath interference can be investigated by varying the distance between the antennas. If multipath effects are present, then the receiver connected to the receive antenna will show an output that resembles a sine wave as the distance between the antennas is increased (or decreased). This is due to the phase differences between the path lengths of the direct and reflected waves; when this path difference is an odd multiple of half wavelengths the receiver's output has its minimum value, and when this path difference is a whole number of wavelengths the output has its maximum value. The direct wave (AB) and a single wave that has undergone one reflection (AG + GB) is depicted in Figure 7.1. When the distance is R, this difference is $\lambda/2$, as shown in Figure 7.1(a), and destructive interference occurs. When this distance is increased to R_1 the difference is λ as shown in Figure 7.1(b), and constructive interference occurs. In practice, there will be more than one reflected wave, and some reflected waves will undergo more than one reflection. The measurement also requires that both the transmit and receive antennas be maintained at boresight (i.e., the positions of maximum radiation of the two antennas are collinear).

7.1.2 Purcell's Method

This method is sometimes called the mirror method and involves the use of just one antenna to perform the calibration. The set-up consists of a transmitting antenna such as a pyramidal horn (as shown in Figure 7.3) placed in front of a square metal reflecting sheet as shown in Figure 7.4, and the power is received by the same antenna.

The sheet reflects power P_m that can be assumed to come from an ''image'' antenna placed as far behind the sheet as the real antenna is in front of it. The normal far-field distance criterion of $R = 2D^2/\lambda$ should be maintained between the antenna and its ''image'' so that the distance between the antenna and the sheet is $R/2 = D^2/\lambda$, where D is the largest dimension of the horn. The sheet should be flat to at least one-sixteenth of a wavelength (see Figure 7.4(c)), and it must be planar in order to avoid errors in the position and dimensions of the image antenna. The sheet should be large so that diffraction effects at its edges are negligible, and its side h should also be large enough to intercept the main beam [4, p. 585] of width 2θ (= $2\lambda/D$) radians between nulls. This results in the following condition, which is shown in Figure 7.4(b).

$$\frac{h/2}{R/2} \geq \frac{\lambda}{D} \tag{7.17}$$

where h is the length of one side of the sheet, R is the distance between the horn and its image, and D is the largest dimension of the horn. This gives the lower limit for h of

$$h \geq \frac{R\lambda}{D} \tag{7.18}$$

For single-frequency measurements using a pyramidal horn, the set-up of Figure 7.4(a) applies. The horn is connected to the signal generator via a *standing wave detector* (SWD) and a stub tuner. When the transmitted and reflected waves interfere, a standing wave results. The ratio of the maximum-to-minimum voltage of the standing wave is the *voltage standing wave ratio* (VSWR). The SWD consists of a probe that is inserted through a longitudinal slot along the center of the broadwall of the rectangular waveguide. The probe can be moved along the longitudinal axis of the waveguide, and its depth into the waveguide can be varied. The probe is connected to a detector (usually a diode), and the detector output is measured on a meter. The probe carriage is moved longitudinally along the axis of the waveguide to record the maximum and minimum values of the standing wave in the waveguide. If there is a very small standing wave, then the ratio of

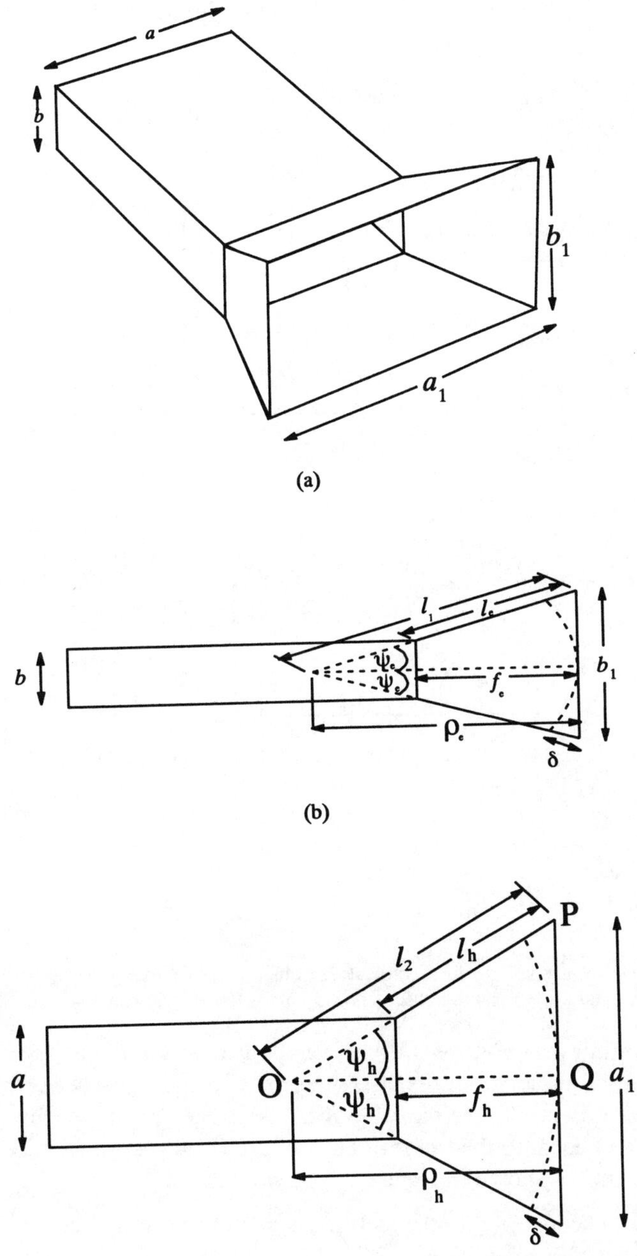

Figure 7.3 Pyramidal horn: (a) perspective view, (b) side elevation—*E* plane, and (c) plan view—*H* plane.

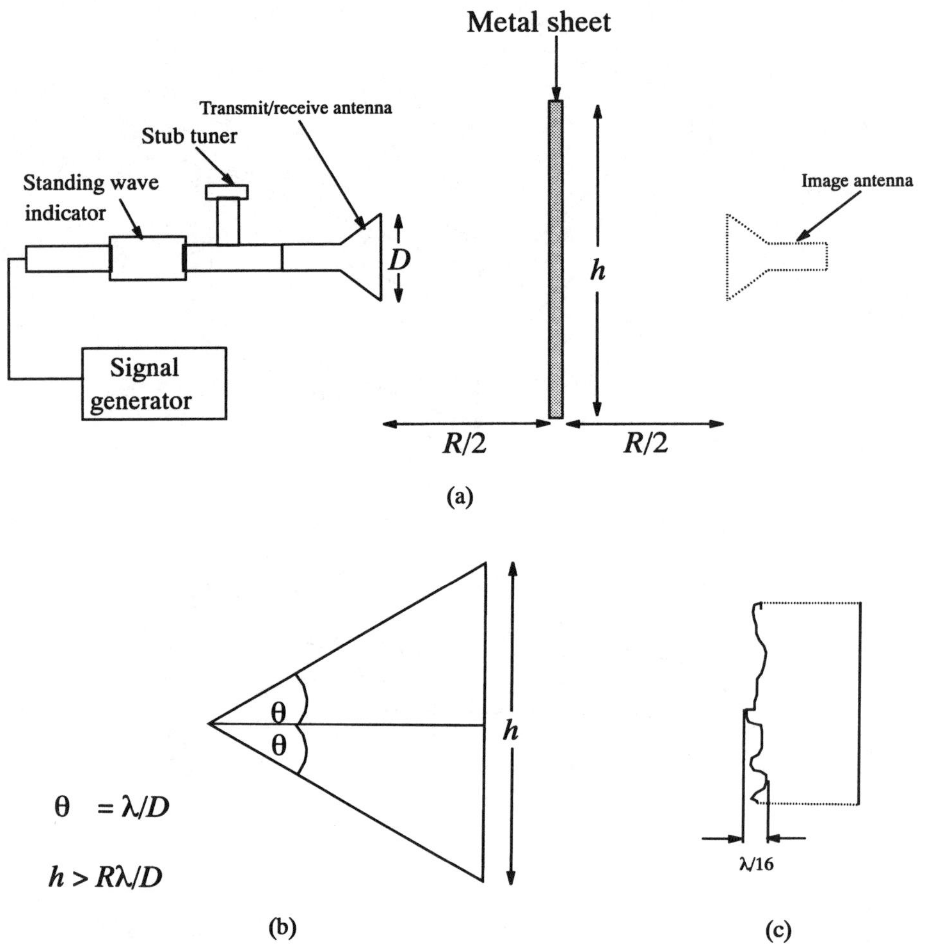

Figure 7.4 Purcell's method for measurement of gain: (a) measurement set-up for Purcell's method, (b) minimum size for side of sheet, and (c) enlarged view showing smoothness of sheet.

maximum-to-minimum voltage will be small, and this ratio is one when there is no reflected wave. This condition exists when the waveguide is perfectly matched. When there is a large reflected wave the VSWR is also large and approaches infinity when the reflected wave is equal to the transmitted wave. The procedure for measuring the gain of a pyramidal horn, as shown in Figure 7.3, is now given.

1. The horn is connected to the stub tuner, SWD, and signal generator and allowed to radiate into free space.
2. The horn is matched to the waveguide run by adjusting the stub tuner (usually a three stub tuner) so that the SWD indicates that the VSWR is very small.

3. The metal sheet is then placed at a distance d (= $R/2$) in front of the horn such that d is $\geq D^2/\lambda$.
4. The VSWR is measured using the SWD. In Friis's transmission formula the distance R between the transmitting and receive horns is now the distance between the horn and its image. The distance between the horn and the metal sheet in Figure 7.4 is d and thus the distance between the transmit horn and its image is $2d$. Thus, $R = 2d$ and (7.7) can be rewritten as

$$P_r = P_t G_r G_t \left(\frac{\lambda}{8\pi d}\right)^2 \tag{7.19}$$

Since the same horn is used for transmitting and receiving, the gains of G_r and G_t can both be replaced by the common value of G. Equation (7.19) thus becomes

$$P_r = P_t G^2 \left(\frac{\lambda}{8\pi d}\right)^2 \tag{7.20}$$

The square root of the ratio of P_r to P_t is the magnitude of the reflection coefficient $|\tau|$ and is given by

$$\sqrt{\frac{P_r}{P_t}} = |\tau| = \frac{G\lambda}{8\pi d} \tag{7.21}$$

which can be rewritten in the form

$$\frac{1}{|\tau|} = \frac{8\pi d}{G\lambda} \tag{7.21b}$$

The magnitude of the reflection coefficient τ is related to the VSWR S by the equation

$$|\tau| = \frac{S-1}{S+1} \tag{7.22}$$

The distance d is varied, the VSWR is measured in each case, and the magnitude of the reflection coefficient is calculated. A plot of $|\tau|$ against $1/d$ is shown in Figure 7.5. The plot should give us a straight line, and the slope yields the value of the constant ($8\pi/G\lambda$) on the right-hand side of (7.21b), from which G can be calculated at each value of λ. In practice, a straight line is not obtained because of the edge effects, measurement errors, and the fact that the transmitting horn reradiates some of the power reflected from the sheet. With the measured results shown in Figure 7.5, the best straight line gives a slope of approximately 18. In this case the frequency at which the measurements were performed was 10 GHz, so $\lambda = 0.03$m. The value of linear gain G calculated is thus 45.6, which is equal to a gain of 16.7 dB.

In the case of swept frequency measurements the SWD and meter are replaced by a dual directional coupler and connected to a network analyzer. The return loss (from which the VSWR can be calculated) is measured over the whole frequency range of interest, but the principle of the measurement is the same as in the single frequency measurements described above.

7.1.3 Two Antenna Method

Friis's formula of (7.7) is used to determine the gain of antennas by employing two identical antennas. The Friis formula can be rewritten for the two antennas as

$$G = \sqrt{\left[\frac{P_t}{P_r}\left(\frac{4\pi R}{\lambda}\right)^2\right]} \tag{7.23}$$

where G is the linear (or numeric) gain of either antenna, P_r is the power transmitted by one antenna, and

P_t is the power received by the other antenna. Since the gain is usually required in

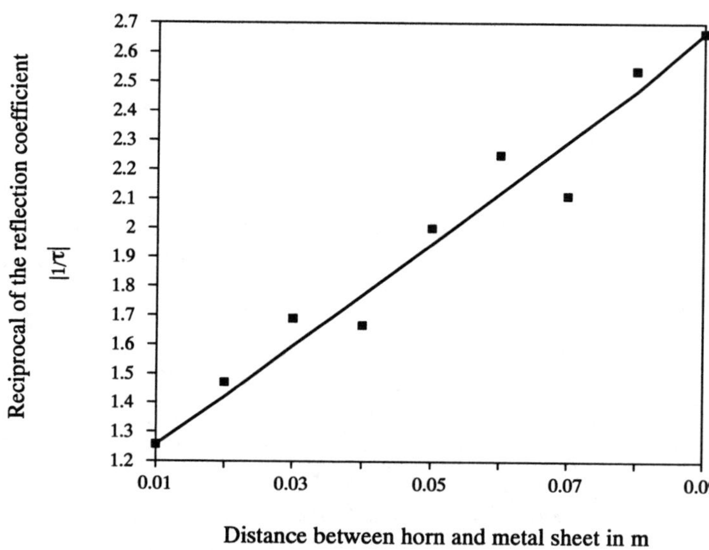

Figure 7.5 Purcell's method measured results.

decibels and the transmitted and received powers are also usually measured in decibels, (7.23) is more conveniently given in decibels in the form

$$g_{dB} = \frac{1}{2}\left[20 \log_{10}\left(\frac{4\pi R}{\lambda}\right) - 10 \log_{10}\left(\frac{P_t}{P_r}\right)\right] \tag{7.24}$$

Alternatively (7.24)can be rewritten in the form

$$g_{dB} = \frac{1}{2}\left[20 \log_{10}\left(\frac{4\pi R}{\lambda}\right) + 10 \log_{10}\left(\frac{P_r}{P_t}\right)\right]$$

Note that the ratio of P_t to P_r is inverted, and the sign of its log is changed since

$$-\log_{10}\left(\frac{P_t}{P_r}\right) = +\log_{10}\left(\frac{P_r}{P_t}\right)$$

The received and transmitted powers P_r and P_t are measured, and the correction terms of (7.15) are applied to obtain the corrected values. The principle of the measurement set-up is shown in Figure 7.1, and the checks to ensure that multipath effects are minimized (described in section 7.1.1) should be undertaken. The distance R between the two antennas is varied to give several values for the received and transmitted powers at each frequency. The average value for the gain can then be calculated from (7.24).

7.1.4 Three Antenna Method

If two identical antennas are not available, then the gain measurements can be performed by using three antennas. The same measurements as described in Section 7.1.3 are performed with two of the antennas each time. Three sets of measurements will be required altogether to cover all the combinations. This will result in three simultaneous equations, which are given by

$$g_1 + g_2 = 20 \log_{10}\left(\frac{4\pi R}{\lambda}\right) + 10 \log_{10}\left(\frac{P_r}{P_t}\right)_{1,2} \tag{7.25}$$

with antenna 1 used for transmitting and antenna 2 used for receiving,

$$g_1 + g_3 = 20 \log_{10}\left(\frac{4\pi R}{\lambda}\right) + 10 \log_{10}\left(\frac{P_r}{P_t}\right)_{1,3} \tag{7.26}$$

with antenna 1 used for transmitting and antenna 3 used for receiving, and

$$g_2 + g_3 = 20 \log_{10}\left(\frac{4\pi R}{\lambda}\right) + 10 \log_{10}\left(\frac{P_r}{P_t}\right)_{2,3} \tag{7.27}$$

with antenna 2 used for transmitting and antenna 3 used for receiving. In (7.25) to (7.27), g_1, g_2, and g_3 are the gains in decibels of the first, second, and third antennas, respectively. The measurement involving each pair of antennas is the same as in the case of the two-antenna method of Section 7.1.3. The precautions and correction terms described in Section 7.1.1 also apply in this case.

7.2 CALCULATION OF GAIN

The rectangular pyramidal horn is the most commonly used standard horn; thus, the method of calculation of its gain alone is included here. A pyramidal horn, such as the one shown in Figure 7.3, is fed by a standard rectangular waveguide and the walls of the guide flare in both the E and H planes. The angle of flares and length of the flared section determine the final dimensions of the open mouth of the horn, which in turn determine the gain of the horn.

The gain of a horn depends on

1. the dimensions a and b of the rectangular waveguide feeding the horn;
2. the dimensions a_1 and b_1 of the open end of the horn;
3. the slant lengths l_e and l_h in E and H planes of the flared section;
4. the flare angles $2\psi_e$ and $2\psi_h$ in the E and H planes, respectively.

The formulas are given for calculating the total gain of the pyramidal horn. This gain is dependent on the individual gains G_e and G_h that would be obtained by having E- and H-plane sectoral horns with the same flare angles. As long as the aperture of the pyramidal horn in each plane is at least one wavelength, the pattern in one plane is substantially independent of the aperture in the other plane [5, p. 648]. In order to compute the gain of a pyramidal horn we may therefore calculate the gains of individual E- and H-plane sectoral horns, and the product of these linear (or numeric) gains together with the efficiency will give us the gain of the pyramidal horn.

7.2.1 *E*-Plane Sectoral Horn

In the E-plane sectoral horn depicted in Figure 7.6 we can see that the broad walls of the waveguide are flared in the direction parallel to the electric field vector, and the flare angle is $2\psi_e$. The narrow dimension of the guide is increased from b to b_1 at the open mouth of the horn, whereas the broad dimension a remains at the same value. The flared walls have

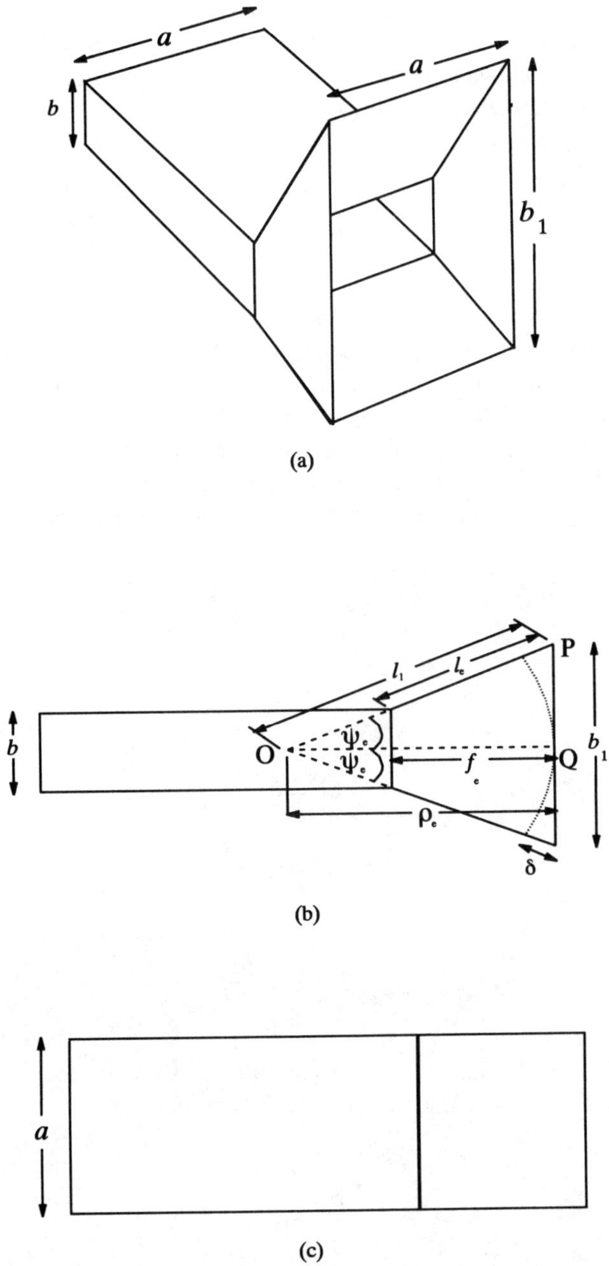

Figure 7.6 *E*-plane sectoral horn: (a) perspective view, (b) side elevation, and (c) plan view.

a virtual apex and the virtual lengths l_1 to the open end are longer than the length ρ_e (the virtual perpendicular distance) by δ. This difference causes the spherical waves emanating from the virtual apex to have a phase difference at the edges of the horn (compared with the wavefronts at the center of the horn).

This phase difference (Fermat's principle) is given by

$$\phi = \frac{2\pi\delta}{\lambda} \tag{7.28}$$

where ϕ is the phase difference in radians and λ is the wavelength. Both δ and λ are required to be in the same units. When the phase difference is π the waves from the edges are in antiphase with those from the center; hence, they destructively interfere, resulting in a decrease in the gain and an increase in the sidelobe level.

In order to obtain universal relationships between the phase deviation δ and the loss in gain associated with it, it is convenient to express δ in terms of wavelengths. Consider triangle OPQ of Figure 7.6(b). Using Pythagoras's theorem we can see that

$$\rho_e^2 = l_1^2 - \left(\frac{b_1}{2}\right)^2 \tag{7.29}$$

But since ρ_e is equal to l_1 - δ,

$$(l_1 - \delta)^2 = l_1^2 - \frac{b_1^2}{4} \tag{7.30}$$

$$l_1^2 - 2l_1\delta + \delta^2 = l_1^2 - \frac{b_1^2}{4}$$

$$2l_1\delta - \delta^2 = \frac{b_1^2}{4} \tag{7.31}$$

However since δ is much smaller than the length l_1, δ^2 is negligible compared with $2l_1\delta$ and can therefore be ignored. Equation (7.31) thus reduces to

$$\delta = \frac{b_1^2}{8l_1} \tag{7.32}$$

where b_1 is the height of the open mouth of the horn and l_1 is the virtual length of the flared section. The phase deviation or error T_e in terms of wavelength is given by

$$T_e = \frac{\delta}{\lambda} = \frac{b_1^2}{8l_1\lambda} \tag{7.33}$$

7.2.1.1 *Loss Due to Phase Error*

The efficiency of the horn is reduced by the phase error at the open mouth so that the effective aperture is only about 50% of the effective aperture. This means that efficiencies as low as 50% are to be expected. Thus, the gain could be only half the magnitude of the directivity or 3 dB less when they are both expressed in decibels.

Dielectric loading of the horn can be used to correct the phase error so that an uniform phase is obtained across the aperture. Efficiencies as high as 90% are achievable with zero-phase taper at the mouth of the horn.

The loss in gain due to the phase error is plotted in Figure 7.7 for an *E*-plane sectoral horn. The loss L_e (in decibels) is shown as a function of T_e; and T_e has been given in terms of the virtual slant length l_1 of the flare [1, p. 10–8] as

$$T_e = \frac{\delta}{\lambda} = \frac{(b_1)^2}{8\lambda l_1} \tag{7.34}$$

However, it is more useful to have T_e in terms of the physical slant length l_e of the flared section, which can be easily measured.

Referring to Figure 7.6(b) we can see that the flare semiangle ψ_e is given by

$$\sin \psi_e = \frac{b_1/2}{l_1} = \frac{(b_1 - b)/2}{l_e} \tag{7.35}$$

$$\frac{b_1}{l_1} = \frac{b_1 - b}{l_e}$$

Thus T_e can be rewritten in terms of l_e as

$$T_e = \frac{b_1(b_1 - b)}{8\lambda l_e} \tag{7.36}$$

The maximum gain of a horn is directly proportional to the size of its aperture. We can increase the aperture by having a large flare angle and/or by having a long flare length. However, in order to restrict the phase variation at the open end, we must have a long horn with a small flare angle. Whilst this can be attained at high (millimeter wave) frequencies with relatively small horns, the horns become impracticably large at lower frequencies. Thus some compromise must be sought between the flare angle and the length f_e of the

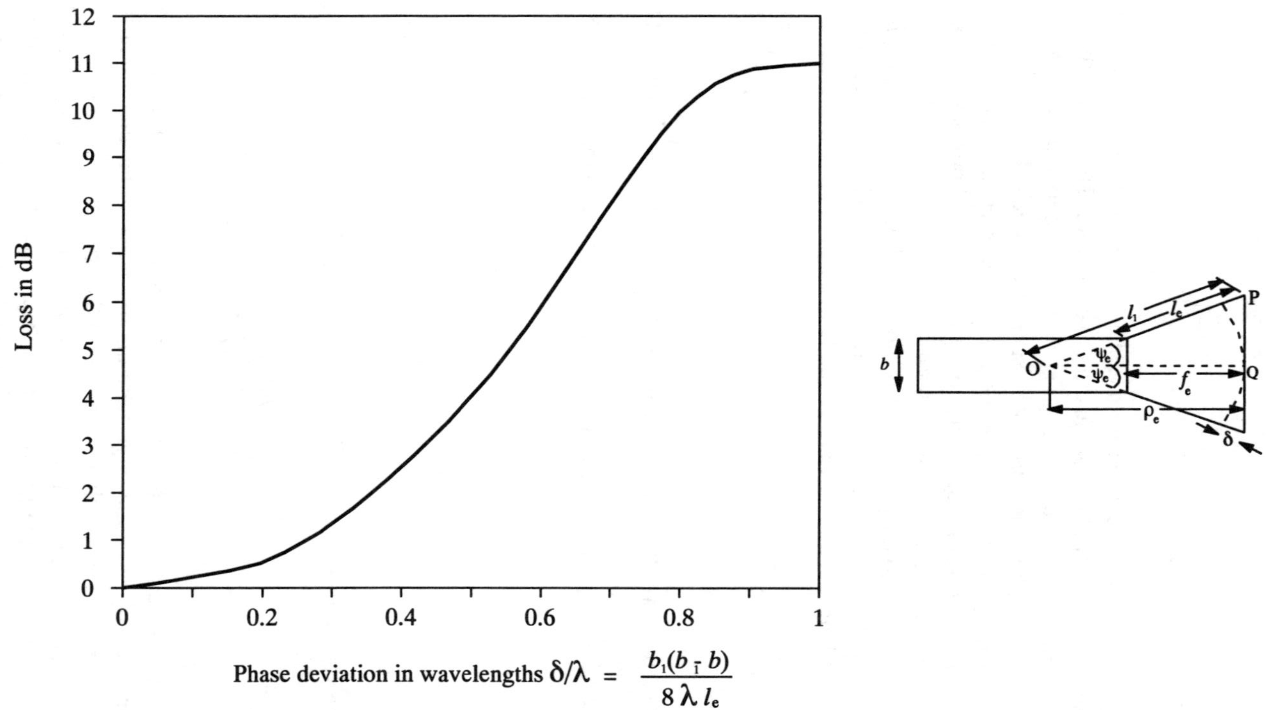

Figure 7.7 Variation of loss with phase deviation—*E*-plane sectoral horn.

flared section for a particular value of δ. We can see from Figure 7.6(b) that the relationship between the flare semiangle ψ_e and the virtual length l_1 is given by

$$\cos \psi_e = \frac{\rho_e}{l_1} \qquad \text{or} \qquad \cos \psi_e = \frac{\rho_e}{\rho_e + \delta}$$

$$\rho_e = (\rho_e + \delta)\cos \psi_e$$

$$\rho_e(1 - \cos \psi_e) = \delta \cos \psi_e$$

$$\rho_e = \frac{\delta \cos \psi_e}{(1 - \cos \psi_e)} \tag{7.37}$$

Thus for a specific value of δ we can relate the flare angle to the flare length ρ_e. The value of δ that can be tolerated depends on the type of sectoral horn. In the case of the *E*-plane horn the maximum value of δ is a quarter of a wavelength at the highest frequency of operation. This gives a phase difference of 90 deg or $\pi/2$ radians.

Figure 7.8 shows the variation of the virtual flare length ρ_e with flare semiangle for a phase error of 90 deg at 100 MHz, 1 GHz, and 10 GHz. Note that for a given flare semiangle the permissible maximum flare length is much smaller at high frequencies than at lower frequencies. For instance, for a flare semiangle of 30 deg the maximum permissible flare length at 10 GHz is only about 5 cm, whereas it is 4.8m at 100 MHz.

Since the virtual length ρ_e is not easy to measure, it is more convenient to derive the variation of the flare semiangle ψ_e in terms of the physical lengths l_e of the flared section and the heights b and b_1 (of the waveguide and horn, respectively). Referring to Figure 7.6(b) we can see that

$$\tan \psi_e = \frac{b_1}{2\rho_e} = \frac{b_1 - b}{2f_e}$$

$$\rho_e = \left(\frac{b_1}{b_1 - b}\right) f_e$$

and

$$f_e = \sqrt{l_e^2 - \left(\frac{b_1 - b}{2}\right)^2}$$

Thus

$$\rho_e = \left(\frac{b_1}{b_1 - b}\right) \sqrt{l_e^2 - \left(\frac{b_1 - b}{2}\right)^2} \tag{7.38}$$

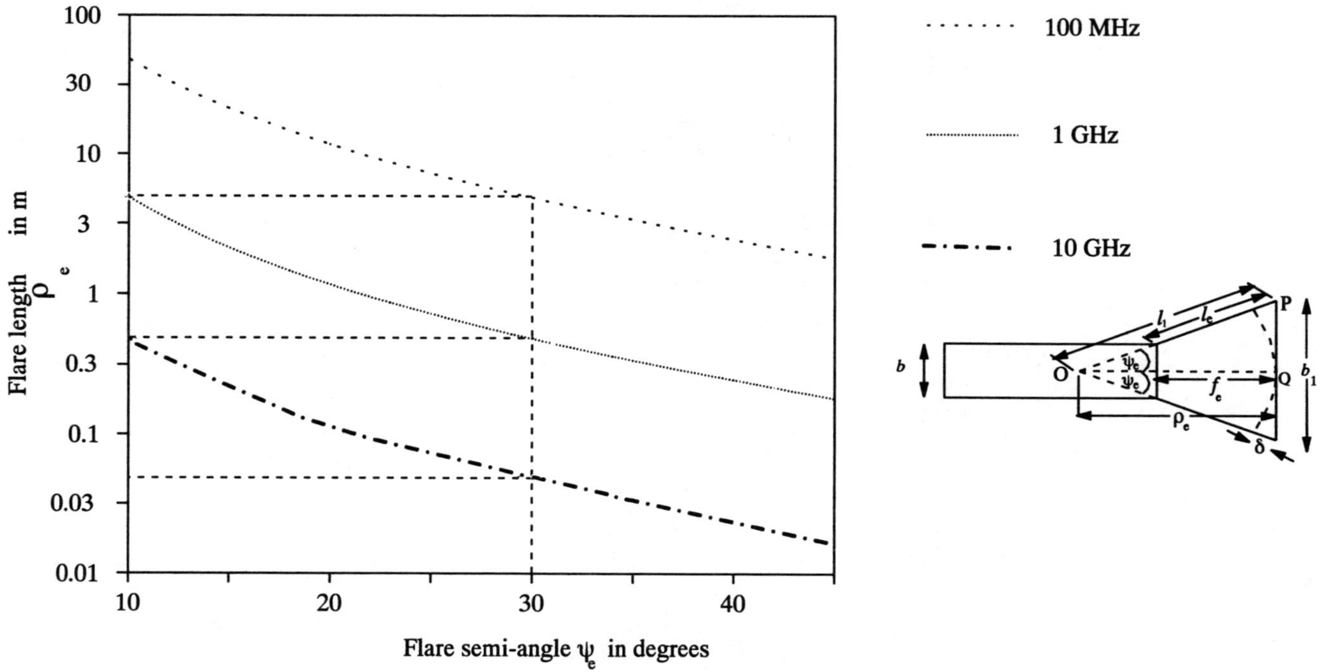

Figure 7.8 Variation of flare length with flare semiangle for an E-plane horn.

7.2.1.2 *Gain of an E-Plane Horn*

The numeric or linear gain G_e of an *E*-plane horn is given by the formula [9, p. 546]

$$G_e = \frac{64\, a\, \rho_e}{\pi \lambda b_1}\left[C^2\left(\frac{b_1}{\sqrt{2\lambda\rho_e}}\right) + S^2\left(\frac{b_1}{\sqrt{2\lambda\rho_e}}\right)\right] \tag{7.39}$$

where a is the width of the waveguide feeding the horn, ρ_e is the virtual length of the horn, b_1 is the height of the open mouth of the horn, C is the Fresnel cosine integral, and S is the Fresnel sine integral. The Fresnel cosine and sine integrals are defined by the relationships

$$C(x) = \int_0^x \cos\left(\frac{\pi}{2}x^2\right) dx$$

$$S(x) = \int_0^x \sin\left(\frac{\pi}{2}x^2\right) dx$$

The values of the Fresnel cosine and sine integrals are shown in Figure 7.9.

The gain of an *E*-plane horn has been calculated using (7.39) for different values of ρ_e to give the normalized curves of Figure 7.10, which show the gains divided by the

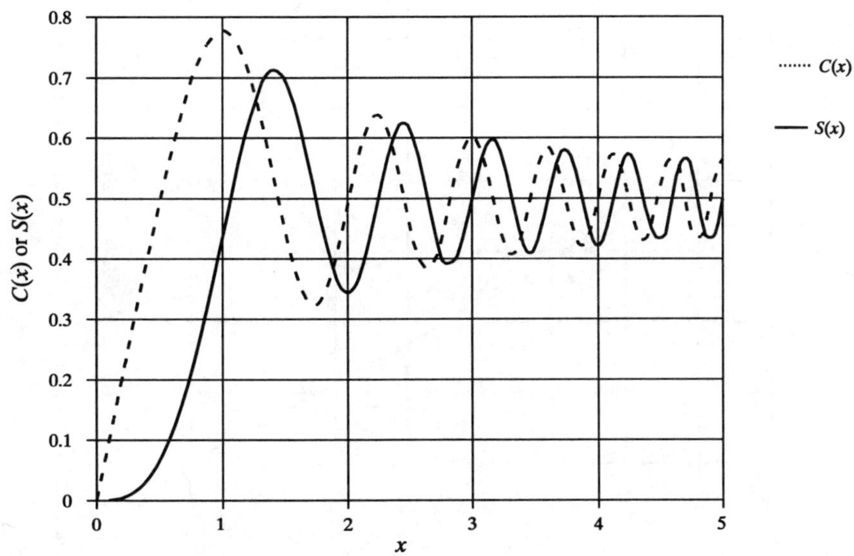

Figure 7.9 Fresnel cosine and sine integrals.

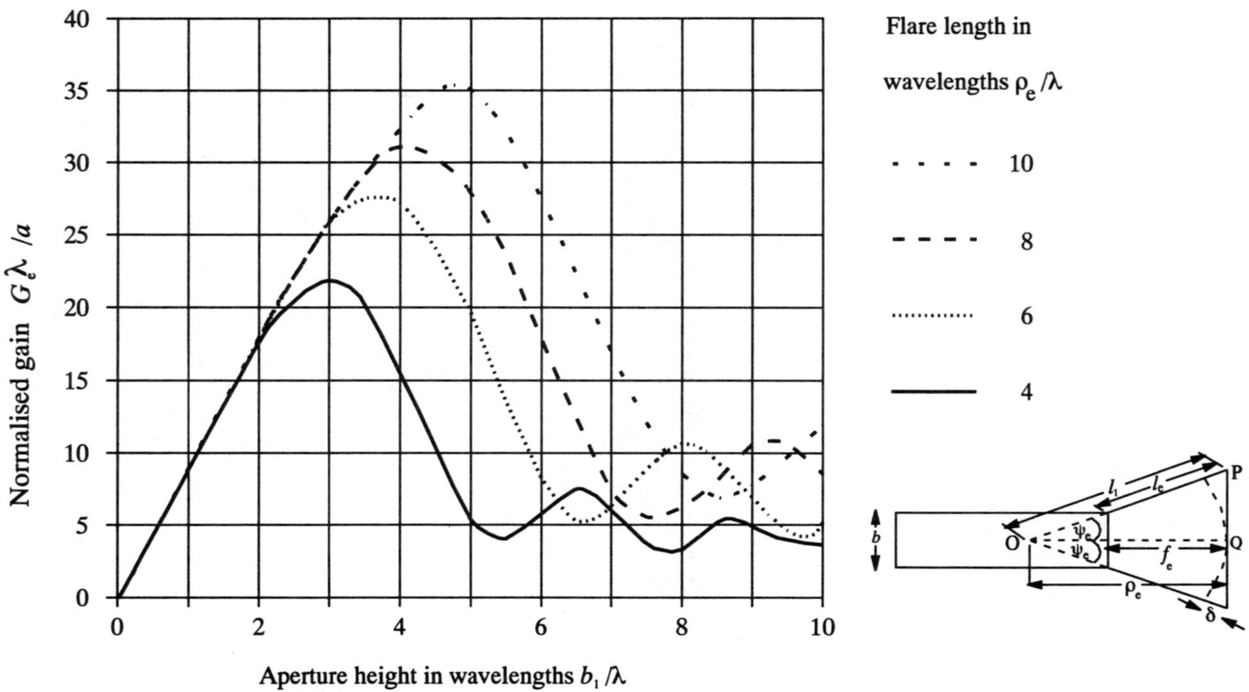

Figure 7.10 Normalized gain for an *E*-plane sectoral horn.

width a (in terms of the wavelength). It should be noted that the gains increase to a maximum in each case as b_1/λ is varied and then start to decrease again. Closer examinations of these curves show that the value of the phase deviation at these maxima is 0.25λ. For instance, in the case of a horn of length 8λ, the maximum occurs at $b_1/\lambda = 4$, resulting in a value of T_e equal to 0.25, and thus $\delta = 0.25\lambda$. This corresponds to a phase error of 90 deg or $\pi/2$ rad, which is the maximum phase error that can be tolerated for an E-plane horn.

7.2.2 *H*-Plane Sectoral Horn

In the H-plane sectoral horn depicted in Figure 7.11 we can see that the narrow walls of the waveguide are flared in the plane parallel to the magnetic field vector, and the flare angle is $2\psi_h$. The broad dimension of the guide is increased from a to a_1 at the open mouth of the horn, whereas the narrow dimension or height b remains at the same value. The flared walls have a virtual apex, and the virtual lengths l_2 to the open end are longer than the length ρ_h (the virtual perpendicular distance) by δ. This difference causes the spherical waves emanating from the virtual apex to have a phase difference at the edges of the horn (compared with the waves from the center of the horn). This phase difference results in a loss in gain as in the case of the E-plane horn described in Section 7.2.1 above; however, in this case a larger phase difference can be tolerated. Instead of 90 deg, a maximum phase difference of 135 deg can be tolerated for the H-plane sectoral horn when the electric field vector is perpendicular to the H-plane. This phase difference corresponds to a value of d equal to $3\lambda/8$.

The phase deviation or error in terms of wavelengths, T_h, can be derived in a manner similar to the case of the E-plane horn and is given by

$$T_h = \frac{\delta}{\lambda} = \frac{a_1^2}{8l_2\lambda} \tag{7.40}$$

where a_1 is the width of the aperture of the horn and l_2 is the virtual slant length of the flared section.

7.2.2.1 Loss Due to Phase Error

The efficiency of the horn is reduced by the phase error at the open mouth, and the effective aperture is only about 50% of the physical aperture. Thus the gain could be only half the magnitude of the directivity.

Dielectric loading of the horn can be used to correct the phase error so that an uniform phase is obtained across the aperture. Efficiencies as high as 90% are achievable with zero-phase taper at the mouth of the horn.

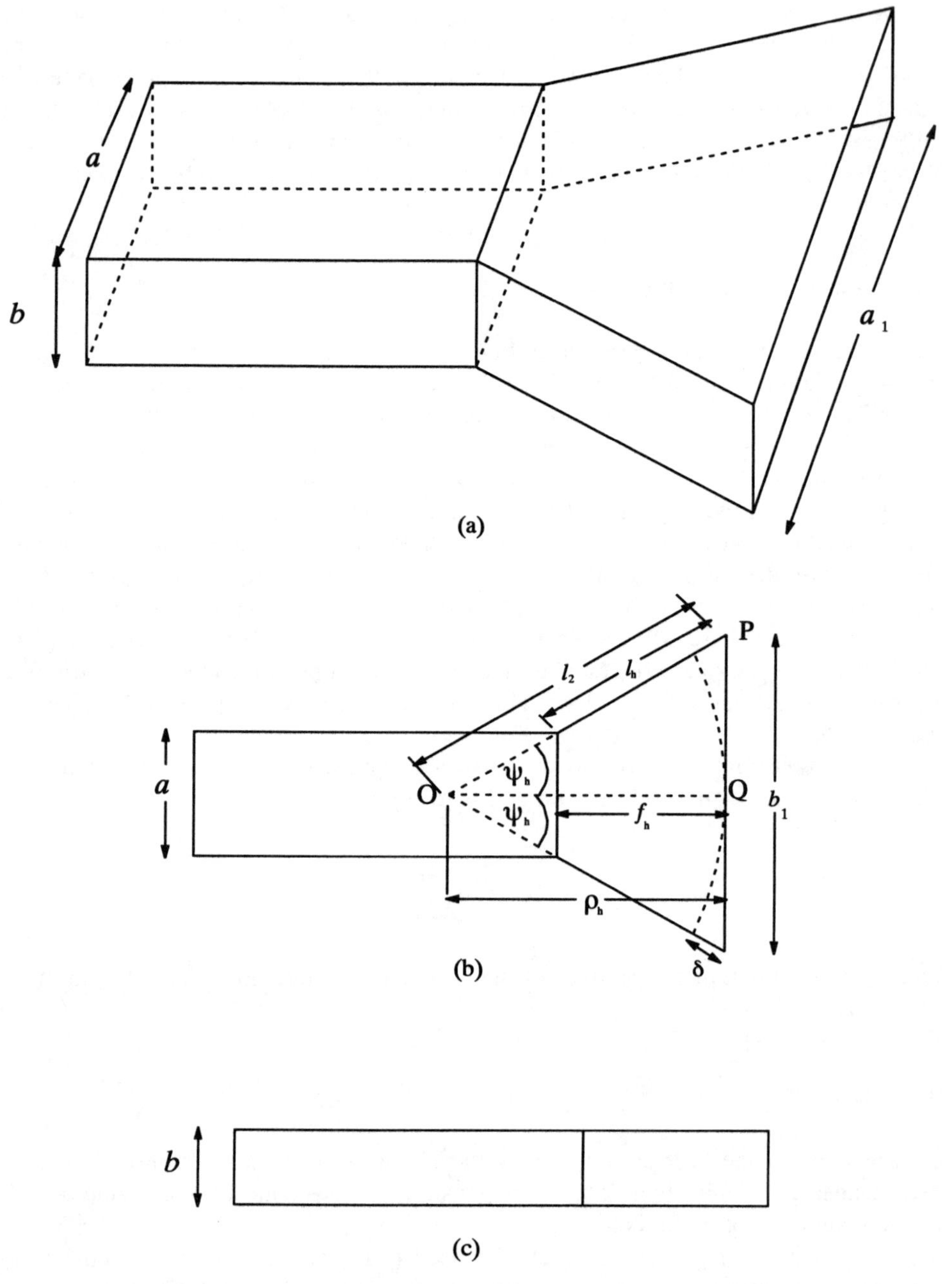

Figure 7.11 *H*-plane sectoral horn: (a) perspective view, (b) plan view, and (c) side view.

The loss in gain due to the phase error is plotted in Figure 7.12 for an *H*-plane sectoral horn. The loss L_h in decibels, against T_h, has been given in terms of the virtual slant length l_2 [1, p. 10–8], which is the virtual length of the flared section. However, it is more useful to have T_h in terms of the physical slant length l_h of the flared section, which can be easily measured.

Referring to Figure 7.11(b) we can see that the flare semiangle ψ_e is given by

$$\sin \psi_1 = \frac{a_1/2}{l_2} = \frac{(a_1 - a)/2}{l_h} \tag{7.41}$$

Thus T_h can be rewritten in terms of l_h as

$$T_h - \frac{a_1(a_1 - a)}{8\lambda l_h} \tag{7.42}$$

The maximum gain of a horn is directly proportional to the size of its aperture. We can increase the aperture by having a large flare angle and/or by having a long flare length. However, in order to restrict the phase variation at the open end, we must have a long horn with a small flare angle. Whilst this can be attained at high (millimeter wave) frequencies with relatively small horns, the horns become impracticably large at lower frequencies. Thus, some compromise must be sought between the flare angle and the length f_h of the flared section for a particular value of δ. We can see from Figure 7.11(b) that the relationship between the flare semiangle ψ_h and the virtual length l_2 is given by

$$\rho_h = (\rho_h + \delta)\cos \psi_h$$

$$\rho_h(1 - \cos \psi_h) = \delta \cos \psi_h \qquad \rho_h = \frac{\delta \cos \psi_h}{(1 - \cos \psi_h)} \tag{7.43}$$

where ψ_h is the flare semiangle in degrees and ρ_h is the virtual length of the flare. Thus for a specific value of δ we can relate the flare angle to the flare length l_2. The value of δ that can be tolerated depends on the type of sectoral horn. In the case of the *H*-plane horn the maximum value of δ is three-eighths of a wavelength at the highest frequency of operation. This gives a phase difference of 135 deg or $3\pi/4$ rad.

Figure 7.13 shows the variation of the virtual flare length ρ_h with flare semiangle for a phase error of 135 deg at 100 MHz, 1 GHz, and 10 GHz. Note that for a given flare semiangle the maximum permissible flare length is much smaller at high frequencies than at lower frequencies. For instance, for a flare semiangle of 30 deg the maximum permissible flare length at 10 GHz is only about 7.3 cm, whereas it is 7.3m at 100 MHz. However, these lengths are about 67% larger than those for an *E*-plane sectoral horn with

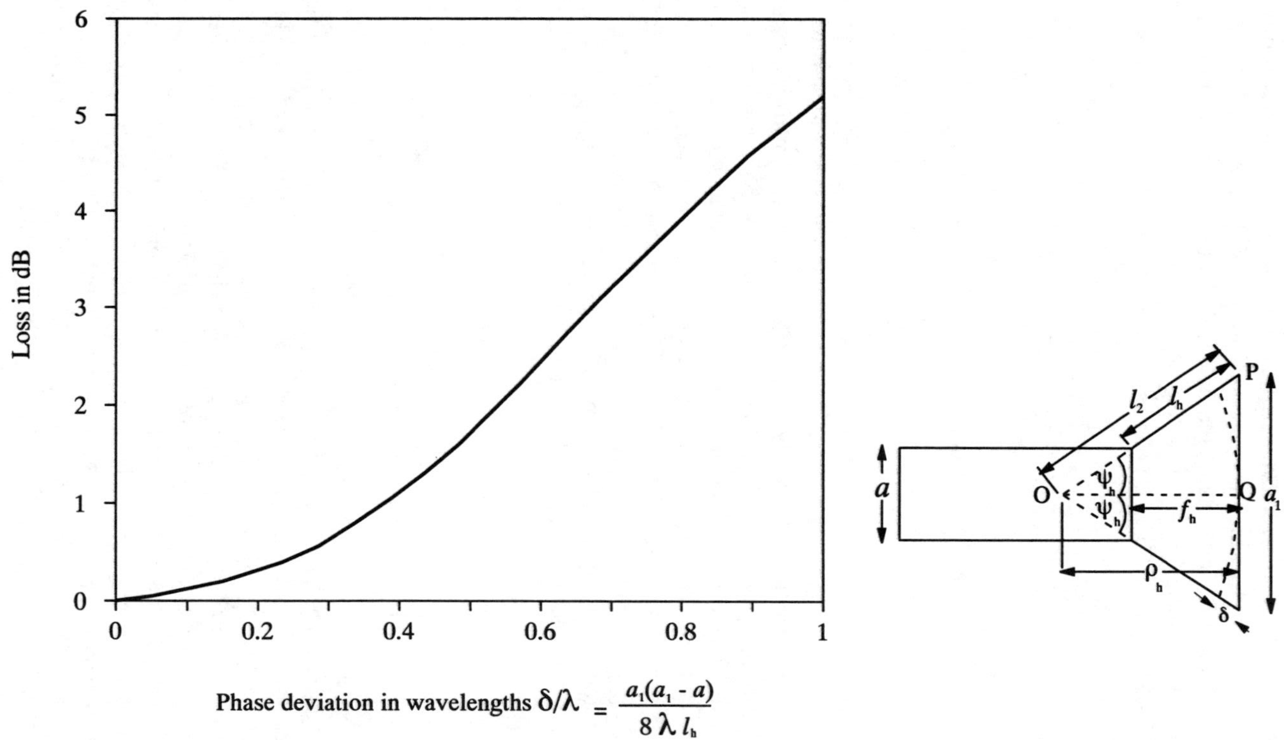

Figure 7.12 Variation of loss with phase deviation—*H*-plane sectoral horn.

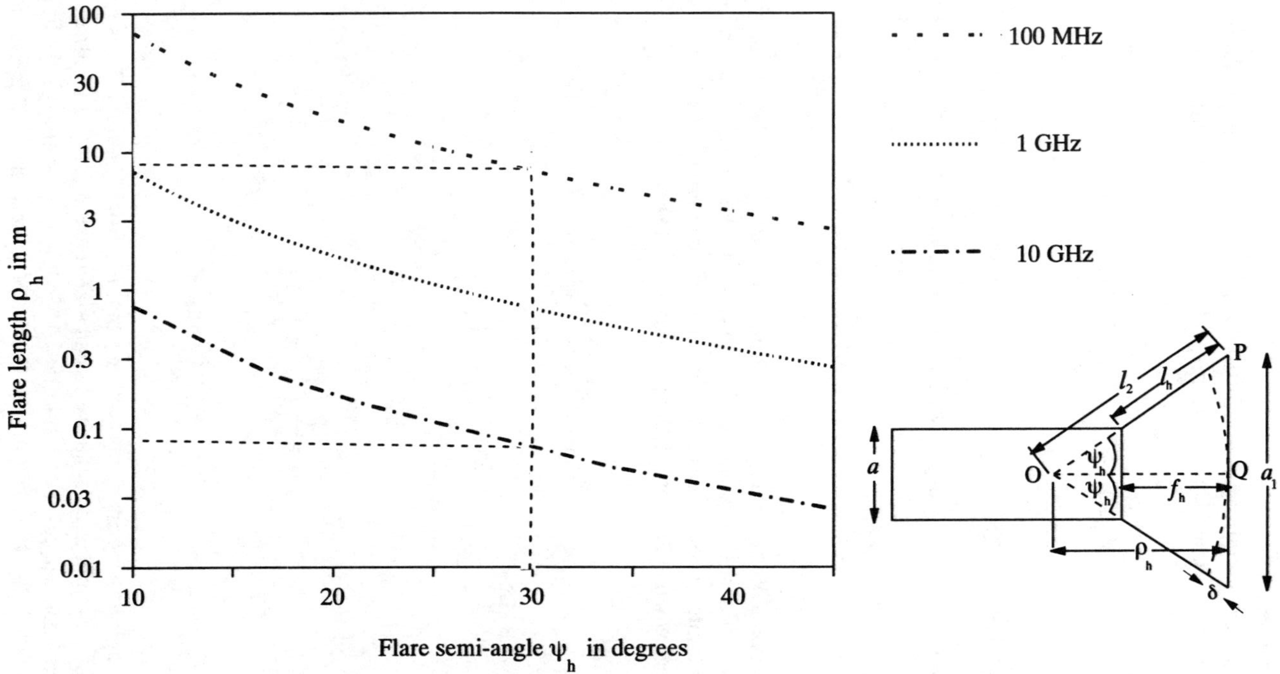

Figure 7.13 Variation of flare length with flare semiangle for an *H*-plane horn.

the same flare angles. Thus, a longer flare length can be tolerated in the H-plane horn than in the E-plane sectoral horn.

The virtual flare length ρ_h is related to the physical flare length l_h, and the widths a and a_1 of the waveguide and the horn, respectively, by the expression

$$\rho_h = \left(\frac{a_1}{a_1 - a}\right) \sqrt{(l_h)^2 - \left(\frac{a_1 - a}{2}\right)^2} \tag{7.44}$$

7.2.2.2 Gain of an H-Plane Sectoral Horn

The linear or numeric gain G_h of an H-plane sectoral horn [9, p. 561] is given by

$$G_h = \frac{4\pi b \rho_h}{a_1 \lambda} \{[C(u) - C(v)]^2 + [S(u) - S(v)]^2\} \tag{7.45}$$

where b is the height of the waveguide feeding the horn, ρ_h is the virtual length of the horn, a_1 is the width of the horn aperture, $C(u)$ is the Fresnel cosine integral of u, $C(v)$ is the Fresnel cosine integral of v, $S(u)$ is the Fresnel sine integral of u, and $S(v)$ is the Fresnel sine integral of v, where the Fresnel cosine and sine integrals are given by

$$C(x) = \int_0^x \cos\left(\frac{\pi}{2}x^2\right) dx$$

$$S(x) = \int_0^x \sin\left(\frac{\pi}{2}x^2\right) dx$$

and u and v are given by

$$u = \frac{1}{\sqrt{2}}\left(\frac{\sqrt{\lambda\rho_h}}{a_1} + \frac{a_1}{\sqrt{\lambda\rho_h}}\right)$$

$$v = \frac{1}{\sqrt{2}}\left(\frac{\sqrt{\lambda\rho_h}}{a_1} - \frac{a_1}{\sqrt{\lambda\rho_h}}\right)$$

The Fresnel cosine and sine integrals can be obtained from Figure 7.9.

The gain of an H-plane horn has been calculated using (7.45) for different values of ρ_h to give the normalized curves of Figure 7.14, which show the gains divided by the width b (in terms of the wavelength). It should be noted that the gains increase to a maximum in each case as a_1/λ is varied and then start to decrease again. Closer exami-

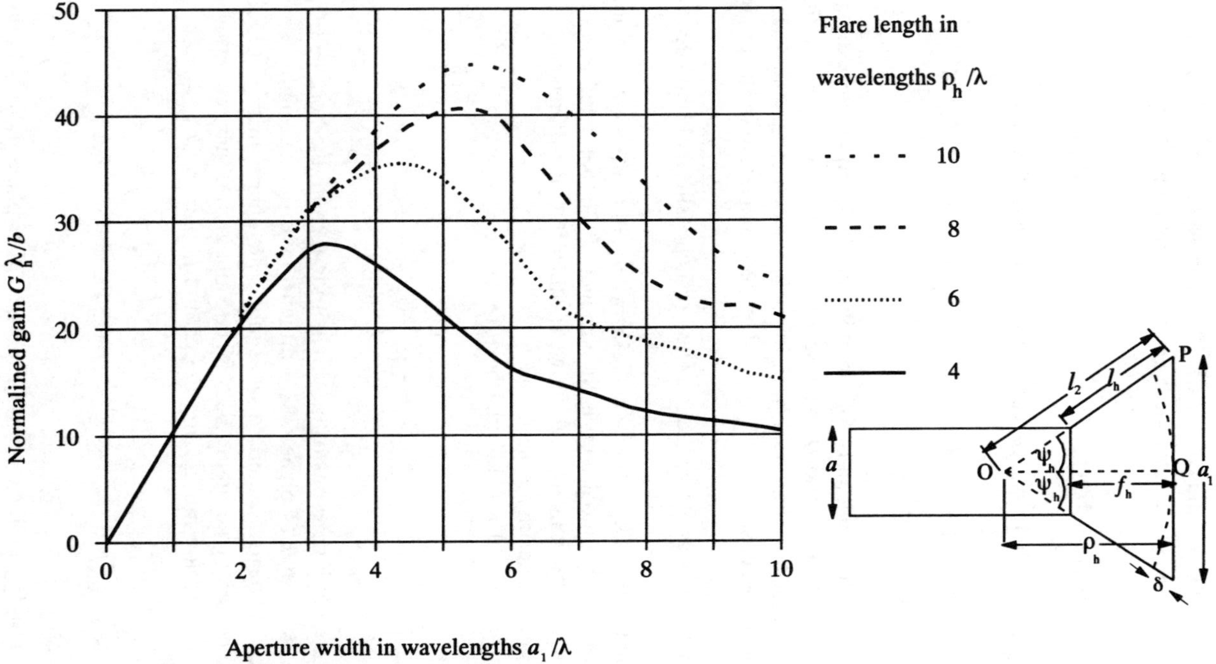

Figure 7.14 Normalized gain for an *H*-plane sectoral horn.

nations of these curves show that the value of the phase deviation at these maxima is $3\lambda/8$. For instance, in the case of a horn of length 8λ, the maximum occurs at $a_1/\lambda \simeq 5$, resulting in a value of $T = 0.390$, and thus $\delta \simeq 3\lambda/8$. This corresponds to a phase error of 135 deg or $3\pi/4$ rad, which is the maximum phase error that can be tolerated for an H-plane horn.

7.2.3 Gain of a Pyramidal Horn

The gain of a pyramidal horn can be calculated in three different ways depending on the degree of accuracy required.

7.2.3.1 Accurate Calculation

This method involves the calculation of the gains in the E and H planes by assuming that their gains are the same as E- and H-plane sectoral horns with equivalent dimensions. As long as the aperture of the pyramidal horn in each plane is at least one wavelength, the pattern in one plane is substantially independent of the aperture in the other plane [5, p. 648].

The E-plane gain is given by (7.39). The Fresnel cosine and sine integrals are shown in Figure 7.9.

The gain in the H plane is given by (7.45) where all terms are as previously defined.

The gain of the pyramidal horn is the product of the individual gains of the E- and H-plane horns divided by a factor that allows for the effect the dimensions of the feeding waveguide have on the resultant gain.

The linear gain G_p of a pyramidal horn [8, p. 16.8] is given by

$$G_p = \frac{\pi\lambda^2}{32\,ab} G_e G_h \tag{7.46}$$

where a and b are the dimensions of the waveguide feeding the horn and G_e, G_h are the linear gains in the E and H planes. The dimensions in wavelengths for the midband frequency in the case of most standard waveguides are approximately 0.75 and 0.375 for a/λ and b/λ, respectively (see Figure 6.25). The value of $\pi/32$ is approximately 0.098, so the value of the expression multiplying the E- and H-plane gains is given by

$$\frac{\pi\lambda^2}{32ab} = 0.349 \tag{7.47}$$

Thus we must multiply the linear gains of the E- and H-plane sectoral horns by 0.349 to obtain the gain for a pyramidal horn at the midband frequency.

In decibels the expression given by (7.47) is equal to -4.57. Thus, after adding the gains in decibels for the E- and H-plane sectoral horns, we must subtract 4.57 dB from the result.

The gain in decibels g_p for a pyramidal horn is given by

$$g_p = g_e + g_h - 4.57 \tag{7.48}$$

where g_e and g_h are the gains in decibels for the E- and H-plane sectoral horns.

7.2.3.2 Semiaccurate Method

In this case the E- and H-plane gains (divided by a and b in wavelengths, respectively) are obtained from Figures 7.10 and 7.14 for given values of ρ_e and ρ_h. For intermediate values of ρ_e and ρ_h the values of the gains can be interpolated. The linear gain of the pyramidal horn G_p can calculated from the formula

$$G_p = \frac{\pi}{32}\left(G_e\frac{\lambda}{a}\right)\left(G_h\frac{\lambda}{b}\right) \tag{7.49}$$

where a is the width of the waveguide feeding the horn and b is the height of the waveguide feeding the horn. The values for the expressions in brackets are obtained directly from the curves of Figures 7.10 and 7.14, and multiplying the product by the $\pi/32$ gives the gain of the pyramidal horn.

Note that (7.49) is exactly the same as (7.46). If we want to express the gains in decibels, then the gain of a pyramidal horn g_p is given by

$$g_p = 10 \log_{10} (G_e\lambda/a) + 10 \log_{10} (G_h\lambda/b) - 10.08 \tag{7.50}$$

since $10 \log_{10}(\pi/32)$ is equal to -10.08.

7.2.3.3 Approximate Method

The gain g_p of a pyramidal horn in decibels can be calculated [9, p. 570] from the approximate formula

$$g_p = 10 \log\left(\frac{32\, a_1 b_1}{\pi\lambda^2}\right) - L_e - L_h \tag{7.51}$$

where a_1 is the width of the horn aperture, b_1 is the height of the horn aperture, L_e is the loss in decibels for an E-plane horn, and L_h is the loss in decibels for an H-plane horn. The values of L_e and L_h can be obtained from Figures 7.7 and 7.12, respectively.

7.3 EXAMPLE

A standard gain horn has aperture dimensions of 16.5 cm and 7.25 cm, and the lengths of the flare in the *E* and *H* planes are 17.75 cm and 17 cm, respectively. The dimensions of the waveguide feeding the horn are 2.29 cm and 1 cm. Calculate the gain at 10 GHz by the three methods described above.

Solution. The lengths in the *E* and *H* planes are the those denoted by l_e and l_h in Figures 7.6 and 7.11. In order to use the formulas in Section 7.2.3 we first must calculate the values of ρ_e and ρ_h. We can calculate these values using (7.38) and (7.44).

Using (7.38) and inserting the above values for b_1, b, and l_e we get ρ_e = 18.9 cm. Similarly, using (7.44) we get a value for ρ_h of 18.9 cm.

We should expect ρ_e and ρ_h to have the same values, since the two lengths are coincident and represent the perpendicular distance from the virtual apex of the horn to its open end.

We shall now apply the three methods of calculation described above. We need to calculate the dimensions in terms of wavelengths.

At 10 GHz, λ = 3 cm; b_1 = 16.5 cm, giving b_1/λ = 5.5; a_1 = 8.25 cm, giving a_1/λ = 2.75; l_e = 17.75 cm, giving l_e/λ = 5.92; and l_h = 17 cm, giving l_e/λ = 5.67.

7.3.1 Accurate Method

7.3.1.1 E-*Plane Gain*

The value of $b_1/\sqrt{2\lambda\rho_e}$ is calculated as 0.7994. The Fresnel cosine and sine integrals for 0.8 can be read from Figure 7.9 as 0.72 and 0.25, respectively. After squaring these we should insert them into formula (7.39) to yield a calculated value of the *E*-plane gain of 19.56.

7.3.1.2 H-*Plane Gain*

The values of *u* and *v* must first be calculated using the formulas

$$u = \frac{1}{\sqrt{2}}\left(\frac{\sqrt{\lambda\rho_h}}{a_1} + \frac{a_1}{\sqrt{\lambda\rho_h}}\right)$$

$$v = \frac{1}{\sqrt{2}}\left(\frac{\sqrt{\lambda\rho_h}}{a_1} - \frac{a_1}{\sqrt{\lambda\rho_h}}\right)$$

These formulas yield values of u = 1.87 and v = 1.23. The cosine and sine integrals for u can be read from Figure 7.9 as 0.36 and 0.41, respectively; and those for v are 0.72 and 0.63, respectively. Inserting these values into (7.45) yields a calculated value of the H-plane, G_h, of 10.6.

The linear gain G_p of the pyramidal horn can now be calculated from (7.46).

The value of the fraction, $\pi\lambda^2/32ab$, is 0.385, and multiplying it by the values for G_e and G_h (19.56 and 10.6, respectively) gives a result of 79.83. In decibels we therefore have g_p = 19.02 dB.

7.3.2 Semiaccurate Method

The magnitude of ρ_e/λ(and ρ_h/λ) is 6.31, but the nearest values to this are the 6λ curves of Figures 7.10 and 7.14. From Figure 7.10 at $b_1/\lambda = 2.75$, we have a value of $G_e a/\lambda = 5$. In decibels this equals 13.98.

From Figure 7.14 at $a_1/\lambda = 5.5$, we have a value of $G_h b/\lambda = 31$. In decibels this is equal to 14.91.

The linear gain G_p of the pyramidal horn is given by (7.49). Inserting the values for G_e and G_h in (7.49) gives a value G_p = 76.1, which gives a value of g_p = 18.8 dB.

7.3.3 Approximate Method

The approximate gain g_p in decibels can be calculated from (7.51). In order to find the values of L_e and L_h from Figures 7.7 and 7.12 we have to calculate the values of T_e and T_h from (7.36) and (7.42), respectively. Inserting the values for b, b_1, λ, and l_e in (7.36) gives a value of T_e = 0.147. From Figure 7.7 we get a value of 0.03 dB for L_e. Inserting the values for a, a_1, λ, and l_h in (7.42) gives a value of T_h = 0.589. From Figure 7.12 we get a value of 2.2 dB for L_h.

The value of the first term in (7.51) is 21.87 dB. Inserting this value and the values for L_e and L_h into (7.51) gives a resultant value of g_p = 19.37 dB.

Thus we can see that the values for g_p (19.02 dB, 18.8 dB, and 19.37 dB) obtained by the three methods only differ by about 0.6 dB.

7.4 ANTENNA CORRECTION FACTOR

This is commonly known as the antenna factor. It is a factor that is used to convert the magnitude of the voltage (in V, mV, μV, dBV, dBmV, or dBμV) obtained at the terminals of the receiver to the electric field intensity present at the position of the antenna. Readers not familiar with this topic should refer to Section 2.2.8 first. Thus the electric field E at the antenna is given by

$$E = V \times (AF) \tag{7.52}$$

where E is in volts per meter, V is in volts, and AF is the antenna factor. If we examine (7.52) dimensionally we can see that since E is in volts per meter and V is in volts, the antenna factor AF must be in m^{-1}. In a 50-Ω system AF can be taken as approximately given by the relation [10]

$$AF = \frac{2.76}{\lambda\sqrt{G}} \tag{7.53}$$

where G is the linear or numeric gain and λ is the wavelength at the frequency of operation.

The Society of Automotive Engineers (SAE) also defines an antenna factor (AF_2), which is called the apparent gain. This is the difference obtained in the receiver reading when (1) the transmit and receive antennas are placed at least one meter apart and (2) the antennas are disconnected and the cables from the signal generator and receiver are connected directly together. Thus the losses in the balun and tuning or matching networks are measured.

In order to calculate the antenna factor [6] we should take into account (1) the losses in the balun and matching network and (2) the losses due to the mismatch between the antenna and receiver.

The power density is defined as the vector cross product (see Chapter 3) of the electric and magnetic field vectors (***E*** and ***H***). In the far field the ***E*** and ***H*** fields are perpendicular to each other and to the direction of propagation. Hence, the magnitude of the power density is the product of the magnitudes of the ***E*** and ***H*** fields, and is given by

$$P_d = E \times H \tag{7.54}$$

where P_d is in watts per square meter, E is the electric field intensity in volts per meter, and H is the magnetic field intensity in amperes per meter. The ratio of E to H is known as the intrinsic impedance of the wave ξ_0 and, in the case of a plane wave in free space, has a value of 120π. Thus (7.54) can be written as

$$P_d = E^2/\xi_0 \tag{7.55}$$

If the effective area of the antenna is A, then the power P in watts available at the antenna is given by

$$P = \frac{E^2 A}{\xi_0} \tag{7.56}$$

If a receiver of impedance Z_r is connected to the antenna, then by Thevenin's theorem the antenna is represented by an EMF V and the current I is given by

$$I = \frac{V}{(Z_a + Z_r)} \tag{7.57}$$

When the antenna and receiver are purely resistive (of resistances R_a and R_r) the power transferred is given by

$$P = I^2 R_r = \frac{V^2 R_r}{(R_a + R_r)^2} \tag{7.58}$$

Figure 7.15 shows that if the receiver resistance R_r is varied between values of $0.2R_a$ to $2.4R_a$ the power transferred for an input voltage of 1 V varies between 0.002W and 0.00348W for an antenna resistance of 72Ω. (This is the approximate resistance of a resonant dipole; see Section 5.4). The maximum power of 0.0035Ω is transferred when the receiver resistance R_r is the same as that of the antenna R_a. The receiver is considered to be matched when this condition exists. The magnitude of the maximum power P_{max} given by (7.58) can rewritten as

$$P_{max} = \frac{V^2}{4R_r} \tag{7.59}$$

since $R_r = R_a$.

If the effective height of the antenna is h_e, then (since the voltage V is equal to the product of the electric field E and the effective height), the maximum power available at the antenna is given by

$$P_{max} = \frac{E^2 h_e^2}{4R_a} \tag{7.60}$$

This assumes that the antenna is directly connected to the receiver and that they both have the same resistance. However, this is not the case in practice since the antenna has a higher resistance than the connecting cable and receiver. A dipole, for instance, has a resistance of 72/73Ω whereas the receiver and connecting cable normally have resistances of 50Ω. A balun is also required to convert the balanced output of the dipole to the unbalanced coaxial cable. In order to transfer the maximum power a *balun and matching* (B/M) network is used to provide a conjugate match to the antenna. There are conjugate mismatch losses associated with this type of matching as well as dissipative or ohmic losses in the B/M network and connecting cables. The correct antenna factor should take into account all these power losses. The total of these losses [6] can be represented by the ratio

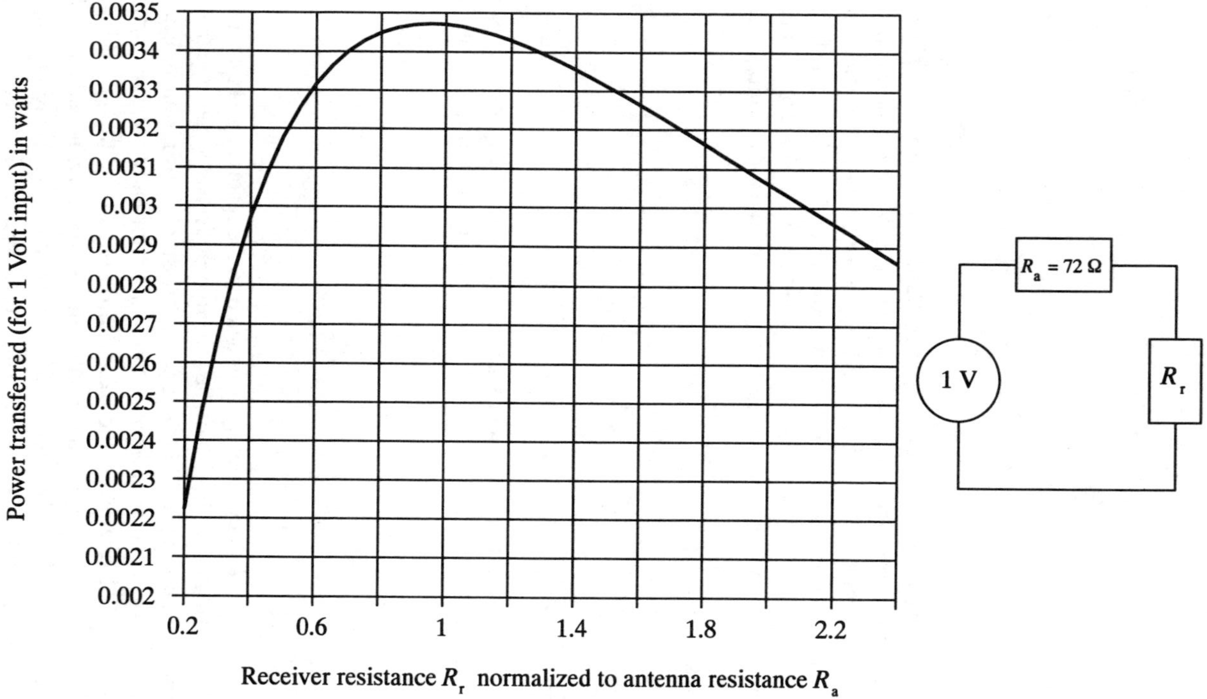

Figure 7.15 Power transfer between an antenna and a receiver.

$$\frac{\text{power available from the antenna}}{\text{power delivered to the receiver}} = M_{ab}^2 \, K_{bu}^2 \, M_{ur}^2 \, e^{2\alpha l} \tag{7.61}$$

where

$$M_{ab}^2 = \frac{\text{power available from the antenna}}{\text{power delivered to the B/M network}}$$

$$K_{bu}^2 = \frac{\text{power delivered to the B/M network}}{\text{power available from the B/M network}}$$

$$M_{ur}^2 = \frac{\text{power available from the B/M network}}{\text{power delivered to the receiver cable}}$$

$$e^{2\alpha l} = \frac{\text{power delivered to the receiver cable}}{\text{power delivered to the receiver input}}$$

α is the attenuation factor of the cable, and l is the length of cable from the unbalanced output of the B/M network to the input of the receiver. Note that the M_{ab}^2 and M_{ur}^2 are not the same as the correction terms of Section 7.1.1.

Referring to Figure 7.16, which shows the simplified equivalent circuit of the measuring system, the maximum power available from the antenna to the B/M network (assuming the B/M network is matched to the antenna) is given by

$$P_{\max} = \frac{V^2}{4R_a} \tag{7.62}$$

where V is the voltage due the electric field E and h_e (the effective height of the antenna) and R_a is the resistance of the antenna.

The power transferred to the B/M network is given by

$$P_{\max} = \frac{V^2 R_b}{|Z + Z_b|^2} \tag{7.63}$$

Where Z_b is the impedance presented to the balun and matching network, and R_b is the resistance.

Combining (7.62) and (7.63) gives the expression for M_{ab}^2 as

$$M_{ab}^2 = \frac{V^2}{4R_a} \frac{|Z_a + Z_b|^2}{V^2 \, R_b} \tag{7.64}$$

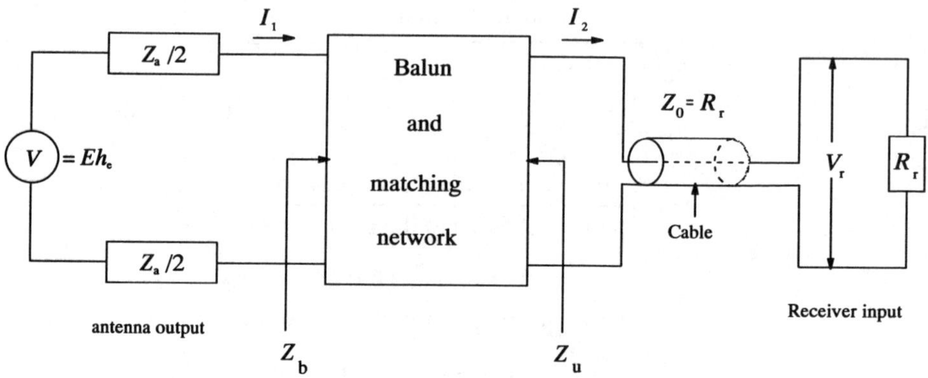

Figure 7.16 Simplified equivalent circuit of the measurement system.

which reduces to

$$M_{\mathrm{ab}} = \frac{|Z_{\mathrm{a}} + Z_{\mathrm{b}}|^2}{4R_{\mathrm{a}}\, R_{\mathrm{b}}} \tag{7.65}$$

Similarly the expression for M_{ur}^2 can be given by

$$M_{\mathrm{ur}}^2 = \frac{|Z_{\mathrm{u}} + R_{\mathrm{r}}|^2}{4R_{\mathrm{a}}\, R_{\mathrm{r}}} \tag{7.66}$$

The expressions for M_{ab}^2 and M_{ur}^2 represent the conjugate mismatch losses at the input and output of the B/M network [6].

The equivalent circuit of the B/M network is shown in Figure 7.17. It can be shown that

$$Z_{\mathrm{b}} = Z_{11} - \frac{Z_{12}^2}{Z_{11} + R_{\mathrm{r}}} \tag{7.67}$$

and

$$Z_{\mathrm{u}} = Z_{22} - \frac{Z_{12}^2}{Z_{11} + Z_{\mathrm{a}}} \tag{7.68}$$

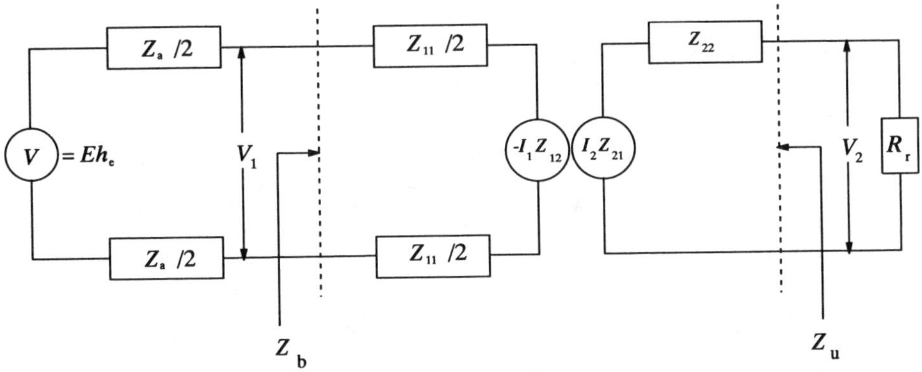

Figure 7.17 Open-circuit two-port network equivalent of the balun and matching network, with source and load connected.

The expression for K_{bu}^2 is given by

$$K_{bu}^2 = \frac{4R_b R_u}{|Z_{12}|^2} \frac{|(Z_{11} + Z_a)\,(Z_{22} + R_r)|^2}{|(Z_{11} + Z_a)\,(Z_{22} + R_r) - Z_{12}^2|^2} \tag{7.69}$$

The maximum power available at the antenna is given by (7.60) and the maximum power delivered to the receiver is given by

$$P_{max} = \frac{V_r^2}{R_r}$$

where V_r is the voltage at the receiver input and R_r is the resistance of the receiver. Thus the ratio of the power available at the antenna to the power delivered to the receiver is given by

$$\frac{\text{power available at antenna}}{\text{power delivered to the receiver}} = \frac{E^2 h_e^2 R_r}{4R_a\, V_r^2} \tag{7.70}$$

Equating (7.61) and (7.70) gives

$$\frac{E^2 h_e^2\, R_r}{4\, R_a\, V_r^2} = M_{ab}^2\, M_{ur}^2\, K_{bu}^2\, e^{2\alpha l}$$

Table 7.1
Comparison Between Calculated and Published Antenna Factors for the Schwarzbeck Dipole

Frequency (MHz)	$2/h_e$ (dB)	$(R_a/R_{ar})^{1/2}$ (dB)	M_{ab} (dB)	K_{bu} (dB)	M_{ur} (dB)	$e^{\alpha l}$	$e^{\alpha l}$ (dB)	E/V_r Calculated	E/V_r Published
30	−4.04	1.65	0.35	9.62	0	1.0040	0.017	7.60	7.45
100	6.42	1.65	0.35	9.62	0	1.0069	0.026	18.07	17.94
300	15.9	1.65	0.35	9.62	0	1.0134	0.058	27.64	27.35
1000	26.42	1.65	0.35	9.62	0	1.029	0.124	38.46	38.03

After [6].

from which we get the ratio of E to V_r, that is, the antenna factor as

$$\frac{E}{V_r} = \frac{2}{h_e} \sqrt{\frac{R_a}{R_r}} \quad M_{ab}\, M_{ur}\, K_{bu}\, e^{\alpha l} \tag{7.71}$$

The values for the antenna factor at four different frequencies was calculated [6] using (7.71) in the case of the Schwarzbeck dipole [7]. It was assumed that the length of the cable connecting the balun to the output connector was 0.5m, and the cable type was RG 214/U. The loss associated with this type of cable is given by

$$e^{2\alpha l} = (0.0015F + 1.11)^{0.03}$$

where F is the frequency in MHz. The calculated individual components as well as the resultant antenna factors are listed in Table 7.1 and compared with the manufacturer's published data [7]. It can be seen that there is very good agreement between the calculated and measured values published by the manufacturer.

REFERENCES

[1] Jasik, H., *Antenna Engineering Handbook*, New York: McGraw-Hill, 1961.

[2] Slater, J. C., *Microwave Transmission*, New York: McGraw-Hill, New York, 1942, Chapter 6.

[3] Rudge, A. W., K. Milne, A. D. Olver, and P. Knight, *The Handbook of Antenna Design, Vol.* I, Stevenage, UK and New York: Peter Peregrinus Ltd., Chapter 8, pp. 647–651.

[4] Silver, S., *Microwave Antenna Theory and Design*, New York: McGraw-Hill, 1947.

[5] Krauss, J. D., *Antennas*, New York: McGraw-Hill, 1988.

[6] Bennett, W. S., "Properly Applied Antenna Factors," *IEEE Trans. on EMC*, Vol. 28, No. 1, February 1986, pp. 2–6.

[7] Schwarzbeck, C., "Antennenfaktoren u Korrekturen Praz-Dipole VHAP-10dB," *Dipl-Ing Mess-Elektronik*, D-6901 Schonau-Alteneudorf, W. Germany 1981.

[8] Schelkunoff, S. A., and H. T. Friis, *Antennas Theory and Practice*, New York: John Wiley & Sons, Inc., 1052.

[9] Balanis, C. A., *Antenna Theory Analysis and Design*, New York: Harper and Row, 1982.

[10] Aerospace Recommended Practice ARP 958, ''Broadband Electromagnetic Interference Measurement Antennas: Standard Calibration Requirements and Methods,'' New York: Society of Automotive Engineers, Inc., 3/1/68.

Appendix A
List of Acronyms and Abbreviations

0TLP	zero transmission level point
AAS	advanced antenna system
ABET	Accreditation Board for Engineering and Technology
AC	alternating current
ACA	American Communications Association
ACARS	ARINC Communications Addressing Reporting System
ACCESS	automatic computer controlled electronic scanning system
ACE	Association for Cooperation in Engineering
ACET	Advisory Committee for Electronics and Telecommunications
ACF	antenna correction factor
ACGIH	American Conference of Governmental Industrial Hygienists
ACI	automatic card identification
ACIL	American Council of Independent Laboratories
ACIS	Association for Computing and Information Sciences
ACM/GAMM	Association for Computing Machinery/German Association for Applied Mathematics and Mechanics
ACOS	Advisory Committee on Safety
ACPDS	Advisory Committee on Personal Dosimetry services
AEA	American Electronics Association
AEA	American Engineering Association
AEA	Association of Engineers and Associates
AEC	American Engineering Council
AEI	Associazio Elettrotecnica Italiana
AEM	Association of Electronic Manufacturers
AEPSC	Atomic Energy Plant Safety Committee
AERE	Atomic Energy Research Establishment (UK)
AF	antenna factor

AF	audio frequency
AFCCE	Association of Federal Communication Consulting Engineers
AFNOR	Association Francaise des Normes (France)
AGARD	Advisory Group for Aerospace Research and Development (NATO)
AGED	Advisory Group on Electron Devices
AGEP	Advisory Group on Electronic Parts
AGET	Advisory Group on Electronic Tubes
AHAM	Association of Home Appliance Manufacturers
ALC	absorber lined chamber
A/m	ampere per meter
AM	amplitude modulation
ANSI	American National Standards Institute
ANZAAS	Australian and New Zealand Association for the Advancement of Science
aps	assumed perfect site
AR	axial ratio
ARCG	American Research Committee on Grounding
ARINC	Aeronautical Radio Incorporated
ARMMS	Automated RF and Microwave Measurement Society
ARMS	Aerial Radiological Measurements and Survey
ARMS	Amateur Radio Mobile Society
ARP	aerospace recommended practice
ARTE	Admiralty Research Test Establishment (UK)
ASA	American Standards Association (now ANSI)
ASAC	Asian Standards Advisory Committee
ASCII	American standard code of information interchange
ASEE	Association of Supervisory Electrical Engineers
ASET	American Society for Engineering Technology
ASI	American Standards Institute
ASM	American Society for Materials
ASTM	American Society for Testing and Materials
AT	ampere-turn
BABT	British Approvals Board for Telecommunications
BALUN	balanced to unbalanced
BAPT	German postal service (for EMC laws)
BBC	broadband conducted
BC	back-conducted
BCI	bulk current injection
BCI	broadcast interference
BEAME	British Electrical and Allied Manufacturing Engineers
BECTO	British Electric Cable Testing Organization

BEMS	Bioelectromagnetics Society
B/M	balun and matching (network)
BS	British Standard
BW	bandwidth
BW	beamwidth
CAA	Civil Aviation Authority (UK)
CAMESA	Canadian Military Electronics Standards Agency
CARS	community antenna relay service
CATV	community antenna television
CB	citizen's band (radio)
CBEMA	Computer and Business Equipment Manufacturer's Association
CCA	CENELEC Certification Agreement
CCB	Common Carrier Bureau (FCC)
CCI	combined form of CCIR and CCIT
CCIR	International Radio Consultative Committee
CCITT	Comite Consultatif International des Telegraphique et Telephonique
CCITT	International Telegraph and Telephone Consultative Committee
CDM	charged device model (ESD)
CE	Commite European
CE	conducted emissions
CE	compromising emanation (TEMPEST)
CECC	CENELEC Electronics Components Committee
CEI	Council of Engineering Institutions
CEM	computational electromagnetics
CEN	Commite European de Normalisation
CEN	European Standard Coordinating Committee
CENELEC	Comite European de Normalisation Electrotechnique (European Committee for Electrotechnical Standardization)
CERL	Central Electricity Research Laboratory
CGA	color graphics adapter (320×200 pixels)
CGM	computer graphics metafile
CGS	centimetre gramme second (system of units) or centimeter-gram-second
CI	conducted interference
CIA	Computer Industry Association
CIP	current injection probe
CISPR	Center for International Systems Research (Department of State)
CISPR	Comite International Special des Perturbation Radio Electronique
CMR	common-mode rejection
CMRR	common-mode rejection ratio
CMV	common-mode voltage
CNET	Centre National d'Etudes des Telecommunication (France)

COMAR	Committee on Man and Radiation (IEEE)
COMSEC	communications security
CONELRAD	control of electromagnetic radiation
COSHH	control of substances hazardous to health
CP	circularly polarized
CPEM	Conference on Precision Electromagnetic Measurement
CPSE	counterpoise
CRT	cathode ray tube
CS	conducted susceptibility
CSA	Canadian Standards Association
CT2	cordless telephone (second generation)
CUI	called unit identify
CW	carrier wave
CW	continuous wave
CYMB	cyan yellow magneta black
CYMK	cyan yellow magneta black
dB	decibel
dBA	decibel referenced to 1 amp
dB(A/m^2)	decibel referenced to 1 amp per square meter
dBa	decibel adjusted
DBAO	digital block AND-OR
dBa0	noise power on dBa referred to or measured at 0TLP
dBc	decibels relative to the carrier
dBi	decibel referenced to an isotropic (antenna)
dBj	relative RF signal levels (j = Jerrold Electronics)
dBk	decibels referred to one kilowatt
dBm	decibels referred to one milliwatt
dBm/m^2	dB referred to one milliwatt per square meter
dBm/m^2/MHz	dB referred to one milliwatt per square meter per megahertz
dBm(PSOPH)	noise power in dBm referred measured by a set with psophometric weighting
dBμA	dB referred to one microamp
dBμA/m	dB referred to one microamp per meter
dBμV	dB referred to one microvolt
dBμV/m	dB referred to one microvolt per meter
dBRN	decibels above reference noise
DBS	direct broadcast by satellite
dBV	decibels referred to 1 Volt
dBW	decibels referred to 1 Watt
dBx	decibels above the reference coupling
DC	direct current

DCI	direct current injection
DFT	discrete Fourier transform
DIN	Deutsches Institut fûr Normung (German Standards Organization)
div	divergence
DOC	Department of Communications (Canada)
DOD	Department of Defense (USA)
DOT	Department of Transportation
DRA	Defence Research Agency (UK)
DSB	double sideband
DSSC	double sideband suppressed carrier (modulation)
DTC	dual TEM cell
DTSC	digital termination shielded chamber
DUT	device under test
EBU	European Broadcasting Union
EC	Engineering Council
EC	European Communities
ECAC	Electromagnetic Analysis Center
ECC	Electronic Calibration Center (NBS)
ECE	Economic Commission for Europe (CEE Geneve)
ECMA	Electronic Computer Manufacturers Association
ECMA	European Computer Manufacturers Association
ECMWF	European Center for Medium-Range Weather Forecasting
ECP	electromagnetic compatibility program
EED	electroexplosive device
EED	electromagnetic energy density
E^3	electromagnetic environmental effects
EEM	electric equipment monitoring
EEPA	Electromagnetic Energy Policy Alliance
EFT	electrical fast transients
EGA	enhanced graphics array (640×320 pixels)
EIA	Electronic Industries Association
EIAJ	Electronic Industries Association of Japan
EIRP	equivalent isotropically radiated power
EITB	Engineering Industry Training Board
EM	electromagnetic
EMC	electromagnetic compatibility
EMC	electromagnetically coupled
EMC	electromagnetic coupling
EMCAB	Electromagnetic Compatibility Advisory Board
EMCOM	emission control
EMCP	electromagnetic compatibility program

EMCS	electromagnetic compatibility standardization (program)
EMCTP	electromagnetic compatibility test plan
EME	electromagnetic energy
EME	electromagnetic environmental effects
EMETF	electromagnetic environmental test facility
EMF	electromotive force
EMH	electromagnetic health
EMI	electromagnetic interference
EMICP	electromagnetic interference control plan
EMINT	electromagnetic intelligence
EMP	electromagnetic pulse (radiation)
EMR	electromagnetic radiation
EMS	electromagnetic susceptibility
EMSEC	emanation security (TEMPEST)
EMU	electromagnetic unit
EMV	Gesetz (German EMC law)
EN	Europaische Norm (European standard)
EOTC	European Organisation for Testing and Certification
EPA	Environment Protection Agency
ERP	effective radiated power
ERP	equivalent radiated power
ERTZ	equipment radiation TEMPEST zone
ESA	electrostatic attraction
ESD	electrostatic discharge
ESDS	electrostatic discharge sensitive
ESTLA	electrically short-tuned loop antenna
ETL	Electrotechnical Laboratory (Japan)
ETSI	European Telecommunications Standards Institute
EUT	equipment under test
FAA	Federal Aviation Administration (USA)
F/B	front-to-back ratio
FC	ferrite core
FCC	Federal Communications Commission (USA)
FEA	Federal Energy Administration
FEANI	Federation Europeane d'Associations Nationales d'Ingenieurs (France)
FERROD	ferrite-rod antenna
FFT	fast Fourier transform
FG	function generator
FI	frequency independent
FM	frequency modulation

FMCW	frequency modulated continuous wave
FMCW	frequency modulated carrier wave
FOB	Field Operations Bureau (FCC)
FOL	fiber optic link
FS	field strength
FSM	field strength meter
FT	Fourier transform
FTC	Federal Trade Commission
FTZ	Fernmelde Technischer Zentralamt (Germany)
GSM	group special mobile
G/T	antenna gain-to-noise temperature
GTD	geometrical theory of diffraction
HBM	human body model (ESD)
HCP	horizontal coupling plane (for ESD testing)
HEMP	high-altitude electromagnetic pulse
HERF	hazards of electromagnetic radiation to flight
HERF	hazards of electromagnetic radiation to fuel
HERO	hazards of electromagnetic radiation to ordnance
HERP	hazards of electromagnetic radiation to personnel
HIRF	high-intensity radiated fields
HP	horizontal polarization
HPBW	half-power beamwidth
HPM	high-power microwaves
HSI	hardness surveillance illuminator
IAC	Industry Advisory Committee
IBS	Institute for Basic Standards
ICAO	International Civil Aviation Organization
ICP	inherently conductive polymers
IDP	inherently dissipative polymers
IEC	International Electrotechnical Commission
IEE	Institution of Electrical Engineers (UK)
IEEE	Institute of Electrical and Electronics Engineers (USA)
IEEJ	Institute of Electrical and Engineers (Japan)
IEPS	International Electronics Packaging Society
IES	Institute of Environmental Sciences
IICIT	International Institute of Connector and Interconnection Technology
INIRC	International Non-ionising Radiation Committee
IPC	Institute for Interconnecting and Packaging Electronics Circuits
IRPA	International Radiation Protection Association

IRR	International Radio Regulations
IRT	Institute fur Rundfunktechnik GmHB (Germany)
ISDN	integrated services digital network
ISHM	International Society for Hybrid Microelectronics
ISM	industrial scientific and medical
ISO	International Standards Organization
ITE	information technology equipment
ITE	Institute of Telecommunications Engineers
ITEM	Interference Technology Engineers' Master
ITU	International Telecommunication Union (UNO)
JEDEC	Joint Electron Device Engineering Council
kWh	kilowatt-hour
LA	lighting arrestor
LASSY	Licence Administration Support System
LCR	inductance-capacitance-resistance (network)
LE	antenna effective length (for electric field antennas)
LEM	antenna effective length (for magnetic field antennas)
LEMP	electromagnetic pulse generated by a nearby lightning strike
LHCP	left-hand circular polarization or circularly polarized
LISN	line impedance stabilization network
LLD	low-level detector
LLSC	low-level sweep coupling
ln	Naperian or natural logarithm (to the base e)
log	logarithm (usually to base 10)
LORAN	long-range navigation
LORAP	long range aerial photography
LPDA	log-periodic dipole antennas or array
LSB	lower sideband
LSN	line stabilization network
LUF	lowest usable frequency
MAGGI	Million Ampere Generator (Aldermaston, UK)
Maser	microwave amplification by stimulated emission of radiation
MCS	Mobile Communications Systems (IEEE vehicular Technology Technical Committee)
MCW	modulated continuous wave
MDS	minimum detectable (or discernible) signal
MIC	minimum ignition current
MIC	mutual interference chart

MIL	Defense Logistics Agency (USA)
MIL-I	military specification on interference
MIL-STD	military standard (book)
mil	one-thousandth of an inch
MKS	meter-kilogram-second
MM	machine model (ESD)
MMB	Mass Media Bureau (FCC)
MMF	magnetomotive force
MoD	Ministry of Defence (UK)
MOF	maximum operating frequency
MP-4	an FCC measurement procedure
MPE	maximum permissible exposure
MPL	maximum permissible level
MPW	modified plane wave
MS	military standard sheet, prefix to numbered series issued by DOD (USA)
MSC	mode stirred cavity
MTF	modulation transfer function
MUB	maximum usable bandwidth
MUF	maximum usable frequency
MUT	material under test
NAB	National Association of Broadcasters
NACSEM	National COMSEC/EMSEC Information Memorandum (TEMPEST)
NACSI	National COMSEC Instruction (TEMPEST)
NACSIM	National COMSEC Instruction Memorandum (TEMPEST)
NAMAS	National Measurement Accreditation Service (UK)
NARTE	National Association of Radio and Telecommunications Engineers
NASA	National Aeronautics Space Administration
NATA	North American Telecommunications Association
NATLAS	National Testing Laboratory Accreditation Scheme (UK)
NBC	narrowband conducted
NBC	nuclear biological and chemical
NBS	National Bureau of Standards (USA) now NIST
NBSFS	NBS frequency standard
NEC	numerical electromagnetics code
NEMA	National Electrical Manufacturer's Association
NEMP	nuclear electromagnetic pulse
NEPA	National Environment Policy Act
NFP	near-field probe
NFPA	National Fire Protection Association
NIH	National Institute of Health

NIST	National Institute of Standards and Technology (USA)
NMR	normal-mode rejection
NPL	National Physical Laboratory (UK)
NRC	National Research Council
NRC	Nuclear Regulatory Commission
NRL	National Research Laboratory (USA)
NRPB	National Radiological Protection Board (UK)
NRTL	nationally recognized testing laboratory
NSA	normalized site attenuation
NSA	National Security Agency
NTG	Nachrichtentechnische Gesellschaft (German Communications Society)
NTIA	National Telecommunications and Information Administration (US Department of Commerce)
NTISSI	National Telecommunications and Information Systems Security Instruction
NTP	normal temperature and pressure
NTZ	Nachrichtechnische Zeitschrift (Germany)
NTSC	National Television System Committee (US)
NVLAP	National Voluntary Laboratory Accreditation Program
OATS	open area test site
OBR	optical bar code reader
oe	oersted (ampere per meter)
OEM	original equipment manufacturer
OET	Office of Engineering and Technology (FCC)
OSHA	Occupational Health and Safety Administration
OST	Office of Science and Technology
P-P	peak to peak
PBS	Public Broadcasting Service
PCZ	physical control zone (TEMPEST)
pd	potential difference
PDS	power density spectra
PF	power factor
PLC	power-line carrier
PM	phase modulation
PM	pulse modulation
PadB	perceived noise level expressed in decibels
POE	point of entry (TEMPEST)
PRB	Private Radio Bureau (FCC)
PRF	pulse repetition frequency

PRI	pulse repetition interval
PRR	pulse repetition rate
PSD	power spectral density
PSOPH	psophometric weighting
PTFE	polytetrafluouroethylene
PTM	pulse time modulation
PW	plane wave
PWS	plane wave spectrum
QP	quasi peak
QPA	quasi-peak adapter
QRM	manmade interference
RAC	Representative Advisory Committee
radar	radio assisted detection and ranging
RADFAC	radiation facility
RADHAZ	radiation hazards
RADIMT	radiation intelligence
radome	radar dome
RAE	Royal Aeronautical Establishment (now DRA)
RAM	radar absorbing material
RBE	radiation biological effectiveness
RD	radiation detector
RE	radiated emission
REC	radiant energy conversion
REG	radio frequency environment generator
REG	radar environment generator
RF	radio frequency
RFA	recticulated radio absorber
RFC	radio-frequency compatibility program
RFI	radio-frequency interference
RFPG	radio frequency protection guidelines
RH	radiological health
RHE	radiation hazard effects
RI	radio interference
RM	radio monitoring
RMS	radiation monitoring system
RMS	root mean square
RRSWG	random repetitive square wave pulse generator
RS	radiated susceptibility
RSS	root sum square
RTMA	Radio Television Manufacturers' Association

SA	spectrum analyzer
SAE	Society of Automotive Engineers
SAMPE	Society for the Advancement of Material and Process Engineering
SAR	specific absorption rate
SB	sideband
s/c	signal-to-clutter ratio
SCSI	small computer system interface
SE	shielding or screening effectiveness
SELDS	shielded enclosure leak detection system
SEV	Schweizerische Elektrotechnischer Verein (Swiss Electrotechnical Association
SI	Systeme Internationale d'Unites (International system of units)
SIR	signal-to-interference ratio
SMC	stirred mode cavity
SNI	signal-to-noise improvement
SNR	signal-to-noise ratio
SPEHS	single-point excitation for hardness surveillance
SPG	single-point ground
SS	Swedish Standard
SSB	single sideband
SSBSC	single-sideband suppressed carrier
SSSC	single-switched suppressed carrier
SSPM	single-sideband pulse modulation
stat	electrostatic units in CGS system (prefix)
SUT	site under test
SWR	standing wave ratio
TE	transverse electric (field)
TEM	transverse electromagnetic
TEMPEST	U.S. government radiated emissions security program
TM	transverse magnetic (field)
TSB	twin sideband
TSI	threshold signal-to-interference ratio
TVAR	TEMPEST Vulnerability Assessment Request
TVS	transient voltage suppressor
TWSB	twin sideband
TWT	traveling-wave tube
TWTA	traveling-wave tube amplifier
UL	Underwriter's Laboratory
UNO	United Nations Organization
UPS	uninterruptible power supply

USB	upper sideband
UUT	unit under test
UWB	ultra wideband
VCCI	Voluntary Control Commission for Interference (Japan)
VCP	vertical coupling plane (for ESD testing)
VDE	Verband Deutsche Electrotechniker—Society of German Electrical Engineers
Vfg	Verfugung (FTZ regulations that affect EDP equipment)
VGA	video graphics array (640 × 480 pixels)
V/Hz	volts per hertz
VP	vertical polarization
VSWR	voltage standing wave ratio
WARC	World Administrative Radio Conference
WG	waveguide
WG	Working Group (CISPR)
XPD	cross polar discrimination
XT	cross talk

Appendix B
Preferred Scientific Prefixes

Prefix	*Abbreviation*	*Decimal Equivalent*
Exa	E	10^{18}
Peta	P	10^{15}
Tera	T	10^{12}
Giga	G	10^{9}
Mega	M	10^{6}
Kilo	k	10^{3}
Centi	c	10^{-2}
Milli	m	10^{-3}
Micro	μ	10^{-6}
Nano	n	10^{-9}
Pico	p	10^{-12}
Femto	f	10^{-15}
Atto	a	10^{-18}

Notes: Upper case K is in common usage to denote memory as kilobytes in computing technology. However, a kilobyte is not strictly 1,000 bytes but 1024 or 2^{10} bytes. The lower case k stands for 1,000 whereas upper case K stands for 1024.
Optical wavelengths are often quoted in Angstrom units.
Angstrom (A°) is 10^{-10} m.

Appendix C
List of Scientific Constants

Avogadro's constant, $N_a = 6.02 \times 10^{23}$ mol^{-1}
Boltzmann's constant, $k = 1.38 \times 10^{-23}$ J K^{-1}
Base of the Naperian or natural system of logarithms, $e = 2.718$
Intrinsic impedance of a plane wave in free space, $Z_0 = 120\pi$ or 377 Ω
Electronic charge, $e = 1.602 \times 10^{-19}$ C
Faraday's constant, $F = 9.65 \times 10^4$ C mol^{-1}
Gravitational acceleration, $g = 9.81$ m s^{-2}
Gravitational constant, $G = 6.67 \times 10^{-11}$ m^3 s^{-2} kg^{-1}
Molar gas constant, $R = 8.3$ J mol^{-1} K^{-1}
Neper is equal to 8.68 dB, {20 log(e)}.
Permeability of free space = $4\pi \times 10^{-7}$ H m^{-1}
Permittivity of free space = $1/(36\pi \times 10^9)$ or 8.85×10^{-12} F m^{-1}
Pi = π = 3.142
Speed of electromagnetic waves in a vacuum, $c = 2.99773 \times 10^8$ m s^{-1} (usually taken as 3×10^8 m s^{-1})
Speed of sound in free space in dry air at STP is 331.5 m/s

Appendix D
Conductivities of Common Metals

Metal	*Conductivity in Siemen/m*	*Conductivity Relative to Cu*	*Multiplier for Skin Depth*
Aluminium	3.538×10^7	0.61	1.28
Beryllium	5.8×10^6	0.10	3.16
Brass (70% Cu)	1.450×10^7	0.25	2.00
Brass (90% Cu)	2.41×10^7	0.42	1.54
Cadmium	1.334×10^7	0.23	2.09
Copper	5.8×10^7	1.00	1.00
Gold	4.06×10^7	0.70	1.20
Iron	9.86×10^7	0.17	2.43
Lead	4.64×10^6	0.08	3.54
Magnesium	2.204×10^7	0.38	1.62
Mu-metal	1.74×10^6	0.03	5.77
Nickel	1.16×10^6	0.20	2.24
Permalloy	1.74×10^6	0.03	5.77
Phosphor-bronze	5.8×10^6	0.18	2.36
Silver	6.09×10^7	1.05	0.98
Stainless steel	1.16×10^6	0.02	7.87
Tin	8.7×10^6	0.15	2.58
Zinc	1.68×10^7	0.29	1.86

Appendix E
Dielectric Constants and Loss Tangents of Common Materials

This is the same as the permittivity of the material relative to the permittivity of a vacuum. The dielectric constant ε_r of a material is made up of a real ε_r' and imaginary part ε_r'' and is given by $\varepsilon_r = \varepsilon_r' + j\varepsilon_r''$. The ratio of the imaginary to real parts ($\varepsilon_r''/\varepsilon_r'$) is the loss tangent, tan δ. The dielectric constant changes with frequency, so the magnitudes are quoted at two frequencies where these were available.

	ε_r		*tan* δ	
Material	*3 GHz*	*10 GHz*	*3 GHz*	*10 GHz*
Alumina (99%) ceramic	8.7	9.7–10.3		0.0004
Corning glass 1990	7.99	7.94	0.00199	0.0032
Teflon (PTFE) glass woven fiber	2.1	2.17–2.55		0.0009–0.0022
Fiberglass BK-164	3.88	3.99	0.01	0.0131
Ice (−12°C)	3.20	3.17	0.0009	0.0007
Mica-glass bonded		7.5		0.0020
Plexiglass	2.60	2.59	0.0057	0.0067
Polyethylene	2.32	2.31	0.0050	0.0044
Polystyrene hydrogenated	2.25	2.25	0.00016	0.00041
Teflon (PTFE) unreinforced	2.1	2.08	0.00015	0.00037
Quartz fused		3.78		0.001
Water (at 25°C)	77	55	0.15	0.54

Appendix F
Conversion Table for dBμV to μV, dBV, dBW, dBm, and Watts

To convert decibel microvolts (dBμV) to μV, the following formula

$$\mu v = 10^{(dB\mu V/10)}$$

should be used.

To convert the power in milliwatts to power in dBm, the formula

$$dBm = 10\log_{10}(mW)$$

is used.

To convert volts to watts using a characteristic impedance of R, $V = \sqrt{PR}$ where V is the voltage in volts, P is the power in watts, and R is the resistance in ohms.

The relation between dBV and dBW is given by

$$20 \log_{10} V = 10 \log_{10} P + 10 \log_{10} R$$

The voltage ratio in dBV is given by $V = 20 \log_{10}(V_1/V_2)$ where V_1 and V_2 are measured in Volts.

The following table assumes an impedance of 50 ohms.

dBμV	*Voltage*	*dBV*	*dBW*	*dBm*	*Power*
0.00	1.00 μV	−120.00	−137.00	−107.00	20.00 fW
1.00	1.12 μV	−119.00	−136.00	−106.00	25.20 fW
2.00	1.26 μV	−118.00	−135.00	−105.00	31.70 fW
3.00	1.41 μV	−117.00	−134.00	−104.00	39.90 fW
4.00	1.58 μV	−116.00	−133.00	−103.00	50.20 fW
5.00	1.78 μV	−115.00	−132.00	−102.00	63.20 fW
6.00	2.00 μV	−114.00	−131.00	−101.00	79.60 fW

(*continued*)

dBμV	Voltage	dBV	dBW	dBm	Power
7.00	2.24 μV	−113.00	−130.00	−100.00	100.20 fW
8.00	2.51 μV	−112.00	−129.00	−99.00	126.20 fW
9.00	2.82 μV	−111.00	−128.00	−98.00	0.00 fW
10.00	3.16 μV	−110.00	−127.00	−97.00	200.00 fW
11.00	3.55 μV	−109.00	−126.00	−96.00	251.80 fW
12.00	3.98 μV	−108.00	−125.00	−95.00	317.00 fW
13.00	4.47 μV	−107.00	−124.00	−94.00	399.10 fW
14.00	5.01 μV	−106.00	−123.00	−93.00	502.40 fW
15.00	5.62 μV	−105.00	−122.00	−92.00	632.50 fW
16.00	6.31 μV	−104.00	−121.00	−91.00	796.20 fW
17.00	7.08 μV	−103.00	−120.00	−90.00	1.00 pW
18.00	7.94 μV	−102.00	−119.00	−89.00	1.26 pW
19.00	8.91 μV	−101.00	−118.00	−88.00	1.59 pW
20.00	10.00 μV	−100.00	−117.00	−87.00	2.00 pW
21.00	11.22 μV	−99.00	−116.00	−86.00	2.52 pW
22.00	12.59 μV	−98.00	−115.00	−85.00	3.17 pW
23.00	14.13 μV	−97.00	−114.00	−84.00	3.99 pW
24.00	15.85 μV	−96.00	−113.00	−83.00	5.02 pW
25.00	17.78 μV	−95.00	−112.00	−82.00	6.32 pW
26.00	20.00 μV	−94.00	−111.00	−81.00	7.96 pW
27.00	22.40 μV	−93.00	−110.00	−80.00	10.02 pW
28.00	25.10 μV	−92.00	−109.00	−79.00	12.62 pW
29.00	28.20 μV	−91.00	−108.00	−78.00	15.89 pW
30.00	31.60 μV	−90.00	−107.00	−77.00	20.00 pW
31.00	35.50 μV	−89.00	−106.00	−76.00	25.18 pW
32.00	39.80 μV	−88.00	−105.00	−75.00	31.70 pW
33.00	44.70 μV	−87.00	−104.00	−74.00	39.91 pW
34.00	50.10 μV	−86.00	−103.00	−73.00	50.24 pW
35.00	56.20 μV	−85.00	−102.00	−72.00	63.25 pW
36.00	63.10 μV	−84.00	−101.00	−71.00	79.62 pW
37.00	70.80 μV	−83.00	−100.00	−70.00	100.20 pW
38.00	79.40 μV	−82.00	−99.00	−69.00	126.20 pW
39.00	89.10 μV	−81.00	−98.00	−68.00	158.90 pW
40.00	100.00 μV	−80.00	−97.00	−67.00	200.00 pW
41.00	112.20 μV	−79.00	−96.00	−66.00	251.80 pW
42.00	125.90 μV	−78.00	−95.00	−65.00	317.00 pW
43.00	141.30 μV	−77.00	−94.00	−64.00	399.10 pW
44.00	158.50 μV	−76.00	−93.00	−63.00	502.40 pW
45.00	177.80 μV	−75.00	−92.00	−62.00	632.50 pW
46.00	199.50 μV	−74.00	−91.00	−61.00	796.20 pW
47.00	223.90 μV	−73.00	−90.00	−60.00	1.00 nW
48.00	251.20 μV	−72.00	−89.00	−59.00	1.26 nW
49.00	281.80 μV	−71.00	−88.00	−58.00	1.59 nW
50.00	316.20 μV	−70.00	−87.00	−57.00	2.00 nW
51.00	354.80 μV	−69.00	−86.00	−56.00	2.52 nW
52.00	398.10 μV	−68.00	−85.00	−55.00	3.17 nW
53.00	446.70 μV	−67.00	−84.00	−54.00	3.99 nW

(continued)

$dB\mu V$	Voltage	dBV	dBW	dBm	Power
54.00	501.20 μV	−66.00	−83.00	−53.00	5.02 nW
55.00	562.30 μV	−65.00	−82.00	−52.00	6.32 nW
56.00	631.00 μV	−64.00	−81.00	−51.00	7.96 nW
57.00	707.90 μV	−63.00	−80.00	−50.00	10.02 nW
58.00	794.30 μV	−62.00	−79.00	−49.00	12.62 nW
59.00	891.30 μV	−61.00	−78.00	−48.00	15.89 nW
60.00	1.00 mV	−60.00	−77.00	−47.00	20.00 nW
61.00	1.12 mV	−59.00	−76.00	−46.00	25.18 nW
62.00	1.26 mV	−58.00	−75.00	−45.00	31.70 nW
63.00	1.41 mV	−57.00	−74.00	−44.00	39.91 nW
64.00	1.58 mV	−56.00	−73.00	−43.00	50.24 nW
65.00	1.78 mV	−55.00	−72.00	−42.00	63.25 nW
66.00	2.00 mV	−54.00	−71.00	−41.00	79.62 nW
67.00	2.24 mV	−53.00	−70.00	−40.00	100.24 nW
68.00	2.51 mV	−52.00	−69.00	0.00	126.19 nW
69.00	2.82 mV	−51.00	−68.00	−38.00	158.87 nW
70.00	3.16 mV	−50.00	−67.00	−37.00	200.00 nW
71.00	3.55 mV	−49.00	−66.00	−36.00	251.79 nW
72.00	3.98 mV	−48.00	−65.00	−35.00	316.98 nW
73.00	4.47 mV	−47.00	−64.00	−34.00	399.05 nW
74.00	5.01 mV	−46.00	−63.00	−33.00	502.38 nW
75.00	5.62 mV	−45.00	−62.00	−32.00	632.46 nW
76.00	6.31 mV	−44.00	−61.00	−31.00	796.21 nW
77.00	7.08 mV	−43.00	−60.00	−30.00	1.00 μW
78.00	7.94 mV	−42.00	−59.00	−29.00	1.26 μW
79.00	8.91 mV	−41.00	−58.00	−28.00	1.59 μW
80.00	10.00 mV	−40.00	−57.00	−27.00	2.00 μW
81.00	11.22 mV	−39.00	−56.00	−26.00	2.52 μW
82.00	12.59 mV	−38.00	−55.00	−25.00	3.17 μW
83.00	14.13 mV	−37.00	−54.00	−24.00	3.99 μW
84.00	15.85 mV	−36.00	−53.00	−23.00	5.02 μW
85.00	17.78 mV	−35.00	−52.00	−22.00	6.32 μW
86.00	19.95 mV	−34.00	−51.00	−21.00	7.96 μW
87.00	22.39 mV	−33.00	−50.00	−20.00	10.02 μW
88.00	25.12 mV	−16.00	−49.00	−19.00	12.62 μW
89.00	28.18 mV	−31.00	−48.00	−18.00	15.89 μW
90.00	31.62 mV	−30.00	−47.00	−17.00	20.00 μW
91.00	35.48 mV	−29.00	−46.00	−16.00	25.18 μW
92.00	39.81 mV	−28.00	−45.00	−15.00	31.70 μW
93.00	44.67 mV	−27.00	−44.00	−14.00	39.91 μW
94.00	50.12 mV	−26.00	−43.00	−13.00	50.24 μW
95.00	56.23 mV	−25.00	−42.00	−12.00	63.25 μW
96.00	63.10 mV	−24.00	−41.00	−11.00	79.62 μW
97.00	70.79 mV	−23.00	−40.00	−10.00	100.20 μW
98.00	79.43 mV	−22.00	−39.00	−9.00	126.20 μW
99.00	89.13 mV	−21.00	−38.00	−8.00	158.90 μW
100.00	100.00 mV	−20.00	−37.00	−7.00	200.00 μW

(*continued*)

dBμV	Voltage	dBV	dBW	dBm	Power
101.00	112.20 mV	−19.00	−36.00	0.00	251.80 μW
102.00	125.90 mV	−18.00	−35.00	−5.00	317.00 μW
103.00	141.30 mV	−17.00	−34.00	−4.00	399.10 μW
104.00	158.50 mV	−16.00	−33.00	−3.00	502.40 μW
105.00	177.80 mV	−15.00	−32.00	−2.00	632.50 μW
106.00	199.50 mV	−14.00	−31.00	−1.00	796.20 μW
107.00	223.90 mV	−13.00	−30.00	0.00	1.00 mW
108.00	251.20 mV	−12.00	−29.00	1.00	1.26 mW
109.00	281.80 mV	−11.00	−28.00	2.00	1.59 mW
110.00	316.20 mV	−10.00	−27.00	3.00	1.00 mW
111.00	354.80 mV	−9.00	−26.00	4.00	2.52 mW
112.00	398.10 mV	−8.00	−25.00	5.00	3.17 mW
113.00	446.70 mV	−7.00	−24.00	6.00	3.99 mW
114.00	501.20 mV	−6.00	−23.00	7.00	5.02 mW
115.00	562.30 mV	−5.00	−22.00	8.00	6.32 mW
116.00	631.00 mV	−4.00	−21.00	9.00	7.96 mW
117.00	707.90 mV	−3.00	−20.00	10.00	10.02 mW
118.00	794.30 mV	−2.00	−19.00	11.00	12.62 mW
119.00	891.30 mV	−1.00	−18.00	12.00	15.89 mW
120.00	1.00 V	0.00	−17.00	13.00	20.00 mW
121.00	1.12 V	1.00	−16.00	14.00	25.18 mW
122.00	1.26 V	1.00	−15.00	15.00	31.70 mW
123.00	1.41 V	3.00	−14.00	16.00	39.91 mW
124.00	1.58 V	4.00	−13.00	17.00	50.24 mW
125.00	1.78 V	5.00	−12.00	18.00	63.25 mW
126.00	2.00 V	6.00	−11.00	19.00	79.62 mW
127.00	2.24 V	7.00	−10.00	20.00	100.20 mW
128.00	2.51 V	4.00	−9.00	21.00	126.20 mW
129.00	2.82 V	9.00	−8.00	22.00	158.90 mW
130.00	3.16 V	10.00	−7.00	23.00	200.00 mW
131.00	3.55 V	11.00	−6.00	24.00	251.80 mW
132.00	3.98 V	12.00	−5.00	25.00	317.00 mW
133.00	4.47 V	13.00	−4.00	26.00	399.10 mW
134.00	5.01 V	14.00	−3.00	27.00	502.40 mW
135.00	5.62 V	15.00	−2.00	28.00	632.50 mW
136.00	6.31 V	16.00	−1.00	29.00	796.20 mW
137.00	7.08 V	17.00	0.00	30.00	1.00 W
138.00	7.94 V	18.00	1.00	31.00	1.26 W
139.00	8.91 V	19.00	2.00	32.00	1.59 W
140.00	10.00 V	20.00	3.00	33.00	2.00 W

Appendix G
The Electromagnetic Spectrum

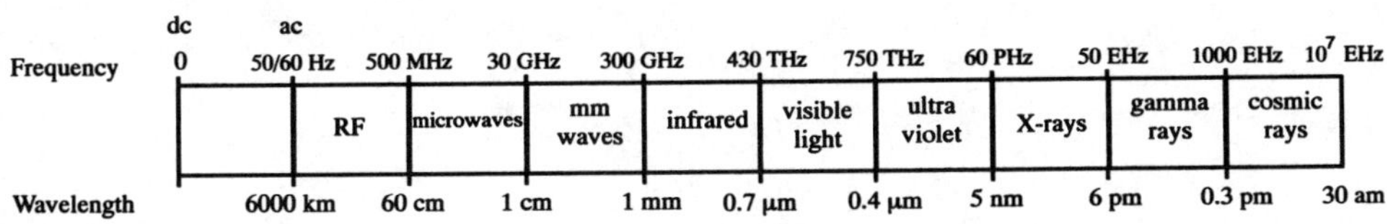

Appendix H
Frequency Band Designations

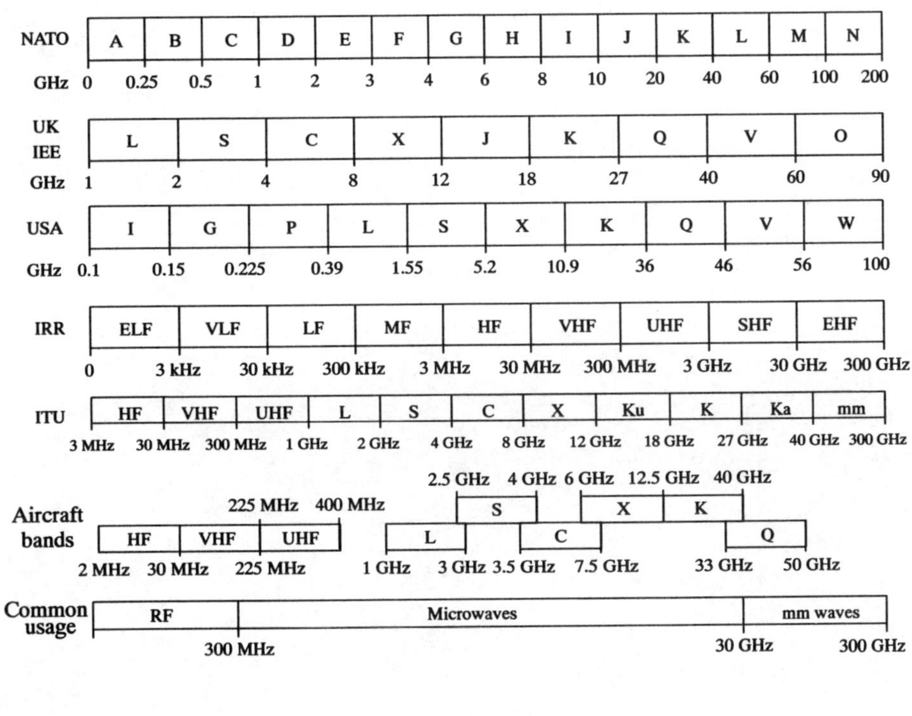

Appendix I

Conversion Between Gain in dB and Gain as a Linear Ratio Gain in dBi to Linear Gain

To calculate intermediate values of gain in dBi multiply linear gains for example for 5.6 dBi multiply the 5 and 0.6 dBi values, i.,e. 3162 × 1.148 = 3.26998.

Gain dBi	−10.0	−9.0	−8.0	−7.0	−6.0	−5.0	−4.0	−3.0	−2.0
Linear gain	0.100	0.126	0.158	0.200	0.251	0.316	0.393	0.501	0.631
Gain dBi	−1.0	−9.0	−8.0	−7.0	−6.0	−5.0	−4.0	−3.0	−2.0
Linear gain	0.794	0.813	0.851	0.871	0.871	0.891	0.912	0.933	0.955
Gain dBi	0.1	0.2	0.3	0.4	0.5	0.6	0.7	0.8	0.9
Linear gain	1.023	1.047	1.072	1.096	1.122	1.148	1.175	1.202	1.230
Gain dBi	1	2	3	4	5	6	7	8	9
Linear gain	1.259	1.585	1.995	2.512	3.162	3.981	5.012	6.310	7.943
Gain dBi	10	20	30	40	50	60	70	80	90.
Linear gain	10	100	1000	10^4	10^5	10^6	10^7	10^8	10^9

Linear gain to gain in dBi

Linear gain	0.1	0.2	0.3	0.4	0.5	0.6	0.7	0.8	0.9
Gain dBi	−10.0	−7.0	−5.2	−4.0	−3.0	−2.2	−1.5	−1.0	−0.5
Linear gain	1	2	3	4	5	6	7	8	9
Gain dBi	0	3.01	4.77	6.02	6.99	7.78	8.45	9.03	9.54
Linear gain	10	20	30	40	50	60	70	80	90
Gain dBi	10.00	13.01	14.77	16.02	16.99	17.78	18.45	19.03	19.54
Linear gain	100	200	300	400	500	600	700	800	900
Gain dBi	20.00	23.01	24.77	26.02	26.99	27.78	28.45	29.03	29.54
Linear gain	1000	2000	3000	4000	5000	6000	7000	8000	9000
Gain dBi	30.00	33.01	34.77	36.02	36.99	37.78	38.45	39.03	39.54
Linear gain	10000	2000	30000	40000	50000	60000	70000	80000	90000
Gain dBi	40.00	43.01	44.77	46.02	46.99	47.78	48.45	49.03	49.54

Appendix J
The Periodic Table Listed Alphabetically by Chemical Symbol

Symbol	*Element*	*Group*	*Atomic No.*	*Atomic Weight*
Ac	Actinium	3A	89	227.0278
Ag	Silver	1B	47	107.868
Al	Aluminium	3B	13	26.981
Am	Americium	3A	95	(243)
Ar	Argon	Gas	18	39.948
As	Arsenic	5B	33	74.9216
At	Astatine	7B	85	(210)
Au	Gold	1B	79	196.9665
B	Boron	3B	5	10.81
Ba	Barium	2A	56	137.33
Be	Beryllium	2A	4	9.01218
Bi	Bismuth	5B	83	208.9804
Bk	Berkelium	3A	97	(247)
Br	Bromine	7B	35	79.904
C	Carbon	4B	6	12.011
Ca	Calcium	2A	20	40.08
Cd	Cadmium	2B	48	112.41
Ce	Cerium	3A	58	130.12
Cf	Californium	3A	98	(251)
Cl	Chlorine	7B	17	35.453
Cm	Curium	3A	96	(247)
Co	Cobalt	8	27	58.9332
Cr	Chromium	6A	24	51.996
Cs	Caesium	1A	55	132.9054

(*continued*)

Symbol	*Element*	*Group*	*Atomic No.*	*Atomic Weight*
Cu	Copper	1B	29	63.54
Dy	Dyprosium	3A	66	162.5
Ee	Einsteinium	3A	99	(252)
Er	Erbium	3A	68	167.2
Eu	Europium	3A	63	151.96
F	Fluorine	7B	9	18.9984
Fe	Iron	8	26	55.84
Fm	Fermium	3A	100	(257)
Fr	Francium	1A	87	(223)
Ga	Gallium	3B	31	69.72
Gd	Gadolinium	3A	64	157.25
Ge	Germanium	4B	32	72.5
H	Hydrogen	1	1	1.00794
He	Helium	7	2	4.0026
Hf	Hafnium	4A	72	178.4
Hg	Mercury	2B	80	200.5
Ho	Holmium	3A	67	164.9304
I	Iodine	7B	53	126.9045
In	Indium	3B	49	114.82
Ir	Irdium	8	77	192.2
K	Potassium	1A	19	39.0983
Kr	Krypton	Gas	36	83.80
La	Lanthanum	3A	57	138.905
Li	Lithium	1A	3	6.94
Lr	Lawrencium	3A	103	(260)
Lu	Lutetium	3A	71	174.967
Md	Mendelevium	3A	101	(258)
Mg	Magnesium	2A	12	24.305
Mn	Manganese	7A	25	54.938
Mo	Molybdenum	6A	42	95.94
N	Nitrogen	5B	7	14.0067
Na	Sodium	1A	11	22.98977
Nb	Niobium	5A	41	92.9064
Nd	Neodynium	3A	60	144.2
Ne	Neon	Gas	10	20.17
Ni	Nickel	8	28	58.69
No	Nobelium	3A	102	(259)
Np	Neptunium	3A	93	237.0482
O	Oxygen	6B	8	15.999
Os	Osmium	8	76	190.2
P	Phosphorus	5B	15	30.973
Pa	Protactinium	3A	91	231.0359
Pb	Lead	4B	82	207.2
Pd	Palladium	8	46	106.42
Pm	Promethium	3A	61	145
Po	Polonium	6B	84	(209)

(*continued*)

Symbol	*Element*	*Group*	*Atomic No.*	*Atomic Weight*
Pr	Praseodymium	3A	59	140.9077
Pt	Platinum	8	78	195.0
Pu	Plutonium	3A	94	(244)
Ra	Radium	2A	88	226.0254
Rb	Rubidium	1A	37	85.467
Rb	Rhodium	8	45	102.99055
Re	Rhenium	7A	75	186.207
Rn	Radon	Gas	86	(222)
Ru	Ruthenium	8	44	101.0
S	Sulphur	6B	16	32.06
Sb	Antimony	5B	51	121.7
Sc	Scandium	3A	21	44.9559
Se	Selenium	6B	34	78.9
Si	Silicon	4B	14	28.085
Sm	Samarium	3A	62	150.3
Sn	Tin	4B	50	118.6
Sr	Strontium	2A	38	87.62
Ta	Tantalum	5A	73	180.9479
Tb	Terbium	3A	65	158.9254
Te	Technetium	7A	43	98
Te	Tellurium	6B	52	127.60
Th	Thorium	3A	90	232.0381
Ti	Titanium	4A	22	47.8
Tl	Thallium	3B	81	204.383
Tm	Thulium	3A	69	168.9342
U	Uranium	3A	92	238.0289
V	Vanadium	5A	23	50.9415
W	Tungsten	6A	74	183.8
Xe	Xenon	Gas	54	131.2
Y	Yttrium	3A	39	88.9059
Yb	Ytterbium	3A	70	173
Zn	Zinc	2B	30	65.38
Zr	Zirconium	4A	40	91.22

Appendix K
Magnetic Permeabilities of Common Metals

Material	*Conductivity Relative to Copper*	*Relative Magnetic Permeability*	*Absolute Permeability (H/m)*
Aluminium	0.61	1	0.00000126
Beryllium	0.1	1	0.00000126
Brass	0.26	1	0.00000126
Cadmium	0.23	1	0.00000126
Copper	1	1	0.00000126
Gold	0.7	1	0.00000126
Hypernick	0.06	80000	0.10053096
Iron	0.17	1000	0.00125664
Lead	0.08	1	0.00000126
Magnesium	0.38	1	0.00000126
Nu-metal	0.03	80000	0.10053096
Nickel	0.2	1	0.00000126
Permalloy	0.03	80000	0.10053096
Phosphor-bronze	0.18	1	0.00000126
Silver	1.05	1	0.00000126
Stainless steel	0.02	1000	0.00125664
Tin	0.15	1	0.00000126
Zinc	0.29	1	0.00000126

Appendix L
Polarization Matching Matrix

Receiver polarization	Transmitter polarization: Horizontal →	Vertical ↑	Right slant ↗	Left slant ↖	RHCP ↻	LHCP ↺
Horizontal →	Yes	No *	Yes but 3 dB loss	Yes but 3 dB loss	Yes but 3 dB loss	Yes but 3 dB loss
Vertical ↑	No *	Yes	Yes but 3 dB loss	Yes but 3 dB loss	Yes but 3 dB loss	Yes but 3 dB loss
Right slant ↗	Yes but 3 dB loss	Yes but 3 dB loss	Yes	No *	Yes but 3 dB loss	Yes but 3 dB loss
Left slant ↖	Yes but 3 dB loss	Yes but 3 dB loss	No *	Yes	Yes but 3 dB loss	Yes but 3 dB loss
RHCP ↻	Yes but 3 dB loss	Yes but 3 dB loss	Yes but 3 dB loss	Yes but 3 dB loss	Yes	No *
LHCP ↺	Yes but 3 dB loss	Yes but 3 dB loss	Yes but 3 dB loss	Yes but 3 dB loss	No *	Yes

* Although cross polarization cannot be detected by perfect antennas, practical antennas have levels of cross polar discrimination of the order of -20 dB. Detection is possible at these levels.

Appendix M
Resistivities of Common Materials

Material	*Resistivity (ohm-m)*	*Resistivity Relative to Copper*
Aluminium	2.83×10^{-8}	1.64
Beryllium	17.24×10^{-8}	10
Brass	6.63×10^{-8}	3.85
Cadmium	7.50×10^{-8}	4.35
Copper	1.72×10^{-8}	1
Gold	2.46×10^{-8}	1.43
Iron	10.14×10^{-8}	5.88
Lead	21.55×10^{-8}	12.5
Magnesium	4.54×10^{-8}	2.63
Nu-metal	57.47×10^{-8}	33.33
Nickel	8.62×10^{-8}	5
Permalloy	57.47×10^{-8}	33.33
Phosphor-bronze	9.58×10^{-8}	5.56
Silver	1.64×10^{-8}	0.95
Stainless steel	86.21×10^{-8}	50
Tin	11.49×10^{-8}	6.67
Zinc	5.95×10^{-8}	3.45

Appendix N
Radio Frequency Protection Guides for Nonionizing Radiation

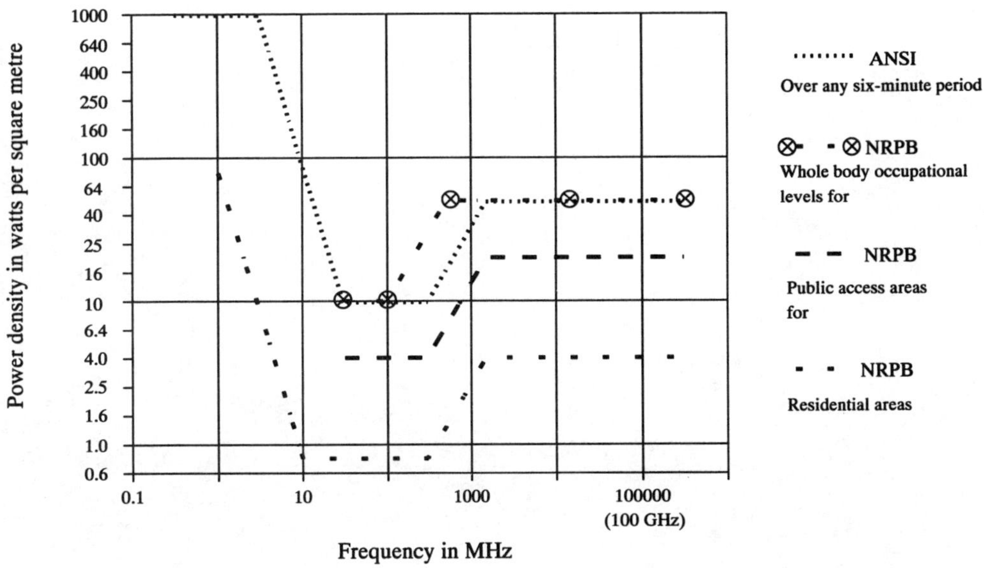

Appendix O

List of Symbols

a	broadside dimension of waveguide side of triangle coefficient of x^2 term of a quadratic radius of a coil expansion coefficient for a planar log spiral antenna
a'	expansion coefficient for an equivalent planar log spiral antenna
a_1	broadside dimension of open end of an H-plane sectoral horn
$\mathbf{a}_r$	unit vector in polar coordinate system
$\mathbf{a}_f$	unit vector in polar coordinate system
$\mathbf{a}_q$	unit vector in polar coordinate system
A	angle number length of vector area of a loop in m^2 or cm^2 height (from apex to opposite side) of one arm of a bowtie antenna
$\boldsymbol{A}$	vector vector field
A_f	antenna correction factor
α	current division factor (folded dipole) angle between the magnetic field vector due to a single coil and the plane containing the coil flare angle of a triangular dipole (bowtie) antenna in deg half-angle of a LPDA angle between the radius vector of a spiral and the x-axis attenuation factor of a cable
b	narrow dimension of waveguide radius of outer conductor of coaxial cable

	radius of loop side of triangle coefficient of x term of a quadratic
b_1	height of open end of an E-plane sectoral horn
B	angle magnetic flux density scalar in tesla (T) number length of vector
$\mathbf{B}$	vector magnetic flux density vector in tesla (T) or picotesla (pT)
B_{ar}	bandwidth of the active region of a conical log spiral
B_s	design bandwidth in hertz (Hz)
B_w	bandwidth in hertz (Hz)
β	half-angle of a supporting section of a trapezoidal tooth LPDA winding or wrap angle of a spiral antenna
c	side of triangle
c_d	cavity diameter for a planar log spiral
C	angle the constant term of a quadratic capacitance in farads (F) circumference of a helical antenna combined correction term of M_g and M_t in dB
C_{MAX}	diameter of the base of the cone of a discone antenna
C_{MIN}	diameter of the top of the cone of a discone antenna
C_p	correction term to allow for the polarization mismatch, in dB
$C(x)$	Fresnel cosine integral
d	diameter of the feed region of a planar log spiral antenna distance between horn and metal sheet in Purcell's Method
dA	area of small loop in m^2
dl	elementary length in m
d_n	distance between the nth and $(n+1)$th element of a LPDA
D	electric flux/displacement density scalar in coulombs per square meter ($C\ m^{-1}$) linear or numeric directivity diameter of a circular aperture diameter of a helical antenna in m

maximum dimension of an aperture antenna
overall diameter of a planar log spiral antenna

D_{dB} directivity in dB
directivity in dB (dB referred to a circularly polarized isotropic source)

D_{dBi} directivity in dBi (dB referred to an isotropic source)

D_{e} effective aperture
diameter of an equivalent reflector antenna

D_m angle in mathematical degrees

D_n and angle in navigational degrees
distance of the *n*th element from the imaginary apex of a LPDA
distance of the back of the *n*th tooth from the imaginary apex of a trapezoidal tooth LPDA

D_{n1} normal component of the electric flux density in medium 1, in volts per meter (V m^{-1})

D_{n2} normal component of the electric flux density in medium 2, in volts per meter (V m^{-1})

D_{tl} tangential component of the electric flux density vector in medium 1, in coulombs per square meter (C m^{-2})

D_{t2} tangential component of the electric flux density vector in medium 2, in coulombs per square meter (C m^{-2})

$\boldsymbol{D}$ electric flux/displacement density vector in coulombs per square meter (C m^{-1})

d skin depth in meters (m)
angular width of the arm of a spiral antenna

E electric field intensity in volts per meter (V m^{-1})

E_{i} incident electric field intensity in volts per meter (V m^{-1})

E_{r} reflected electric field intensity from the edges of a spiral antenna in volts per meter (V m^{-1})
electric field intensity in the radial direction, in volts per meter (V m^{-1})

E_{f} electric field intensity in the azimuth plane, in volts per meter (V m^{-1})

E_{q} electric field intensity in the elevation plane, in volts per meter (V m^{-1})

E_{n1} normal component of the electric field vector in medium 1, in volts per meter (V m^{-1})

E_{n2} normal component of the electric field vector in medium 2, in volts per meter (V m^{-1})

E_{t1} tangential component of the electric field vector in medium 1, in volts per meter (V m^{-1})

E_{t2} tangential component of the electric field vector in medium 2, in volts per meter (V m^{-1})

$\boldsymbol{E}$ electric field vector in volts per meter (V m^{-1})

e permittivity in farads per meter (F m^{-1})

e_1 permittivity in medium 1, farads per meter (F m^{-1})

e_2 permittivity in medium 2, farads per meter (F m^{-1})

h efficiency

η_{dB} efficiency in dB

h_{op} operating efficiency

f frequency in hertz (Hz)

F frequency in megahertz (MHz)

ϕ magnetic flux in webers (Wb)
angle in the azimuth plane
phase difference in radians (rad)

ϕ_1 half power beamwidth in the azimuth plane, in degrees (deg)

ϕ_{p} phase associated with the arm of a spiral antenna, in radians (rad)

g_1 gain of horn 1 in dB

g_2 gain of horn 2 in dB

g_3 gain of horn 3 in dB

g_{dB} gain in dB

g_{p} gain of a pyramid horn in dB

G linear or numeric gain

G_{dBi} gain in decibels referred to isotropic (dBi)

G_{e} linear or numeric gain of an *E*-plane sectoral horn

G_{h} linear or numeric gain of an *H*-plane sectoral horn

G_{p} linear or numeric gain of a pyramid horn

h height of antenna
distance between feed point of a dipole and its center
effective length/height of a half-wave dipole in m

h_{eff} effective length/height of the monopole in m

H magnetic field intensity scalar in amperes per meter (A m^{-1})
half length of a dipole in m

H_0 amplitude of the magnetic field intensity in amperes per meter (A m^{-1})

H_f magnetic field intensity in the azimuth plane in amperes per meter (A m^{-1})

H_r magnetic field intensity in the radial direction in amperes per meter (A m^{-1})

$\boldsymbol{H}_t$ total or resultant magnetic field intensity vector for a coil, in amperes per meter (A m^{-1})

H_q magnetic field intensity in the elevation plane in amperes per meter (A m^{-1})

H_{n1} normal component of the magnetic field vector in medium 1, in amperes per meter (A m^{-1})

H_{n2} normal component of the magnetic vector in medium 2, in amperes per meter (A m^{-1})

H_{t1} tangential component of the magnetic field vector in medium 1, in amperes per meter (A m^{-1})

H_{t2} tangential component of the magnetic field vector in medium 2, in amperes per meter (A m^{-1})

$\boldsymbol{H}$ magnetic field intensity vector in amperes per meter (A m^{-1})

i unit vector along *x*-axis in rectangular coordinate system

I current in amperes (A)

I_1 current in the upper arm of a dipole in amperes (A)
current along the inner conductor of a coaxial cable in amperes (A)

I_2 current in the lower arm of a dipole in amperes (A)
current along the outer conductor of a coaxial cable in amperes (A)

I_3 current along the outside of the outer conductor of a coaxial cable in amperes (A)

$I(z)$ current at a distance *z* along a monopole in amperes (A)

I_{max} maximum current in a monopole

j square root of -1, imaginary axis of Argand diagram

j unit vector along *y*-axis in rectangular coordinate system

$\boldsymbol{J}_d$ displacement current density vector in amperes per meter (A m^{-2})

$\boldsymbol{J}_s$ surface current density vector in amperes per meter (A m^{-2})

k unit vector along *z*-axis in rectangular coordinate system

k phase constant (=2p/l)
constant that determines the start of a spiral antenna

K_2 ratio of the radial distances of the inner and outer edges of the arm of a log spiral antenna

k_3 constant (log spiral antenna)

l length in m
physical length of a monopole in m
length of a transmission line in m

l_e slant length of an *E*-plane sectoral horn

l_h slant length of an *H*-plane sectoral horn

l_n length of the *n*th element of a LPDA

l/d_0 length-to-diameter ratio of the element of a LPDA

L length of a turn of a helix antenna
length of a log spiral antenna
slant height of the cone for a discone antenna

L_a inductance of antenna in ohms (Ω)

L_e loss in dB due to phase error for an *E*-plane sectoral horn

L_h loss in dB due to phase error for an *H*-plane sectoral horn

L_{ext} external inductance of a single turn loop in ohms (Ω)

λ wavelength in m

λ_g guide wavelength in m

λ_c wavelength at the cut-off frequency in m

λ_H wavelength at the highest operating frequency in m

λ_L wavelength at the lowest operating frequency in m

λ_0 free-space wavelength in m

M_t magnitude of the correction term applied to the power output of the generator

M_m magnitude of the correction term applied to the power measured by the power meter

M_t magnitude of the correction term applied to the power measured by the receiver, to allow for the mismatch between the receive antenna and the receiver

m magnetic permeability in henries per meter (H m^{-2})

m_1 magnetic permeability in medium 1, in henries per meter (H m^{-2})

m_2 magnetic permeability in medium 2, in henries per meter (H m^{-2})

$\mathbf{n}$ unit vector normal to a surface

N number of turns of a coil or helical antenna

ω angular frequency in radians per second (rad/s)

P	power in watts (W)
P_1	power in watts (W)
P_2	power in watts (W)
P_d	magnitude of power density in watts per square meter (W m^{-2})
$\boldsymbol{P}_d$	Poynting's power flux density vector in watts per square meter (W m^{-2})
P_g	power received by (or supplied to) generator in watts (W)
P_m	power measured by power meter in watts (W) linear polarization mismatch
P_{max}	maximum power in watts (W)
P_r	received power in watts (W)
P_t	transmitted power in watts (W)
P_{rad}	radiated power in watts (W)
PF_{loss}	loss power factor
PF_{rad}	radiation power factor
r	radial distance from a radiating dipole or loop
r_1	radial distance from center of a spiral
r_2	radial distance from center of a spiral
R	length of position vector distance between a horn and its image, in Purcell's method
$\boldsymbol{R}$	position vector
R_a	radiation resistance of antenna in ohms (Ω)
R_e	effective resistance of antenna in ohms (Ω)
R_h	terminal resistance of a helical antenna in ohms (Ω)
Re	real part of a complex number
R_{int}	internal resistance of a single turn loop in ohms (Ω)
$R(kl)$	function shown in figure 5.4
r_{15^+}	is electrical radius on the base side (of a conical log spiral) at which the magnitude of the near field is 15 dB below the maximum, at the highest frequency of operation
r_{3^+}	is electrical radius on the apex side (of a conical log spiral) at which the magnitude of the near field is 3 dB below the maximum, at the highest frequency of operation
r_n	distance of the front of the *n*th tooth from the imaginary apex of a trapezoidal tooth LPDA

R_c	cone radius of a conical log spiral antenna
R_r	radiation resistance of antenna in ohms (Ω)
r_e	virtual perpendicular length of an *E*-plane sectoral horn
r_h	virtual perpendicalar length of an *H*-plane sectoral horn
ρ_s	surface charge density in coulombs per square meter (C m^{-2})
s	disk-to-cone spacing for a discone antenna
s_f	scaling factor for a LPDA
S	surface spacing between turns of a helical antenna in m voltage standing wave ratio (VSWR)
S_p	spacing parameter for a LPDA
$S(x)$	Fresnel sine integral
y	angle between arrays of LPDAs pitch angle of a helical antenna
y_e	flare angle in the *E*-plane of a horn
y_h	flare angle in the *H*-plane of a horn
t	time in seconds
T_e	phase error/deviation in terms of wavelengths for an *E*-plane sectoral horn
T_h	phase error/deviation in terms of wavelengths for an *H*-plane sectoral horn
T_p	thickness parameter of the wire of a loop
q	angle in the elevation plane angle between vectors
q_l	half power beamwidth in degrees
q_E	half power beamwidth in the *E* plane, in degrees
q_H	half power beamwidth in the *H* plane, in degrees
t_g	complex reflection coefficient of the generator
t_r	complex reflection coefficient of the receive antenna
t_s	complex reflection coefficient of the receiver
t_t	complex reflection coefficient of the transmit antenna
v	velocity of a EM wave in m s^{-1}
v	velocity of a EM wave in medium 1, in m s^{-1}
v_x	velocity in x direction in m s^{-1}

v_y	velocity in x direction in m s^{-1}
v_z	velocity in x direction in m s^{-1}
$\boldsymbol{v}$	velocity vector in m s^{-1}
V	voltage in volts (V)
V_l	voltage across a load in volts (V)
V_1	voltage in volts (V)
V_2	voltage in volts (V)
V_{oc}	open-circuit voltage in volts (V)
w	diameter of the feed conductor for a discone antenna
x	abscissa, coordinate of rectangular Cartesian system variable
X_a	self reactance of antenna in ohms (Ω)
X_e	effective reactance of antenna in ohms (Ω)
$X(kl)$	function shown in figure 5.4
x_0	intrinsic impedance of a free space wave
y	coordinate of rectangular Cartesian system variable
Y_0	susceptance of the waveguide of normal height propagating the TE_{10} mode
Y'_0	susceptance of the waveguide of reduced height
z	coordinate of rectangular Cartesian system
Z	input impedance of a shielded single turn loop in ohms (Ω)
Z_0	characteristic impedance in ohms (Ω) intrinsic impedance of a wave in ohms (Ω)
Z_a	self impedance of antenna in ohms (Ω) total impedance of antenna in ohms (Ω)
Z_d	impedance of a half-wave resonant dipole antenna in ohms (Ω)
Z_e	radiation impedance of antenna in ohms (Ω)
Z_E	characteristic impedance of a waveguide propagating a TE mode in ohms (Ω)
Z_f	impedance of the differential mode on ohms (Ω)
Z_H	characteristic impedance of a waveguide propagating a TM mode in ohms (Ω)
Z_{in}	input impedance of a folded dipole antenna in ohms (Ω)
Z_{int}	internal impedance of a shielded single turn loop in ohms (Ω)

Z_l impedance of a load in ohms (Ω)

Z_r impedance of the common mode in ohms (Ω)

Z_w characteristic impedance of waveguide in ohms (Ω)

$Z_{1/4}$ impedance of half of a resonant half-wave dipole in ohms (Ω)

About the Author

Thereza Macnamara attained her first degree in applied physics and her Master's degree in microwaves at London University. After two years teaching physics up to advanced level, she worked as a microwave engineer at G & E Bradley. She then worked as a research physicist for Morganite Research and Development before returning to work as a microwave engineer working on a standard field facility, waveguide components, thermistor detectors, and calibration at Wayne-Kerr Laboratories, Flann Microwave Instruments and Bradley Electronics. After a short break to have a family, she returned to work as an examiner at the British Patent Office and lecturing in maths and physics while her children were growing up. She then took up a post as a senior RF engineer at ERA Technology, where she worked on antennas, feed networks and as an EMC engineer, before taking up her present post as an antenna systems specialist at British Aerospace.

Index

The Artech House Antenna Library

Helmut E. Schrank, *Series Editor*

Advanced Technology in Satellite Communication Antennas: Electrical and Mechanical Design, Takashi Kitsuregawa

Analysis of Wire Antennas and Scatterers: Software and User's Manual, A. R. Djordjević, M. B. Bazdar, G. M. Bazdar, G. M. Vitosevic, T. K. Sarkar, and R. F. Harrington

Analysis Methods for Electromagnetic Wave Problems, E. Yamashita, ed.

Antenna-Based Signal Processing Techniques for Radar Systems, Alfonso Farina

Braodband Patch Antennas, Jean-François Zürcher and Fred E. Gardiol

CAD for Linear and Planar Antenna Arrays of Various Radiating Elements: Software and User's Manual, Miodrag Mikavica and Aleksandar Nešić

The CG-FFT Method: Application of Signal Processing Techniques to Electromagnetics, Manuel F. Cátedra, Rafael P. Torres, José Basterrechea, Emilio Gago

Electromagnetic Waves in Chiral and Bi-Isotropic Media, I.V. Lindell, S.A. Tretyakov, A.H. Sihvola, A. J. Viitanen

Fixed and Mobile Terminal Antennas, A. Kumar

Generalized Multipole Technique for Computational Electromagnetics, Cristian Hafner

Handbook of Antennas for EMC, Thereza Macnamara

Integral Equation Methods for Electromagnetics, N. Morita, N. Kumagai, and J. Mautz

IONOPROP: Ionospheric Propagation Assessment Program, Version 1.1: Software and User's Manual, by Hernert V. Hitney

Four-Armed Spiral Antennas, Robert G. Corzine and Joseph A. Mosko

Introduction to Electromagnetic Wave Propagation, Paul Rohan

Introduction to the Uniform Geometrical Theory of Diffraction, D. A. McNamara

Microwave Cavity Antennas, A. Kumar and H. D. Hristov

Millimeter-Wave Microstrip and Printed Circuit Antennas, Prakash Bhartia

Mobile Antenna Systems, K. Fujimoto and J. R. James

Modern Methods of Reflector Antenna Analysis and Design, Craig Scott

Moment Methods in Antennas and Scattering, Robert C. Hansen, editor

Monopole Elements on Circular Ground Planes, M. M. Weiner *et al.*

Near-Field Antenna Measurements, D. Slater

Passive Optical Components for Optical Fiber Transmission, Norio Kashima

Phased Array Antenna Handbook, Robert J. Mailloux

Polariztion in Electromagnetic Systems, Warren Stutzman

Practical Phased-Array Antenna Systems, Eli Brookner *et al.*

Practical Simulation of Radar Antennas and Radomes, Herbert L. Hirsch and Douglas C. Grove

Radiowave Propagation and Antennas for Personal Communications, Kazimierz Siwiak

Reflector and Lens Antennas: Software User's Manual and Example Book, Carlyle J. Sletten, editor

Reflector and Lens Antenna Analysis and Design Using Personal Computers, Carlyle J. Sletten, editor

Shipboard Antennas, Second Edition, Preston Law

Small-Aperture Radio Direction-Finding, Herndon Jenkins

Spectral Domain Method in Electromagnetics, Craig Scott

Waveguide Components for Antenna Feed Systems: Theory and CAD, J. Uher, J. Bornemann, and Uwe Rosenberg

For further information on these and other Artech House titles, contact:

Artech House
685 Canton Street
Norwood, MA 02062
617-769-9750
Fax: 617-769-6334
Telex: 951-659
email: artech@world.std.com

Artech House
Portland House - Stag Place
London SW1E 5XA England
+44 (0) 171-973-8077
Fax: +44 (0) 171-630-0166
Telex: 951-659
email: bookco@artech.demon.co.uk